Lecture Notes in Geoinformation and Cartography

Series editors

About the Series

The Lecture Notes in Geoinformation and Cartography series provides a contemporary view of current research and development in Geoinformation and Cartography, including GIS and Geographic Information Science. Publications with associated electronic media examine areas of development and current technology. Editors from multiple continents, in association with national and international organizations and societies bring together the most comprehensive forum for Geoinformation and Cartography.

The scope of Lecture Notes in Geoinformation and Cartography spans the range of interdisciplinary topics in a variety of research and application fields. The type of material published traditionally includes:

- proceedings that are peer-reviewed and published in association with a conference;
- post-proceedings consisting of thoroughly revised final papers; and
- research monographs that may be based on individual research projects.

The Lecture Notes in Geoinformation and Cartography series also includes various other publications, including:

- tutorials or collections of lectures for advanced courses;
- contemporary surveys that offer an objective summary of a current topic of interest; and
- emerging areas of research directed at a broad community of practitioners.

More information about this series at http://www.springer.com/series/7418

Kateřina Růžičková · Tomáš Inspektor
Editors

Surface Models for Geosciences

Springer

Editors
Kateřina Růžičková
Institute of Geoinformatics
VŠB—Technical University of Ostrava
Ostrava
Czech Republic

Tomáš Inspektor
Institute of Geoinformatics
VŠB—Technical University of Ostrava
Ostrava
Czech Republic

ISSN 1863-2246 ISSN 1863-2351 (electronic)
Lecture Notes in Geoinformation and Cartography
ISBN 978-3-319-18406-7 ISBN 978-3-319-18407-4 (eBook)
DOI 10.1007/978-3-319-18407-4

Library of Congress Control Number: 2015939156

Springer Cham Heidelberg New York Dordrecht London

Printed on acid-free paper

Springer International Publishing AG Switzerland is part of Springer Science+Business Media
(www.springer.com)

Preface

This book was published as conference GIS Ostrava 2015 compilation—Surface Models for Geosciences. The main aim of the conference was to present and discuss new methods, issues and challenges encountered in all steps of gradual development, maintenance and utilization of digital surface models and their integration with various spatial data and models. Almost 200 attendees from 15 countries discussed current research directions and applications of geoinformatics in these fields at Technical University of Ostrava, The Czech Republic. All contributions in this book have undergone proper double-blind peer-reviewed process of journal scientific standards.

Digital surface models are used in almost all geoinformatics applications, since the shape of the earth's surface affects many natural processes that are happening on the Earth. The applicability of height data is not only influenced by method of their acquisition, but also by methods of their processing.

Articles in this compilation present the latest technical equipment for acquiring of elevation data using unmanned aerial systems (e.g. RPAS—remotely piloted aircraft systems) brought the need to discuss ways and possibilities of their use.

There is also introduced the use of data from INSAR techniques, or the possibility of using data from laser scanning, e.g., for rockfall monitoring, or estimation of solar radiation.

A large number of articles focuses on the use of DSM in hydrology, where correcting elevation model directly influences the hydrological behaviour analysis, whether it is a flood modelling, or erosion modelling.

Evaluation of DSM and morphological formations accuracy plays a large part in further contributions.

In this book, you can also learn about geoinformatic applications, which are focused on the areas of geology, planned development, population distribution, visibility analysis and engineering. Possibilities of final presentation of models represent the article about 3D models printing.

Acknowledgments

Behalf of editorial team, I would like to thank all contributors for their willingness to share their new knowledge in the processing and use of DSM. I would also like to thank reviewers, who selected and helped to fine-tune the most valuable articles for you.

Kateřina Růžičková

Programme Committee

Kateřina Růžičková	VŠB—Technical University of Ostrava, Czech Republic—Chairman
Martin Adamec	University of Ostrava, Czech Republic
Temenoujka Bandrova	University of Architecture, Civil Engineering and Geodesy, Bulgaria
Farès Belhadj	University of Paris 8, France
Jan Brus	Palacky University in Olomouc, Czech Republic
Jan Caha	VŠB—Technical University of Ostrava, Czech Republic
Lena Halounová	Czech Technical University, Prague
Tomáš Hlásný	National Forest Centre Zvolen, Slovakia
Jaroslav Hofierka	Pavol Josef Safarik University in Kosice, Slovakia
Jiří Horák	VŠB—Technical University of Ostrava, Czech Republic
Michal Kačmařík	VŠB—Technical University of Ostrava, Czech Republic
Mikhail Kanevski	University of Lausanne, Switzerland
Martin Klimánek	Mendel Univerzity in Brno, Czech Republic
Gotffried Konecny	Leibniz University Hannover, Germany
Milan Lazecký	VŠB—Technical University of Ostrava, Czech Republic
Xiaoye Liu	University of Southern Queensland, Australia
Juraj Papčo	Slovak University of Technology in Bratislava, Slovakia
Norbert Pfeifer	Vienna Technical University, Austria
Petter Pilesjö	Lund University GIS Centre, Sweden
Petr Rapant	VŠB—Technical University of Ostrava, Czech Republic
Jan Růžička	VŠB—Technical University of Ostrava, Czech Republic
Joaquim J. Sousa	University of Trás-os-Montes e Alto Douro, Portugal
Jan Šrejber	Czech Hydrometeorological Institute, Czech Republic
Jan Unucka	VŠB—Technical University of Ostrava & Czech Hydrometeorological Institute, Czech Republic
Tomáš Vaníček	Czech Technical University in Prague, Czech Republic
Vít Voženílek	Palacky University in Olomouc, Czech Republic
John P. Wilson	University of Southern California, United States of America
Tian-Xiang Yue	Chinese Academy of Sciences, China
Qiming Zhou	Hong Kong Baptist University, China

Auspices

January 26–28 2015

AGILE—Association of Geographic Information Laboratories for Europe
EuroSDR—European Spatial Data Research Organisation
ISPRS—International Society for Photogrammetry and Remote Sensing
CAGI—Czech Association for Geoinformation
SAGI—Slovak Association for Geoinformatics
Miroslav Novák, President of the Moravian-Silesian Region
Ing. Tomáš Macura, MBA, Mayor of the City of Ostrava
Prof. Ing. Ivo Vondrák, CSc., Rector of VŠB—Technical University of Ostrava

Contents

Modelling and Visualising Landscape and Terrain Impacts of Planned Developments 1
Gábor Barton, Katalin Bódis and Róbert Géczi

Morphological Analyses of the Central Slovakia on Base of the DEM 13
Dalibor Bartoněk, Jiří Bureš, Lubomil Pospíšil and Otakar Švábenský

Surface Classification Above Gas Pipeline Facilities 27
Dalibor Bartoněk, Irena Opatřilová and Jiří Bureš

Rockfall Monitoring Based on Surface Models 37
Snežana Bogdanović, Miloš Marjanović, Biljana Abolmasov, Uroš Đurić and Irena Basarić

Coping with Integrating Low-Cost 3D Printing and Surface Models: A Case Study on Prusa i3 45
Jan Brus and Radek Barvíř

Line-of-Sight Derived Indices: Viewing Angle Difference to a Local Horizon and the Difference of Viewing Angle and the Slope of Line of Sight 61
Jan Caha and Alexandra Rášová

Initial Results of a Surface Deformation by Using InSAR Techniques: Case Study of Babadağ (Denizli), Turkey 73
Fatma Canaslan Çomut, Şule Gürboğa and Aydın Üstün

Results Comparison of the Flow Direction and Accumulation Algorithms Together with Distributed Rainfall-Runoff Models in Czech Switzerland National Park 87
Vladimír Fárek and Jan Unucka

Surveying of Small Water Reservoirs for Water Management Purposes 99
Jakub Fuska and Viliam Bárek

Detailed Digital Terrain Models in the Research of Deserted Settlement: Hydrological Modelling and Environment of Settlement Areas 113
Lukáš Holata and Radek Světlík

DTM Impact on the Results of Dam Break Simulation in 1D Hydraulic Models 125
Andrea Jančíková and Jan Unucka

Dasymetric Mapping of Population Distribution in Serbia Based on Soil Sealing Degrees Layer 137
Nikola Krunić, Branislav Bajat and Milan Kilibarda

Geographic Information System for Flood Hazard Analysis and Early Warning Using Numerical Weather Predictions 151
Marcin Kulawiak, Andrzej Chybicki, Lukasz Markiewicz and Marlena Jankowska

Effect of DEM Inaccuracy on Precision of Satellite InSAR Results 165
Milan Lazecký, Matúš Bakoň and Ivana Hlaváčová

Creating of DSM Based on RPAS Measurement and Accuracy Testing 173
Karel Pavelka and Jaroslav Šedina

The Use of Real and Fictitious Surfaces for Territorial Distribution Assessment of Given Geographic Phenomenon 189
Aleš Ruda and Jaromír Kolejka

Impact of GDAL JPEG 2000 Lossy Compression to a Digital Elevation Model 205
Jan Růžička and Kateřina Růžičková

Design of Test Base for Determine Volume Accuracy 215
Václav Šafář

New Automatic Accuracy Evaluation of Altimetry Data: DTM 5G Compared with ZABAGED® Altimetry 225
Jakub Šilhavý and Václav Čada

Hydraulic Erosion Modeling on a Triangular Mesh 237
Věra Skorkovská, Ivana Kolingerová and Bedrich Benes

Registration of Optical and Radar Satellite Images Using Local Features and Non-rigid Geometric Transformations 249
Mohamed Tahoun, Aboul Ella Hassanien and Ralf Reulke

Experimental Comparison of Sequential and Global Modelling Methods of Subsurface Horizon Sequence 263
Ladislav Vizi and Mária Benčoková

Implementation of Poisson Faulting Algorithm with Circular Fault Function for Hardness Data Synthesis 281
Korneliusz K. Warszawski and Sławomir S. Nikiel

Photogrammetric Point Clouds for GIS-Based High-Resolution Estimation of Solar Radiation for Roof-Top Solar Systems 293
Roland Zink, Luis Ramirez Camargo, Patrick Reidelstürz and Wolfgang Dorner

Author Index 307

Abbreviations

ABS	Acrylonitrile–butadiene–styrene
ADCP	Acoustic Doppler current profiler
ALS	Airborne laser scanning
AOI	Area of interest
ARW	Advanced Research WRF
BCE	Before the Common/Current/Christian Era
CHMI	Czech Hydrometeorological Institute
CLC	Corine land cover
CPU	Central processing unit
CR	Czech Republic
ČÚZK	Czech Office for Surveying, Mapping and Cadastre
DCT	Discrete cosine transform
DEM	Digital elevation model
DInSAR	Differential SAR interferometry
DLP	Digital light projection
DPI	Dots per inch
DSM	Digital surface models
DTM	Digital terrain model
EEA	European Environment Agency
ESA	European Space Agency
FDM	Fused deposit modelling
FPL	First pulse points
GCP	Ground control points
GDEM	Global digital elevation model
GIS	Geographic information system
GKU	Geodetic and Cartographic Institute Bratislava
GNSS	Global navigation satellite system
GPS	Global positioning system
GPW	Gridded population of the world
GRUMP	Global rural–urban mapping project
GSD	Ground sample distance

HD	Hydraulic modelling
HPPS ČR	Operational flood forecasting service of Czech Republic
HTML	Hypertext Markup Language
IDW	Inverse distance weighting
IMU	Inertial measurement unit
InSAR	Interferometric synthetic aperture radar
LAU	Local administrative units
LIDAR	Light detection and ranging
LOM	Laminated object manufacturing
LoS	Line of sight
LPL	Last pulse points
LSGPD	LandScan Global Population Database
LZW	Lempel-Ziv-Welch
MPD	Mid-point displacement
NASA	National Aeronautics and Space Administration
NCAR	National Center for Atmospheric Research
NDVI	Normalized difference vegetation index
NMEA	National Marine Electronics Association
NN	Nearest neighbour
NNR	National Natural Reserve
NUTS	Nomenclature of units for territorial statistics
NWP	Numerical weather prediction
ORNL	Oak Ridge National Laboratory
PCI	Population change index
PDF	Portable document format
PLA	Polylactic acid
PSInSAR	Persistent scatterer interferometric synthetic aperture radar
RMGC	Roşia Montana Gold Corporation
RMSE	Root-mean-square error
RPAS	Remotely piloted aircraft systems
RST	Rotation, scaling and translation
RTK	Real-time kinematic
SAR	Synthetic aperture radar
SCS-CN	Soil conservation service curve number
SIFT	Scale invariant feature transform
SLA	Stereolithography
SLC	Single look complex
SLS	Selective laser sintering
SPH	Smoothed-particle hydrodynamics
SRES	Special report on emission scenarios
SRTM	Shuttle radar topography mission
SSD	Soil sealing database
STL	Standard template library

SURF	Speeded-up robust features
TIFF	Tag image file format
TIN	Triangulated irregular network
TLS	Terrestrial laser scanning
UAS	Unmanned aircraft systems
UAV	Unmanned aircraft vehicle
USB	Universal serial bus
USGS	United States Geological Survey
UTC	Coordinated universal time
UWB	University of West Bohemia in Pilsen
VUGTK	Research Institute of Geodesy, Topography and Cartography
WRF	Weather research and forecast
ZABAGED	Fundamental base of geographic data

Contributors

Biljana Abolmasov Department of Geotechnics, Faculty of Mining and Geology, University of Belgrade, Belgrade, Republic of Serbia

Branislav Bajat Faculty of Civil Engineering, Institute for Geodesy and Geoinformatics, University of Belgrade, Belgrade, Serbia

Matúš Bakoň STU Bratislava, Bratislava, Slovakia; UTAD Vila Real, Vila Real, Portugal

Viliam Bárek Department of Landscape Enginering, Faculty of Horticulture and Landscape Engineering, Slovak University of Agriculture, Nitra, Slovakia

Gábor Barton GIS Developer BartonGeo.Com, Pécs, Hungary

Dalibor Bartoněk Faculty of Civil Engineering, Institute of Geodesy, Brno University of Technology, Brno, Czech Republic; European Polytechnic Institute, Kunovice, Czech Republic

Radek Barvíř Department of Geoinformatics, Faculty of Science, Palacký University, Olomouc, Czech Republic

Irena Basarić Department of Geotechnics, Faculty of Mining and Geology, University of Belgrade, Belgrade, Republic of Serbia

Mária Benčoková Faculty of BERG, Institute of Geoscience, Technical University of Košice, Košice, Slovak Republic

Bedrich Benes Department of Computer Graphics Technology, Knoy Hall of Technology, West Lafayette, IN, USA

Katalin Bódis Renewable Energy Unit, European Commission—Joint Research Centre (JRC), Institute for Energy and Transport, Ispra, Varese, Italy

Snežana Bogdanović Department of Geotechnics, Faculty of Mining and Geology, University of Belgrade, Belgrade, Republic of Serbia

Jan Brus Department of Geoinformatics, Faculty of Science, Palacký University, Olomouc, Czech Republic

Jiří Bureš Faculty of Civil Engineering, Institute of Geodesy, Brno University of Technology, Brno, Czech Republic

Václav Čada Geomatics Section of Department of Mathematics, Faculty of Applied Sciences, University of West Bohemia in Pilsen, Pilsen, Czech Republic

Jan Caha Institute of Geoinformatics, VŠB—Technical University of Ostrava, Ostrava, Czech Republic

Andrzej Chybicki Department of Geoinformatics, Faculty of Electronics, Telecommunication and Informatics, Gdansk University of Technology, Gdansk, Poland

Fatma Canaslan Çomut Disaster and Emergency Directorate of Denizli (AFAD), Denizli, Turkey

Wolfgang Dorner Applied Energy Research Group, Technologie Campus Freyung, Deggendorf Institute of Technology, Freyung, Germany

Uroš Đurić Department of Geotechnical Engineering, Faculty of Civil Engineering, University of Belgrade, Belgrade, Republic of Serbia

Vladimír Fárek Czech Hydrometeorogical Institute, Ústí nad Labem, Czech Republic

Jakub Fuska Department of Landscape Enginering, Faculty of Horticulture and Landscape Engineering, Slovak University of Agriculture, Nitra, Slovakia

Róbert Géczi John Wesley Theological College, Budapest, Hungary

Şule Gürboğa Department of Geological Engineering, Akdeniz University, Antalya, Turkey

Aboul Ella Hassanien Department of Information Technology, FCI, Cairo University, Giza, Cairo, Egypt

Ivana Hlaváčová CVUT Prague, Praha 6, Czech Republic; UTAD Vila Real, Vila Real, Portugal

Lukáš Holata Department of Interdisciplinary Activities, New Technologies Research Centre, University of West Bohemia, Pilsen, Czech Republic

Andrea Jančíková Katastrální úřad, Olomouc, Czech Republic

Marlena Jankowska Department of Civil Law and Private International Law, Faculty of Law and Administration, University of Silesia, Katowice, Poland

Milan Kilibarda Faculty of Civil Engineering, Institute for Geodesy and Geoinformatics, University of Belgrade, Belgrade, Serbia

Jaromír Kolejka Department of Geography, Faculty of Education, Masaryk University, Brno, Czech Republic

Ivana Kolingerová NTIS—New Technologies for Information Society, University of West Bohemia, Plzeň, Czech Republic; Department of Computer Science and Engineering, Faculty of Applied Sciences, University of West Bohemia, Plzeň, Czech Republic

Nikola Krunić Institute of Architecture and Urban and Spatial Planning of Serbia, Belgrade, Serbia

Marcin Kulawiak Department of Geoinformatics, Faculty of Electronics, Telecommunication and Informatics, Gdansk University of Technology, Gdansk, Poland

Milan Lazecký IT4Innovations, VŠB—Technical University of Ostrava, Ostrava-Poruba, Czech Republic; UTAD Vila Real, Vila Real, Portugal

Miloš Marjanović Department of Geotechnics, Faculty of Mining and Geology, University of Belgrade, Belgrade, Republic of Serbia; Chair of Landslide Research, Faculty of Civil, Geo and Environmental Engineering, Technische Universität München, Munich, Germany

Lukasz Markiewicz Department of Geoinformatics, Faculty of Electronics, Telecommunication and Informatics, Gdansk University of Technology, Gdansk, Poland

Sławomir S. Nikiel Faculty of Economics and Management, University of Zielona Góra, Zielona Góra, Poland

Irena Opatřilová Faculty of Civil Engineering, Institute of Geodesy, Brno University of Technology, Brno, Czech Republic

Karel Pavelka Department of Geomatics, Faculty of Civil Engineering, Czech Technical University in Prague, Prague 6, Czech Republic

Lubomil Pospíšil Faculty of Civil Engineering, Institute of Geodesy, Brno University of Technology, Brno, Czech Republic

Luis Ramirez Camargo Applied Energy Research Group, Technologie Campus Freyung, Deggendorf Institute of Technology, Freyung, Germany; Institute of Spatial Planning and Rural Development, University of Natural Resources and Life Sciences, Vienna, Austria

Alexandra Rášová Slovak University of Technology in Bratislava, Bratislava, Slovakia

Patrick Reidelstürz Applied Energy Research Group, Technologie Campus Freyung, Deggendorf Institute of Technology, Freyung, Germany

Ralf Reulke Department of Computer Science, Humboldt University in Berlin, Berlin, Germany

Aleš Ruda Department of Regional Development and Public Administration, Faculty of Regional Development and International Studies, Mendel University in Brno, Brno, Czech Republic

Jan Růžička Institute of Geoinformatics, HGF, VŠB—Technical University of Ostrava, Ostrava, Czech Republic

Kateřina Růžičková Institute of Geoinformatics, HGF, VŠB—Technical University of Ostrava, Ostrava, Czech Republic

Václav Šafář The Research Institute of Geodesy, Topography and Cartography, Zdiby, Czech Republic

Jaroslav Šedina Department of Geomatics, Faculty of Civil Engineering, Czech Technical University in Prague, Prague 6, Czech Republic

Jakub Šilhavý Geomatics Section of Department of Mathematics, Faculty of Applied Sciences, University of West Bohemia in Pilsen, Pilsen, Czech Republic

Věra Skorkovská NTIS—New Technologies for Information Society, University of West Bohemia, Plzeň, Czech Republic; Department of Computer Science and Engineering, Faculty of Applied Sciences, University of West Bohemia, Plzeň, Czech Republic

Otakar Švábenský Faculty of Civil Engineering, Institute of Geodesy, Brno University of Technology, Brno, Czech Republic

Radek Světlík Department of Archaeology, Faculty of Philosophy and Arts, University of West Bohemia, Pilsen, Czech Republic

Mohamed Tahoun Department of Computer Science, Humboldt University in Berlin, Berlin, Germany

Jan Unucka Czech Hydrometeorogical Institute, Ostrava-Poruba, Czech Republic; Faculty of Mining and Geology, VŠB—Technical University of Ostrava, Ostrava-Poruba, Czech Republic

Aydın Üstün Department of Geomatics Engineering, Kocaeli University, Izmit, Turkey

Ladislav Vizi Faculty of BERG, Institute of Geoscience, Technical University of Košice, Košice, Slovak Republic

Korneliusz K. Warszawski Faculty of Electrical Engineering, Computer Science and Telecommunications, University of Zielona Góra, Zielona Góra, Poland

Roland Zink Applied Energy Research Group, Technologie Campus Freyung, Deggendorf Institute of Technology, Freyung, Germany

Modelling and Visualising Landscape and Terrain Impacts of Planned Developments

Gábor Barton, Katalin Bódis and Róbert Géczi

Abstract While it is true that modern remote sensing based terrain modelling techniques offer a fast, efficient and relatively cheap way to create terrain models, they are all limited to currently existing features. This means that they cannot provide information about future, planned changes to the terrain. In such cases we need to create surface models that can represent both the existing landforms and the planned modifications with appropriate reliability and accuracy. Such simulations can play a significant role in impact assessment studies, public consultation communication as well as offer additional insight into the consequences of a project. This paper will demonstrate some of the methods I have been developing using the example study area of a planned open-cast gold and silver mining operation in Roşia Montana, Romania. It describes the data sources used, the required conversion and harmonisation steps, the modelling of planned mining operations and the resulting visualisations and data products.

Keywords Visualisation · Terrain modelling · Hybrid surface models · Data conversion

---

G. Barton (✉)
GIS Developer BartonGeo.Com, Hunyadi út 73, Pécs 7625, Hungary
e-mail: bartongabor@bartongeo.com

K. Bódis
Renewable Energy Unit, Institute for Energy and Transport, European Commission—Joint Research Centre (JRC), via Enrico Fermi 2749, TP 450, 21027 Ispra, Varese, Italy
e-mail: katalin.bodis@geography.hu

R. Géczi
John Wesley Theological College, Dankó utca 11, Budapest 1086, Hungary
e-mail: robi@earth.geo.u-szeged.hu

K. Růžičková and T. Inspektor (eds.), *Surface Models for Geosciences*,
Lecture Notes in Geoinformation and Cartography,
DOI 10.1007/978-3-319-18407-4_1

1 Introduction

Gold, silver, copper and other metallic ores have been mined in the Apuseni region of Romania for centuries, with archaeological finds dating back to Roman times. With regards to Roşia Montana, the most impressive of the relics is the vast network of mining shafts throughout the current settlement area, some of which is now part of the mining museum and open for visitors (Fig. 1).

Following a long history of subsurface operations, the communist era brought about the open-cast methods which included blasting away the mountains and processing the resulting rock material to extract the ore chemically. The state-owned mining activity was shut down in 2006, partly because of Romania's EU membership and the subsequent more rigorous environmental requirements, which could not have been met using the old technologies. Many of the surface waters have been heavily polluted and the landscape around the town has been altered with no (or very little) regard to environmental impacts.

Following the shutdown of the state mining company a new developer announced plans to open the largest open-cast gold and silver mine in Europe. Roşia Montana Gold Corporation (RMGC) has been lobbying at the Romanian government for support of the project since 2006. The idea met with large scale opposition in the country and its neighbours caused mainly by environmental concerns regarding the planned cyanide-based extraction process. While it is true that this method has been used previously, the new mining company planned to process the ore on site, as opposed to the previous processing location of Gura Roşiei and Baia Aries. These protests and several other activities have led to the 2014 decision of the Romanian

Fig. 1 Roman mining gallery in Roşia Montana (author's photo, 2005)

government to revoke its support for the project. However RMGC is still looking for a way to extract the 300 tons of gold and 1600 tons of silver that, according to their projections are present in the concession area.

2 Location of Study Area

The initial studies conducted by the researchers at the Department of Physical Geography at the University of Szeged focussed on the ecological assessment of the Roşia watershed (Géczi and Bódis 2003), but since the RMGC mining project area extends beyond the basin, the study area demonstrated in this paper has been expanded accordingly. The township of Roşia Montana is about 80 km Northwest from Alba Iulia (Fig. 2), the capital of Alba County, and it includes several smaller villages along the Roşia river, with a total population of 2656 in 2011 (INS 2011).

3 Methods and Data

In order to achieve the visualisation and modelling objectives set out in this paper several methods and tools had to be used, some of which may not seem relevant in the traditional GIS environment. The land cover component was based on Landsat

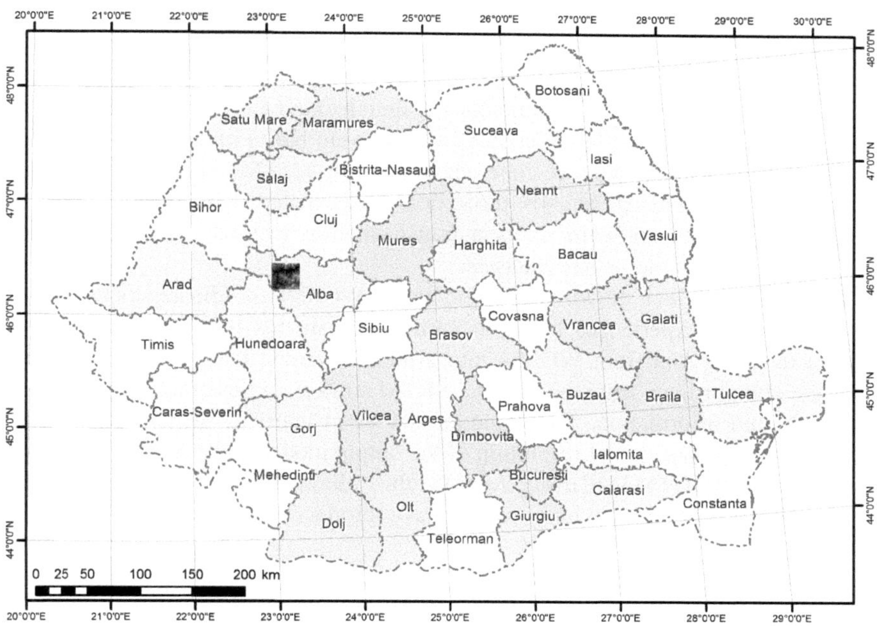

Fig. 2 Location of study area within Romania (ESRI Data and Maps, 2007)

(Chavez et al. 1991) images from 2004 at 25 m resolution. While generally we aim to use as high resolution data as possible, the limits of available computing power and financial resources did not allow the purchase of better terrain and imagery sources. However, the extent of the study area and the planned overview visualisations did not require the highest possible detail and made the processing much more efficient. The original Landsat image was modified using the Open Source GiMP software.

Terrain modelling was done using ESRI ArcGIS, specifically Spatial and 3D Analyst modules. The baseline was the freely available SRTM (Rodriguez et al. 2005) surface model at 90 m/pixel resolution, which had to be refined in order to allow the representation of artificial features. The initial SRTM surface was augmented with the models of the artificial features, whose creation and integration is the main focus of this paper. The source data of the planned surface objects was obtained from the maps found in the Environmental Impact Assessment report pages (RMGC 2014) of the RMGC website.

The research presented several data processing and conversion challenges, namely:

- converting and georeferencing maps in PDF files
- combining TIN and raster based surface models
- modifying and re-integrating satellite imagery

3.1 Surface Model Preprocessing

As a first step a common spatial reference system had to be determined. This is very important, because the different sources of data were based on different coordinate systems and in order to achieve coherent results, we need to have all the input data in a common system. UTM 34N projection was selected, which is usable in the study area and the software offers a transformation method from all the input projections used in the source datasets.

The SRTM surface model was originally in WGS84 coordinate system. Based on the negative experiences found in earlier experiments involving the transformation of raster data from WGS84 into a metric system it was more effective to generate contour lines from the original SRTM surface and use those as the source for the terrain information. The contours had 10 m intervals, which proved sufficient for the planned spatial resolution of the output models. The contour lines were transformed into UTM34N using ArcGIS, thus eliminating the adverse effects of reprojecting raster data (which would have produced significant errors in the UTM34N raster output).

The original SRTM surface had a spatial resolution of 90 m, which does not allow for the representation of smaller artificial features. This was an additional reason to use contour lines as a basis for terrain modelling, since using contours enabled the generation of a model with any resolution as deemed necessary. It is

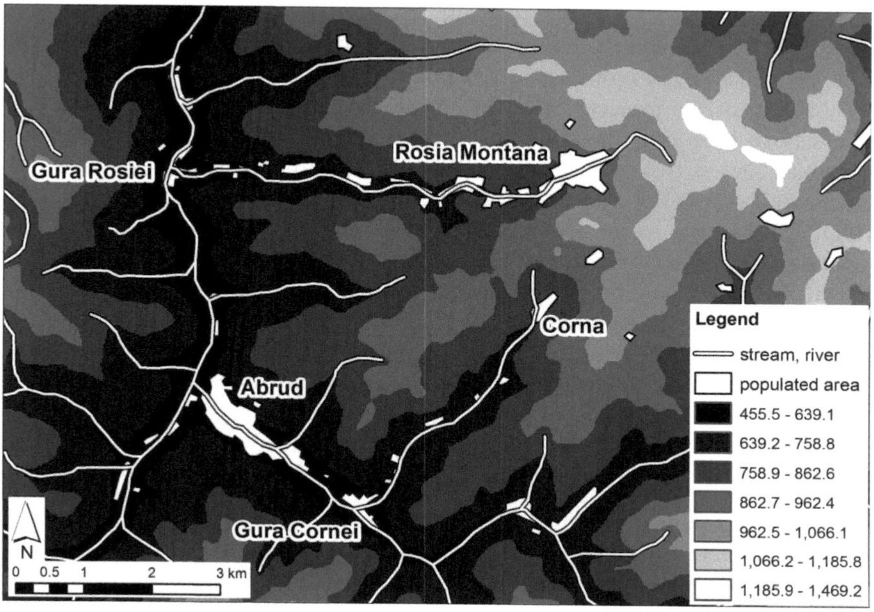

Fig. 3 Elevation, populated areas and stream network in study area

important to remember, that this "higher resolution" terrain does not mean that it offers higher detail than the original SRTM data, merely that the smaller cell size allows for better manipulation and editing that is required in the modelling process. The surface was created at 20 m resolution, which is detailed enough for the inclusion of most artificial features and still does not produce too large a dataset to handle.

One additional enhancement for the surface model was to generate a water flow network to avoid having unnatural surface elements. This stream network was manually digitised based on the morphological characteristics of the area and using the data found in the INSPIRE Hydrography view service (ANCPI 2014) of the Romanian National Agency of Cadastre and Land Registration. The river network was used in the surface interpolation as input to ArcGIS Spatial Analyst's Topo to Raster tool, including the generated 10 m interval contour lines, resulting in a 20 m resolution raster elevation model (Fig. 3).

3.2 *Modelling of Artificial Surface Elements*

The biggest challenge of the research was to find a way to integrate the models of the planned landscape alterations into the terrain model. The source for these features was found in the PDF files available in the Environmental Impact Assessment reports page of the RMGC website. The report contains quite detailed plans and maps of the surroundings, including roads, mining pits, quarries, waste dumps, water and material reservoirs for several points in time (current conditions and years 0, 7, 14, 16 of the project). The map documents also contained the information necessary to determine the coordinate system of the shown data, which was the Romanian Stereo70 system. First the PDF files had to be converted into a format understood by ArcGIS. The easiest solution was to create TIFF files from the map, which also allowed easier georeferencing. In order to reduce the number of conversion steps the coordinate graticule on the PDF maps to determine points of known Stereo70 coordinates, reprojected them directly into UTM34N and used these as reference points to place the TIFF files into their correct geographic position.

The next step was to generate 3D elements using the information found on the map sheets. Fortunately most of the required features had elevation values as labels and the map also contained contour lines at 5–10 m intervals. The elements were grouped into the following categories based on their function and shape (Fig. 4):

- flat surfaces (water, mining terrace)
- accumulation of material (waste dumps, dams)
- extraction (mining pits, quarries)

Modelling the flat surfaces was achieved by creating simple polygon shapes and giving them the elevation value from the map. In order to ensure seamless integration of artificial shapes into the source surface, some of these polygons had to be extended beyond the actual feature they represent (e.g. adding a 50–100 m buffer to the outline of the reservoir).

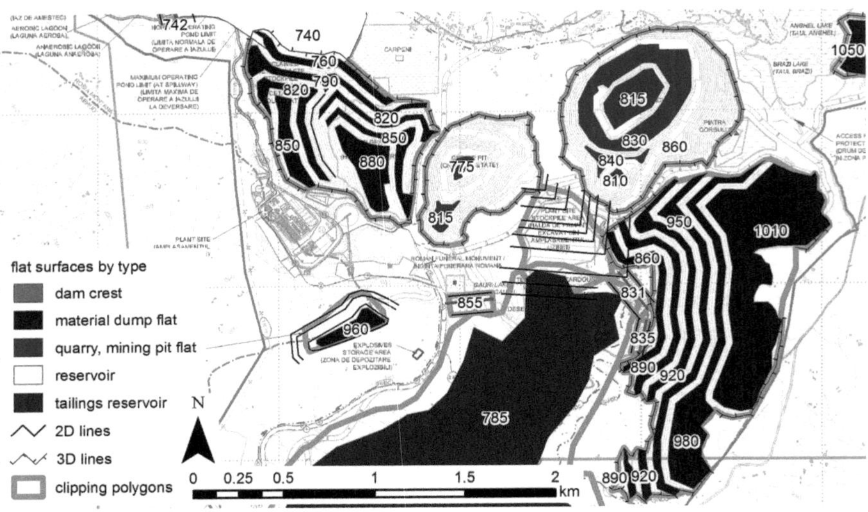

Fig. 4 Categories of artificial features and elevation values from project year 07

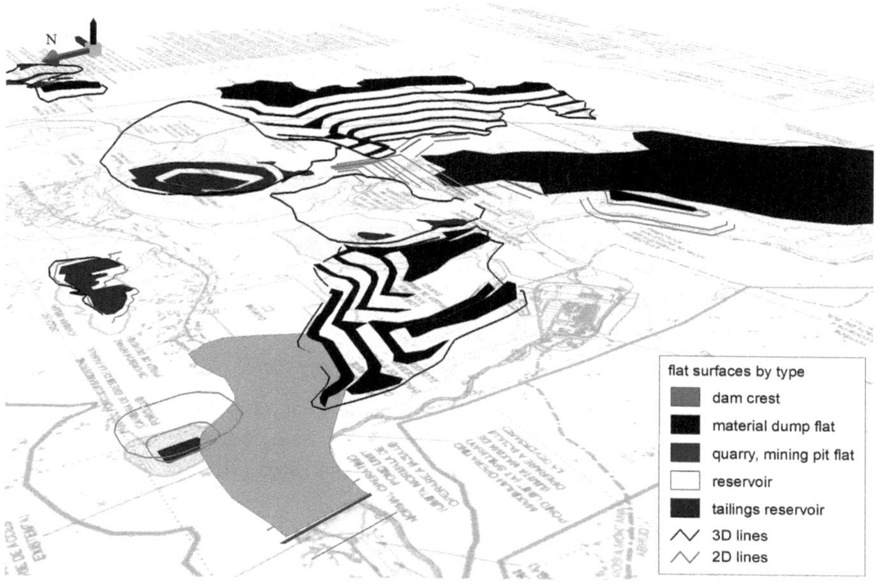

Fig. 5 Artificial terrain components of project year 07 in 3D (ESRI ArcScene)

Smaller dams were modelled by creating the crest as a flat polygon feature and two "contour lines" at the base on each side, setting the elevation value of the polygon to the height of the point where the dam structure meets the original surface. The large dam holding the sludge reservoir in Corna Valley was planned to be built in several stages and each stage had a different cross-section, so this dam had to be modelled for each time segment individually. The sludge reservoir presented an additional task because its surface is planned to be sloping in different directions, so additional contour lines had to be digitised as well.

Waste dumps containing the solid waste products and leftovers from quarrying are planned to be built in stages, resulting in step-like slopes and flat surfaces as the top levels. The most efficient way to represent these forms in 3D was to digitise the flat surfaces of the sides and the top with their corresponding elevation values and fill the gaps among them by simple triangulation (Fig. 5).

In addition to the generic features described above, some smaller size shapes were also required for the modelling which would add details that could not be modelled using the generic methods. Such features include the walls or ridges dividing mining pits, which are necessary to improve the accuracy of the model and the results of area and volume calculations. These were digitised as 3D lines where each vertex would have its specific elevation value based on the contour lines it intersects on the map.

In order to ensure the seamless integration of these surface components into the original terrain a 3D boundary line was created around the modelled features, which would obtain the elevation of its vertices from the original surface model. This is required for the outward slopes around the top level flat element of the waste dumps to reach the natural terrain where they are supposed to instead of creating vertical (or near vertical) walls around them.

The resulting component surfaces were all stored as individual TIN models covering only the area affected by each feature, except where the specific case warranted the use of a larger coverage for better integration. These cases were the dams and the various reservoirs, where using a larger area avoids the occurrence any gaps between the artificial feature and the original terrain surface.

3.3 Integration of Terrain Components

Once the component surfaces were ready they were joined together and merged into the original surface. At this point a choice had to be made between creating a fully vector or raster output surface. Both solutions have their pros and cons, some of which are listed in Table 1.

Based on the final output requirements and the spatial resolution of the available input datasets, it was decided to use raster surfaces in the integration process. This allowed better control of defining the relationships between the partial surfaces and the source terrain model as well as among the individual components themselves.

The integration process involved logical raster mosaic operations based on whether the component(s) represented features above the original surface or below. The artificial forms that overlapped one another also had to be taken into account and to make sure that each of them is mosaicked correctly (Fig. 6). One example was a smaller soil stockpile just south of the large sludge reservoir, which was supposed to cover parts of the dam slope, but not cut into it.

3.4 Satellite Image Manipulation

In order to simulate the visual impact of the various artificial features they also had to be represented in the imagery data. While the draped location map shown above

Table 1 Comparison of raster versus vector surface modelling techniques

Raster surface		Vector surface	
Advantages	Disadvantages	Advantages	Disadvantages
Simple storage logic	Detail level limited by cell size	Higher detail possible	More “edgy” visual output
More “natural” visual output	Larger files sizes	Smaller file size	Difficult to modify
Easy to manipulate with raster arithmetic			Higher display processing requirements

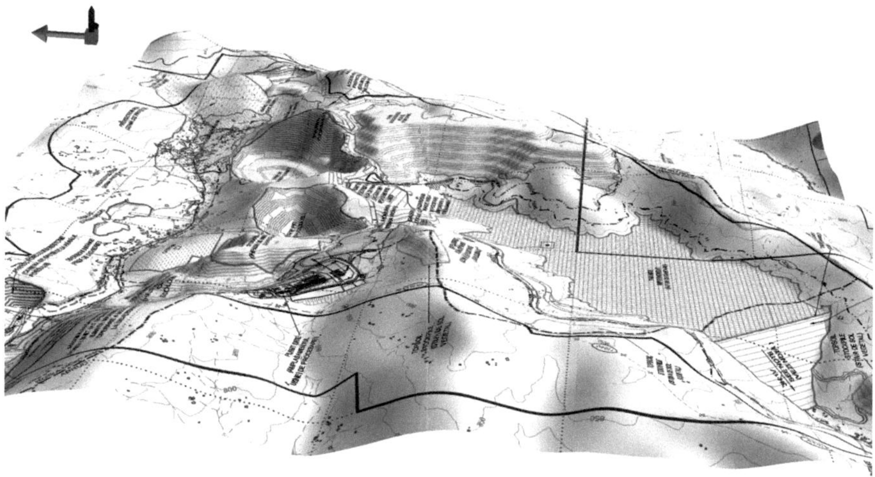

Fig. 6 RMGC map of year 07 draped over merged terrain components (ESRI ArcScene)

is quite informative for a more professional audience, the general public is more interested in the visuals rather than the technical details. The spatial resolution of the input data, the level of detail achieved in the terrain modelling process allowed the use of free Landsat images from 2001 with 25 m pixel size. Aiming to produce a visual simulation of the landscape, the standard RGB bands were used from the image and a simple minimum/maximum stretch with no modifications to the colouring to achieve a more natural result.

The first step was to convert the satellite image from a GIS-friendly format into one that can be understood and edited by standard graphics software (in our case GiMP for Windows). This process presented the risk of losing the georeferencing information stored within the original Landsat GeoTIFF file, because the tags containing geographic data are not picked up by GiMP and are lost in the output image. To solve this issue, the GeoTIFF file was converted into a regular TIFF format with an attached world file storing the georeferencing information required for ArcGIS to locate the image properly. The world file contains the X/Y coordinates of the top left corner and the pixel size in metres, which meant that as long as the original image size and orientation were maintained, the world file would always be able to define the geographic location of the file for GIS applications.

Once the world file was created the image could be opened in GiMP for editing. The first step was to define the textures to be used for the various artificial landscape features (waste dump, sludge reservoir, mining pit, dams, lakes, different stages of reforestation). "Fortunately" the area surrounding Roşia Montana has several existing examples of such features in various sizes so many of these could be used as texture and colour sources. Once the new texture had been selected the Clone tool was used in GiMP to paint the new texture over the area affected by the given artificial surface element. In order to keep the modifications within the boundaries of artificial areas their boundaries were also imported into the image as a separate layer and turned invisible before exporting the modified satellite image.

The original satellite image had to be modified for each modelled time instant, so in the end 5 images were created, each showing a different stage of the planned project. These images could now be draped over the modified surface models depicting the corresponding time period to show a simulated 3D view of the landscape.

4 Results

The two parallel modelling exercises (terrain and imagery) resulted in 5 surface models and 5 modified satellite images for the 5 different project stages used in the process. These datasets were then used to create various 3D animations, images, and also offered an opportunity to perform displacement calculations about the volume of the material that would have been moved during the mining project. Due to the relatively low resolution of the input terrain data (20 m), the accuracy of these volume calculations only allows estimated values and they should not be considered as final, definitive results. With higher resolution input data much better estimates can be achieved, but the during the procedures described in the current paper it was not possible to obtain more accurate input datasets. Figures 7, 8 and Tables 2, 3 will show some of the results.

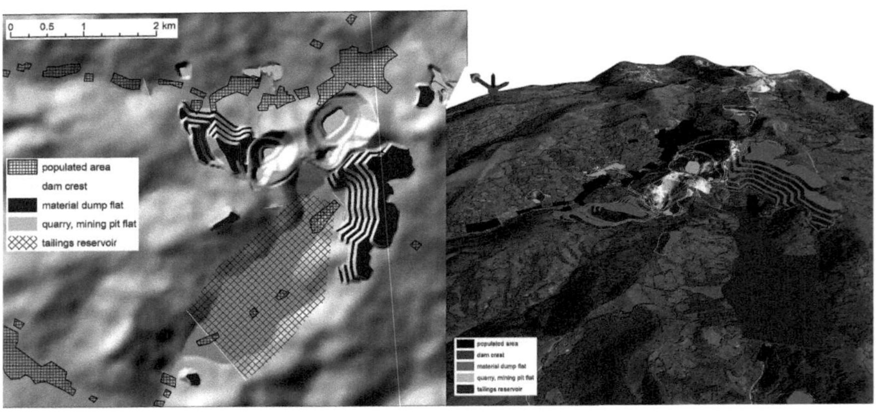

Fig. 7 Modelled features and simulated SPOT image from project year 07

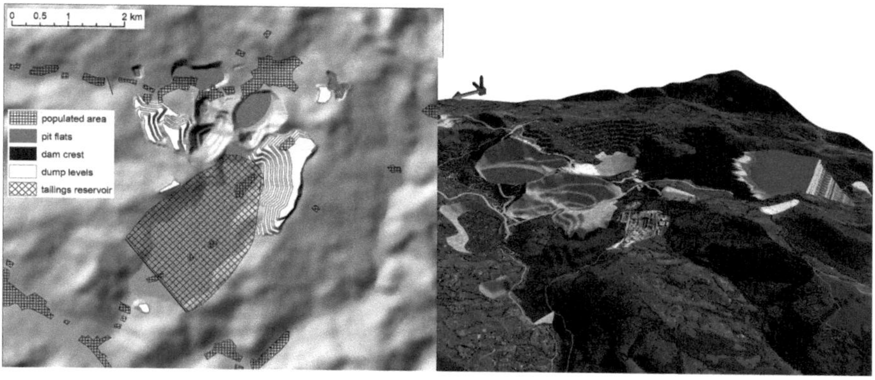

Fig. 8 Modelled features and simulated SPOT image from project year 14

Table 2 Modelled volume and area figures in project year 07 (ESRI 3D Analyst)

Landform	Volume change (×1000 m^3)	Affected area (ha)
Tailings reservoir	+15,150 (since project start)	55

Table 3 Modelled volume and area figures in project year 14 (ESRI 3D Analyst)

Landform	Volume change (×1000 m^3)		Affected area (ha)	
	From year 07	Current volume	From year 07	Current area
Tailings reservoir	+71,448	+139,136	+92	245
Cetate waste dump	0	+26,902	0	62
Cirnic waste dump	+25,885	+65,874	+18	143
Cetate pit	−41,428	−68,826	+21	71
Cirnic pit (lake)	+24,871 (water)	−51,548	0	68
Orlea pit	−24,051	−24,051	+36	36
Jig pit	−6164	−6164	+21	21

5 Conclusion

The present paper demonstrated the use of non-traditional methods and techniques which are required for the modelling and visualisation of landscape and terrain changes caused by future impacts. The main objective was to emphasise the significance and effectiveness of the combination of different data sources and modelling principles when dealing with changes that cannot be captured with remote sensing DTM techniques. The mixed application of vector and raster models offers adequate detail while maintaining the overall seamless visual integration of the resulting surfaces. It also shows the application for the time-series of surfaces in quantifying the volumes and areas of moved materials.

Another result is the procedure for modifying imagery data in order to simulate the landscape change. Converting the GIS images into graphics software compatible format while retaining the geographic information using world files allows the performance of any kind of image editing in the graphics application and easy re-integration into the GIS project for further visualisation stages. Collecting and applying the appropriate textures on modified areas can be achieved with standard graphics tools included in Open Source graphics applications as well.

Acknowledgments The research in this paper was supported by Balassi Institute—Hungarian Scholarship Board Office.

References

ANCPI (2014) http://geoportal.ancpi.ro/arcgis/rest/services/HY/HY_View/MapServer. Accessed 29 Dec 2014

Chavez P, Sides SC, Anderson JA (1991) Comparison of three different methods to merge multiresolution and multispectral data—Landsat TM and SPOT panchromatic. Photogram Eng Remote Sens 57(3):295–303

Géczi R, Bódis K (2003) Környezeti monitoring Verespatak környékén (Environmental monitoring in the region of Roşia Montana). Kriterion, Cluj Napoca, p 72. ISBN 973-26-0734-3, 973-82-3124-8

INS (2011) http://www.recensamantromania.ro/wp-content/uploads/2013/07/sR_Tab_8.xls. Accessed 2 Dec 2014

RMGC (2014) http://en.rmgc.ro/rosia-montana-project/environment/environmental-impact-assessment.html. Accessed 7 Dec 2014

Rodriguez E, Morris CS, Belz JE, Chapin EC, Martin JM, Daffer W, Hensley S (2005) An assessment of the SRTM topographic products. Technical report JPL D-31639, Jet Propulsion Laboratory, Pasadena, California, 143 pp

Morphological Analyses of the Central Slovakia on Base of the DEM

Dalibor Bartoněk, Jiří Bureš, Lubomil Pospíšil and Otakar Švábenský

Abstract Satellite images revealed an extensive deformation on the southern slopes of the Nízké Tatry Mts. Range. This deformation can be ascribed to group of gravity nappes. Geologic survey, however, neither confirmed nor negated its existence. Therefore a detailed analysis of DEM and aerial images was carried out on the scale 1:35,000. Analytical interpretation led to the construction of a special morphotectonic and geodynamic scheme of the area, supported by the various geophysical data. The purpose of contemporary studies on the southern and south-western slopes of Nízke Tatry Mts. and marginal parts of the Velká Fatra Mts. were to verify the important tectonic boundaries and deformation structures interpreted on satellite images by the geophysical, geomorphological and GPS data.

Keywords DEM · Geologic survey · Aerial images · Morphotectonic and geodynamic scheme · Analytical interpretation

D. Bartoněk (✉) · J. Bureš · L. Pospíšil · O. Švábenský
Institute of Geodesy, Faculty of Civil Engineering, Brno University of Technology, Veveří 331/95, 602 00 Brno, Czech Republic
e-mail: bartonek.d@fce.vutbr.cz

J. Bureš
e-mail: bures.j@fce.vutbr.cz

L. Pospíšil
e-mail: pospisil.l@fce.vutbr.cz

O. Švábenský
e-mail: svabensky.o@fce.vutbr.cz

D. Bartoněk
European Polytechnic Institute, Osvobození 899, 686 04 Kunovice, Czech Republic

K. Růžičková and T. Inspektor (eds.), *Surface Models for Geosciences*, Lecture Notes in Geoinformation and Cartography,
DOI 10.1007/978-3-319-18407-4_2

1 Introduction

The complex analysis showed that during the Tertiary the south-western part of the Nízke Tatry Mts., presumably had the character of a horst whose relative fast uplifts had been caused by active deformations within the fundament (Fusán et al. 1979). Complicated faults or fracture systems of great diversity were found, many of them have strike-slip character. Also the expressive dissection of the relief, the chaotic arrangement of planation surfaces and numerous geodynamic features outline the intricate geologic environment and complicated late Alpine development of the whole structure. Interpretation and detailed geological analysis has been previously carried out by Klinec et al. (1985). Most of the anomalous morphologic phenomena detected has been found on base of DEM, satellite and aerial image interpretation. The reason is the vastness and extent of the structure mentioned, which goes beyond the possibilities of field verification. New GNSS measurements at permanent stations, and results of several epoch campaigns that were uniformly processed and evaluated (Hefty et al. 2010) made it possible, in relation to the existing tectonic plan in region of the Western Carpathians, to create the kinematic model of this area (Pospíšil et al. 2012). Overview of the block division of the Western Carpathians territory revealed that the area concerned maintains uniform character from the point of view of velocity vectors orientation, while in its surroundings the vectors are changing in relation to the single active fault systems, mostly of transcurrent character (Pospíšil et al. 2012). Review of all existing data and their interpretation shows that such structure type is closely related to the distinct Hron tectonic system (Pospíšil et al. 1985; Doktór and Graniczny 1983—Fig. 1).

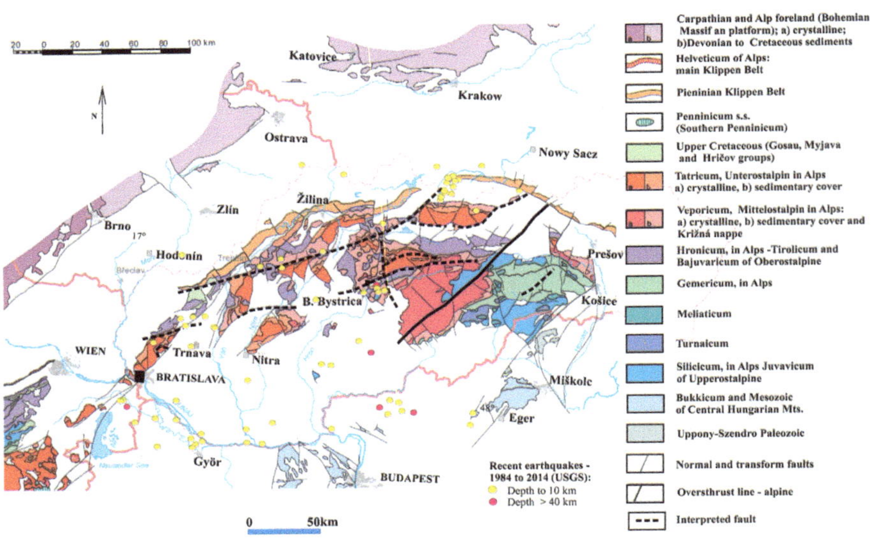

Fig. 1 Simplified geological map of the Western Carpathians supplied by interpreted seismoactive faults and analysed structure of the Nízke Tatry Mts. (NT)

The extent and scale of deformations can be compared with structures of the "foreberg" type, described in detail in Bayasgalan et al. (1999), which appear nearby the reverse faults of the strike slip character type (e.g. the Gurvan Bogd fault system in Mongolia). Unlike the structure mentioned, the interpreted gravitation nappe is not of recent character, but it is an older considerably broken and destroyed structure above an active fault system. The structure was clearly indicated by magnetotelluric (MT) measurement (Varga and Lada 1988), and is continually "resuscitated" by the underlying fault.

2 Methodology

In the interpretation of DEM we proceeded in the following methodological steps:

- evaluation of the tectonics of the area, with regard to orders of faults, fault zones, fracture zones, etc. and to the analysis of the drainage system,
- analysis of the morphological features, reflecting the particularities of the geotectonic development of the area (the relief forms, planation surfaces, etc.),
- distinction of geodynamic processes and other geodynamic phenomena—re-evaluation of the previous and recent findings concerning the geologic-tectonic development,
- evaluation of the latest geophysical and structural—geologic data, correlation of data yielded by the individual research methods applied—compilation of data, construction of maps and cross-sections.

3 Geological Characteristic

The Nízké Tatry Mts. are part of the Central Western Carpathians and belong to the Tatra-Fatra Belt of core mountains, which create Late Tertiary asymmetric horst structures, comprises from bottom to top (Plašienka et al. 1997—Fig. 1):

- the Penninic-Vahic oceanic rock complexes (it is supposed, only);
- the Tatric pre-Alpine crystalline basement and its Late Paleozoic-Mesozoic;
- sedimentary cover;
- the Fatric (Križna) Mesozoic cover nappe system;
- the Hronic (Choč) cover nappe system;
- the Veporicum unit at south;
- the overstepping Late Cretaceous (uncommon) and Tertiary (abundant) post-nappe;
- sedimentary and volcanic cover.

The structure mentioned is located at dividing line of Tatricum and Veporicum. The Tatricum is a crustal sheet present in the Western Carpathians only, but the

Infratatric units may be well correlated with the Lower Austroalpine units (Plašienka et al. 1997); the Veporicum, a large crustal-scale basement wedge best corresponds to the Middle Austroalpine basement units with the typical Cretaceous metamorphic overprint.

4 Active Faults in Tertiary

The basement exceptional activity was noticed in Tertiary, mainly on EW fault systems (Figs. 2 and 3). Primarily it was the Hron fault and its deep tectonic equivalent system. It fits into the system, according to which the southern part of the nappe is subject of progressive stepwise sinking. This principle (sinking) does not involve the part of the nappe that represents piling up of masses in form of bank (Ľubietová anticlinorium Fig. 2).

The Nízke Tatry Mts. fault system had been active during all the Tertiary period. The area of the gravitational nappe is divided by system of primary and secondary faults (Figs. 2 and 3).

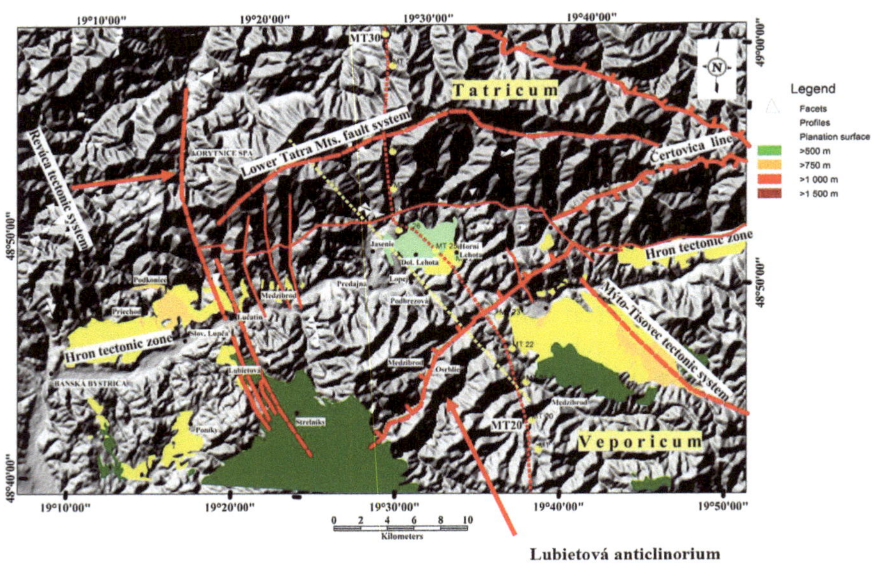

Fig. 2 Explanations to the paleogeographical scheme. (1) the Revúca fault system, (2) the Mýto Tisovec fault system, (3) the Nízke Tatry Mts. fault system, (4) the Hron river fault system, (5) the Čertovica fault system, (6) the Vajsková conglomerates, (7) sediments of motley series (the Polhora development of Paleogene, the Pre-Lutetian age), (8) dolomitic conglomerate facies, (9) sediments of the Podkonice syncline, (10) sandstone—conglomerate facies, (11) claystone lithofacies, (12) regressive facies of the Oligocene (the Upper Hron valley development of Paleogene, Lutetian—Oligocene), (13) gravel (Pliocene)

From overall directing and structural conditioning of relief over the nappe area it is possible to conclude that the movement was in SE direction (roughly perpendicularly to the Čertovica fault system), but there the sliding rock masses collided with NW margin of the Veporic structures. The compress zone, represented in the Čertovica fault system by development of antiform rock composition, is proof of the movement mentioned. Lengthwise the movement was reoriented to S, so the nappe front reached farthest to the south at the Strelník area.

Satellite image of the area of interest indicates possible movement of masses which have in their interior structure style employed in landslides. Its location in bedrock of the fault system mentioned and precondition for development of such phenomena—rapid pulsewise uplift of the area are the bases for analysis of wider area. The analysis is also possible from the point of view of Tertiary sediments development in the area (Fig. 2).

5 Analysis of Morphological Features

Geomorphological research in the area showed the heterogeneous nature of the relief leading to expressive difference in elevations of levelled surfaces in mid mountainous and riverine systems (Fig. 3). The morphological analysis realized on the stereoscopic model confirmed this relief destruction into various individualized blocks and strips, in their heights extraordinarily differentiated. As it is indicated by contemporary investigations are these results of various, not yet strictly

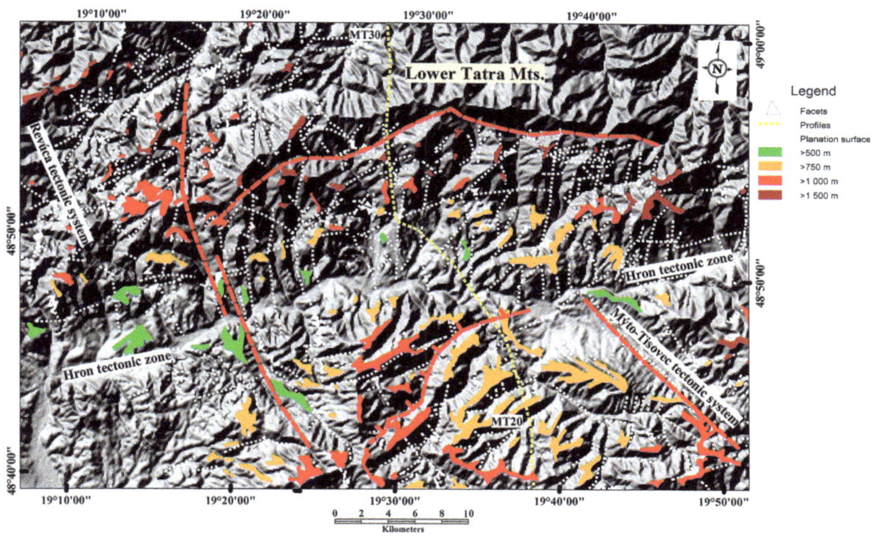

Fig. 3 Map of interpretated linear boundaries and planation surfaces interpreted from DEM and air photos

distinguished movements and of the selective exogenetic destruction of rocks. Neither the extensive gravity movement on the southern slopes of the Nízké Tatry Mts. can be excluded. This assumption is supported by the height differentiation of the mid-mountainous levelled surfaces (in absolute values from 850 to 1400 m a.s.l.) and of the riverine terrace system (100–300 m relatively above the Hron River valley —Fig. 3). Due to the extreme morphological conditions it was possible only to range following groups of levelled surfaces: (1) less than 700 m a.s.l., (2) 700–1000 m a.s. l., (3) 1000–1300 m a.s.l., (4) More than 1300 m a.s.l.

6 Analysis of Geodynamic Processes

Important secondary factors demonstrating the dynamics of the geologic environment, and of all exogenetic processes are geodynamic phenomena.

The following were in the focus of the interest (Fig. 4): (1) Slope deformations (creep, slide and fall gravity deformations, taluses, block disintegration and block fields), (2) Erosion elements (gully and sheet erosion, erosion edges), (3) Surface karst phenomena (sinkholes, areal and linear surface karst elements, (4) Forming of

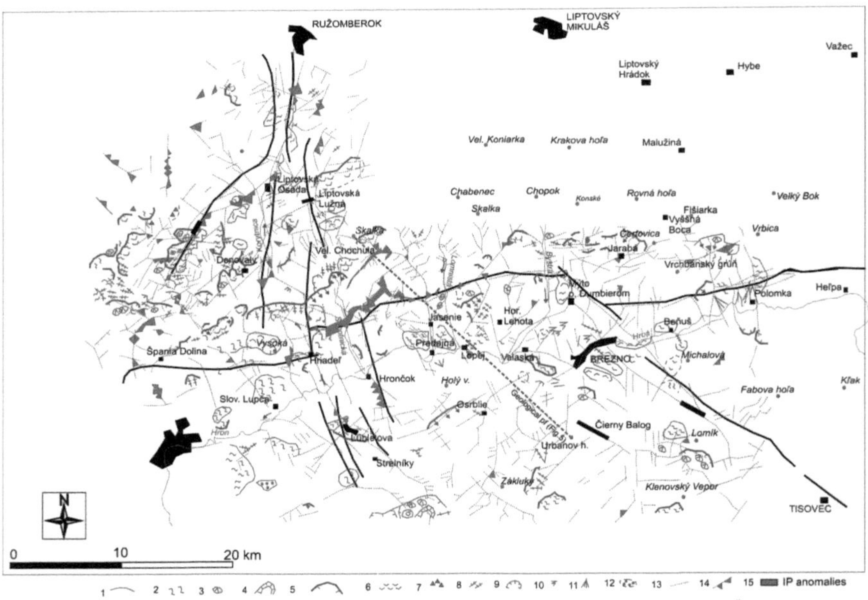

Fig. 4 Analysis of geodynamical phenomena. *Legend 1* slope deformations general, *2* talus creep, *3* block rifts, block fields, *4* individualized blocks, *5* scarps, *6* landslides, *7* alluvial fan, *8* gully erosion, *9* erosion edge, *10* sheet erosion, *11* proluvial cones, *12* sinkholes, karst valleys, *13* linear boundaries (after remote sensing), *14* facets, *15* zone with height IP (induced polarization) anomalies

proluvia (young alluvial cones), (5) Weathering. Slope deformations. The relatively new results of tectonics, engineering-geologic and geophysical investigations show that numerous structure at the Earth's surface so far related to the dynamics and mechanism of tectonic (plicative and disjunctive) movements are actually, connected with gravitational tectonics. In the Velká Fatra Mts. such deformations are represented on top ridges in the vicinity of elevation points Zvolen Mt. (1402 m a.s.l.), Chabenec Mt. (1955 m a.s.l.), Krížna Mt. (1574 m a.s.l.), etc. (Nemčok 1982).

Along the mountain ridges they form a linear succession and may represent a boundary of the active root area of an extensive gravity deformation. The slides could be well identified on Velká Fatra Mts., slopes (Turecká valley), on the Nízké Tatry Mts. slopes (S of the village Dolná Lehota, Prašivá Mt., Kozí chrbát Mt. etc.), but namely on the slopes of intra-mountainous basins, e.g. near Lubietová village (Fig. 4). It cannot be excluded their direct connection with the significant tectonic Revúca zone.

Erosion phenomena. Two types of erosion can be encountered in the area of interest: gully erosion and sheet erosion. Due to a large uplift of the mountain, an intensive down cutting of water flows took place and created a variable erosion relief with various types of the water system. At the present time the size of the erosion down cutting and its extension depend on the gradual disintegration of the rocks and on climatic changes. In places of tectonic or lithological weakening mainly on the steep slopes a considerable progress of back erosion can be observed (Fig. 4). Scours, deep ravines and gorges, and also flat closures of mountain valleys in the top parts of the ridges are formed. In the greater part of the area, back erosion has modelled the top parts into sharply-cut shapes and thus has disintegrated ancient structural planes.

Karst phenomena. Karstification is scarce in the area of interest due to the relatively rare occurrence of soluble karst carbonate rocks. Its surface elements can be encountered only in the south-western part near the village Poníky and Nemecká where the sinkholes form small groups of greater or smaller depressions. Linear arrangement indication occurrences and systems of tectonic lines cannot be observed (Fig. 4).

Other phenomena. Among the best identified phenomena belong proluvial cones. On images they are clearly visible by the tributaries of the Hron River.

The most product of weathering in the diversified relief, remodelled by erosion and denudation activity were transported away.

7 Correlation of Morphological, Regional Geological and Geophysical Data

The position of fault zones, fault lines and other structural features by means of remote sensing, regional geologic and geophysical data in the study area are in a good agreement. First of all the stereoscopic interpretation of various images produces much more continuous picture and complex results) than geologic or other maps on different scales. Evidently, the interpretation criteria make possible an identification of faults also in places where they have not been confirmed by field mapping.

It was impossible to find out on the satellite images which linear elements belong to older evolution stages and which to the Late Alpine development stage. According to the development and diversity of the geomorphological features it can be assumed that, the majority of fault and structural-tectonic features are from the latest evolution stage or are results of several stages. For instance, the linear Hron River valley is presumably based on one ancient fault in partial segments documented by linear, steep and sporadically facetted slopes. The facets in valleys were taken as a criterion for interpretation of faults (e.g. numerous facets along the northern part of the Korytnica River valley indicating the Revúca fault line—Fig. 4).

The fault and fracture zones are best characterized by the configuration of the drainage system. It can be assumed that the river follows the lines or zones of tectonic or lithological "weakening". In this way the principal directions of prevailingly linear tectonic zones, dislocations and shorter fault-fracture elements were distinguished (Figs. 2, 3 and 4):

1. NE-SW, ENE-WSW systems (western branch of the Revúca fault zone, the Hron River valley and the system of lines in the east of the area). In this direction runs the Hron lineament (Pospíšil et al. 1985) interpreted as a transcurrent fault with a large dextral strike-slip.
2. NW-SE, NNW-SSE (the eastern branch of the Revúca fault zone and transverse faults between Lučatín and Brusno, in the east the Rohozná and Čierny Hron River fault).
3. N-S faults forming important valleys in this direction on southern slopes of the Nízké Tatry Mts. as well as the Northern part of Revúca fault system.

All the tectono-structural features are indicated by the forms of valleys abrupt changes of drainage system, spring lines, erosional features, morphological structures, etc. (Fig. 4).

All the identified faults are mostly wide zones of the highest inhomogeneity rate in terms of secondary lithological changes, fragmentation or blockiness. They are all very expressive in the morphology. In this connection of greatest significance is the confirmation of extraordinarily manifested tectonic zone, corresponding partly to the Revúca fault system. First of all the interpreted branching into the two separate systems was of significance. One of them N-S or NNE-SSW system progresses along the Revúca River valley and indicates the continuation of important regional tectonic boundary—so-called Povážany-Ihráč fault system. The second branch in direction NNW-SSW could mark the indication of the west border of the Nízké Tatry Mts. gravity deformation. The analysis of satellite, aerial and radar images confirmed a great relief diversity, frequent changes in its morphological forms, development of gravity elements on slopes, etc. within the entire Revúca system.

The vertical density boundary was based on the gravity data which demonstrate the diversity and frequent tectonic delimitation of structural blocks (Fig. 3).

The interpreted tectonic line often coincide with density boundaries or with places of disrupted correlation. The loss of correlation can be observed along the entire length of the Revúca fault zone, Hron lineament etc. Geophysical data focused our interest to the disruption of the NW-SE branch of the Revúca system. This not exactly limited wide zone is de-formed there and along about 5 km it is bending to the WNW-ESE direction. It may be the crossing point of several tectonic faults with variable orientations.

The tectonic predisposition of the Hron River between Banská Bystrica and Brezno where two distinct NW-SE and NE-SW striking faults have been interpreted, can be supported by an intensive gravity gradient. Many of interpreted structural and tectonic boundaries, e.g. Mýto-Tisovec fault zone or Čertovica line have been confirmed by geo-electric, magnetometric and gravity data.

The broader surroundings of the south slopes in the middle-west part of the mountain has been the focus of investigations for are prospecting. Geophysical measurements and aerial photographs yielded a detail picture of the tectonics of the area. Mapping of the so-called "productive zone" showed equal values of resistivity and polarization to those of the immediate surroundings of the existing deposit (Vybíral et al. 1986). The zone of the anomalous induced polarization values strikes from NE to SW and it is evidently bounded by fault zones. Tectonic lines interpreted on the images disturb this zone in form of shifts and disruptions so that it proves the particular interest from the viewpoint of tectonics, geological structure and mineralization in this area.

Obtained geophysical data from various methodological stages cannot be fully illustrated in the simplified geologic cross-sections, but only in the schematic section (Fig. 5). It is because of an in clarity in the geologic interpretation of the Nízké Tatry Mts. structure itself.

8 Discussion

Interpretation of multispectral space image LANDSAT and aerial B/W images combined with DEM was based on evaluation of the relief and its main elements—hollows and ridges. On base of their structuring the tectonic predispositions were considered. It is therefore possible to state that the evaluation of relief texture in the image of interest It can therefore be concluded that the appreciation of the relief texture to the image significantly contributed to support the geologic and geophysical indices, and also to analysis of the paleo-horst and its supposed gravitational deformation (structurally conditioned relief in Veporicum area, i.e. in supposed accumulation part of the gravitational nappe, cannot be decidedly taken as proof of its existence).

After Tertiary period the SE part of the Nízke Tatry Mts. territory had character of a horst. Its rapid pulsewise movements had been conditioned by bedrock active faults that formed its exterior limitations. From W-SE side it was the Čertovica fault system in section of the Čertovica saddle—Ľubietová crystalline—Strelníky.

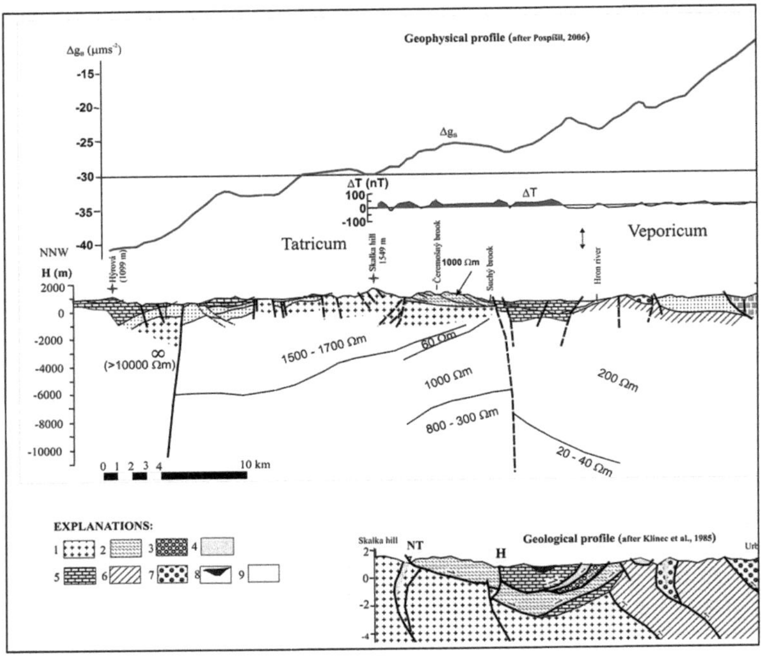

Fig. 5 Geological (*bottom*—Klinec et al. 1985) and geophysical profiles located in area where is interpreted "gravity nappe". *Area of Tatrides 1* granitoids, *2* crystalline schist, *3* sandstones, conglomerates, rhyolites (Permian), *4* quartzites and carbonates of lower tectonic unit, *5* carbonates and melaphyres of the upper tectonic unit. *Area of Veporides 6* mica schists, amphibolites, *7* gneisses, migmatites, *8* the Vajsková conglomerates, *9a* basement of the gravity nappe, *9b* dislocations of higher order, *9c* dislocations of lower order, *NT* the Nízke Tatry Mts. fault system, *H* the Hron river fault system, *Č* the Čertovica line (thrust of the Veporides on the Tatrides). The geophysical profile (Pospíšil 2006) clearly demonstrate different resistivity image of the contact area between Tatrides and the Veporides

From W-SW side it was the Revuca fault system in section Korytnica-kúpele over Medzibrod to Strelníky, and from N the section of Nízke Tatry Mts. fault (E-W) from Sopotnice springs to Čertovica.

If the concept of gravitational nappe is accepted, then it must be of pre-Pannonian era. Reason for such opinion is the existence of considerably reshaped planation levels of pre-Pannonian age (Pospíšil et al. 2006). Indices for this are flat parts of the Nízke Tatry Mts. crevasses of elevations 1200–1400 m (Vanko and Wyrzykowski 1981). In case of the Pannonian gravitational nappe of supposed extent the planation surfaces would be completely destructed.

It is necessary to emphasize that it is altogether impossible to explain the identified gravitational nappe structure in complete space-time development only on base of remote sensing. It is possible to point at some influences which had to exist during formation of such structure. Paleogeographical analysis of the development in Tertiary showed that until Oligocene the whole area had tendency to rise against its surroundings. That movement, even if pulsewise, should have enormously rapid course in final phase. The horizontal motions of Carpathian blocks along the ENE-WSW oriented faults could play the deciding role, how it was pointed out before on base of the satellite image analysis and the detection of linear divides of this type by (Janků et al. 1984—Fig. 6).

One of these divides (the Hron tectonic zone) runs in direction of line Banská Bystrica–Brezno towns and reaches out as far as the area concerned. Early Miocene age of these movements corresponds to dynamic conditions which existed at that time. The gravitational nappe rapid uplift and tearing away may be attributed to the collision of two (or more) blocks moving in different directions with different velocities (Figs. 2 and 6). It cannot be excluded that also other linear divides (clearly observable in satellite images and in illuminated relief)—the Muráň fault, along with its northern parallel—the Divín fault (the Muráň-Malcov system—Pospíšil et al. 1985), controlled similar movements in vicinity of more to the south (Veporic) blocks.

Contributing in some way to understanding and explanation of this mechanism are also some geophysical findings. In the gravity map (Ibrmajer 1964; Pospíšil et al. 1985; Pospíšilová et al. 2012) the both linear ENE-WSW oriented divides

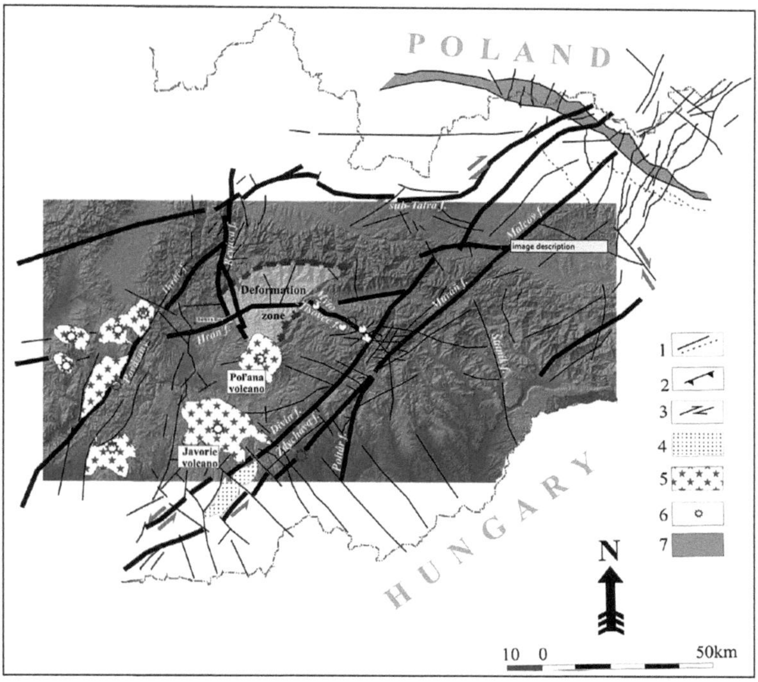

Fig. 6 Regional relationship among the main tectonic zones surrounding of the Nízke Tatry Mts. *Explanations 1* known and supposed faults, *2* thrust, *3* strike slip character, *4* graben structure, *5* Pannonian volcanic, *6* central volcanic zone, *7* Klippen Belt

appear as border lines separating the individual blocks, that are arranged asymmetrically in relation to these lines, but are maintaining constant mutual distance.

In space between the divides mentioned there are preserved the sediments of the Central Carpathians paleogene, partly also of Early Miocene (Fusán et al. 1979; Nemčok 1978—Figs. 1 and 2). If the fault systems (e.g. in areas of intra-mountain hollows) and the timing of sedimentation cycles are accepted, it is possible to assume that the mentioned system of divides plays the role also in opening, resp. in founding of the molasse basins.

It is reasonable that during the horizontal movements there were initiated the extended transversal crevices, resp. faults, which could be of a deep reach. The Revúca fault and the Mýto-Tisovec fault can be considered as such fault types, which controlled not only the formation and extending of the gravitational nappe, but also the ascending of the neogene intermediate magma to the surface.

To proving of the gravitational nappe existence in the area concerned, but also in other areas of Carpathians where the anomalies in stratigraphic sequencing and the zonality destructions are discovered, must be paid increased attention because part of these anomalies could originate from the same genesis. Contemporary means for tracing the subtle Earth surface deformations with help of GNSS monitoring could particularize some lack of clarities of relation to surrounding blocks. Considering the increased seismic hazard in the area it could be utilized for risk studies in the territory concerned.

Because the interior arrangement of the gravitational nappe structure shows the landslide body features, it is possible to apply this distinction also by exploration of accumulations of utility minerals. Inside the gravitational nappe body the W-Au ore is contained, which in the past times was corroborated by exploratory works. Their spacing is, with regard to their character (mobility of elements), originally deformed in relation to the environment, and they have to adapt to the plan of the developing deformation. In accordance with this is the contemporary asymmetric distribution of ores without distinctive zonality in space (section of the gravitational nappe). It is notable that most of Pb, Cu, Sb, W, Au ores in region of the Nízke Tatry Mts. is localised within the gravitational nappe area.

9 Conclusions

The purpose of contemporary studies on the southern and south-western slopes of Nízke Tatry Mts. and marginal parts of the Velká Fatra Mts. were to verify the important tectonic boundaries and deformation structures interpreted on satellite images by the geophysical, geomorphological and GPS data.

This analysis showed their great heterogeneity rate in the area of interest. The relief diversity, existing deep founded and at the same time hidden transcurrent tectonic zones offer clear indications about presence of a seismo-tectonic active risk zone. Supposition about existence of a gravitation nappe are also offered by the secondary influences accompanying alike structures (relicts sediment, exogene dynamics elements, distribution of isolated neovolcanic bodies etc.). The interpreted gravitation nappe structure very closely resembles structures accompanied by numerous thrust-related folds, thrust faults, and elongated low ridges ('forebergs'), sub-parallel to the mountain range, all of which result from the shortening component on the fault zone (Bayasgalann et al. 1999).

Evidence from geomorphology and surface ruptures suggests that they may all serve a common function, which is to broaden the deforming zone by creating new structures that are able to accommodate both the strike-slip and the shortening components of motion.

The migration of faulting away from the main range is likely to be driven by stresses associated with topography, which in turn is a consequence of the shortening component.

Development of such structure in Nízke Tatry Mts. region may be ascribed to the present Hron strike-slip fault system, which has reverse component and which can produce more elements of the "flower structures" that are often described in such oblique-shortening zones (Pospíšil et al. 1989). Particularly yet neglected results of MT measurements along the 2T profile (Varga and Lada 1988) provide conclusive information about significant deformation in the area. While the results of seismic measurements along the same profile are ambiguous (Tomek et al. 1989), the unambiguously different resistance distribution in the area north of the Hron tectonic zone allows to explain also the other yet neglected interpretations—presence of light granitoid masses (with "infinite resistances") under the Nízke Tatry Mts. mountain range (Pospíšil and Filo 1980). This could explain the cause of uplift and instability of the mountain range during Tertiary and up to the recent. The presented results and interpretations show that even at contemporary level of exploration and research of the Western Carpathians it is possible to encounter the essential problems of regional dynamics, which from the recent point of view can lead to some risks in connection with subsequent unexpected earthquake occurrences. From that reason it would be suitable to supplement such areas by a number of additional permanent GNSS stations which could contribute to monitoring of eventual changes and movements of single blocks (Gašinec et al. 2012; Staňková and Černota 2013).

Acknowledgments The research was supported by the Grants Nos. BD 12300008 and BD 12400012 of the Brno University of Technology.

References

Bayasgalan A, Jackson J, Ritz J-F, Carretier S (1999) Forebergs', flower structures, and the development of large intra-continental strike-slip faults: the Gurvan Bogd fault system in Mongolia. J Struct Geol 21(10):1285–1302

Doktór S, Graniczny M (1983) Fotogeologiczna analiza zdjec satelitarnych Karpat. Kwart Geol 27:645–656

Fusán O, Ibrmajer J, Plančár J (1979) Neotectonics blocks' of the West Carpathians. Geodynamics investigations in Czechoslovakia. Final report, Publ. House of SAS Veda, Bratislava, pp 187–192

Gašinec J, Gašincová S, Černota P, Staňková H (2012) Uses of terrestrial laser scanning in monitoring of ground ice within Dobšinska Ice Cave [Zastosowanie naziemnego skaningu laserowego do monitorowania lodu gruntowego w Dobszyńskiej Jaskini Lodowej]. Inzynieria Mineralna 30(2):31–42. ISSN 1640-4920

Hefty J, Hipmanová L, Gerhátová Ľ, Igondová M, Droščák B (2010) Recent geo-kinematics of Slovakia based on homogenized solutions of permanent and epoch GPS network. Acta Geodyn Geomater 7, No. 3(159):1–13

Ibrmajer J (1964) Gravity map of Czechoslovakia, scale 1:200,000. MS Geofyzika Brno (in Czech)

Janků J, Pospíšil L, Vass D (1984) The contribution of the remote sensing to the solution of the geological structure of the West Carpathians. Mineralla Slovaca 16(2):121–137 (in Slovak)

Klinec A, Pospíšil L, Pulec M, Feranec J, Stankovianský M (1985) Identification of a gravity Nappe in the Low Tatras by satellite images. Mineralia Slovaca 17(6):485–499 (in Slovak)

Nemčok J (1978) Deformations of the flysch sediments as a impact of the basement dynamics. Západ Karpaty Ser Geol 35–58 (in Slovak)

Nemčok A (1982) Landslides in the Slovak Carpathians. VEDA, Bratislava, 127 pp (in Slovak)

Plašienka D, Grecula P, Putiš M, Kováč M, Hovorka D (1997) Evolution and structure of the Western Carpathians: an overview. In: Grecula P, Hovorka D, Putiš M (eds) Geological evolution of the Eastern Carpathians. Mineralia Slovaca, Bratislava, pp 1–24

Pospíšil L, Filo M (1980) The West Carpathian central gravity minimum and its interpretation. Mineralia Slovaca 12:149–164

Pospíšil L, Nemčok J, Graniczny M, Doktór S (1985) Contribution of the Remote Sensing to the identification of the strike-slip faults in the West Carpathians. Mineralia Slovaca 18(5): 385–402 (in Slovak)

Pospíšil L, Bezák V, Nemčok J, Feranec J, Vass D, Obernauer D (1989) Muráň tectonic system as example of horizontal displacement in the West Carpathians. Mineralia Slovaca 21(4):305–322 (in Slovak)

Pospíšil L, Ádám A, Bimka A, Bodlak J, Bodoky T, Dövényi P, Granser H, Hegedü E, Joó I, Kendzera A, Lenkey L, Nemčok M, Posgay K, Pylypyshyn B, Sedlák J, Stanley WD, Starodub G, Szalaiová V, Šály B, Šutora A, Várga G, Zsíros T (2006) Crustal and lithospheric structure of the Carpathian—Pannonian region—a geophysical perspective: regional geophysical data on the Carpathian—Pannonian lithosphere. In: Golonka J, Picha FJ (eds) The Carpathians and their foreland: geology and hydrocarbon resources: AAPG Memoir, vol 84, pp 651–697

Pospíšil L, Hefty J, Hipmanová L (2012) Risk and geodynamically active areas of the carpathian lithosphere on the base of geodetical and geophysical data. Acta Geod Geophys Hung 47 (3):287–309. doi:10.1556/AGeod.47.2012.3.2

Pospíšilová L, Pospíšil J, Staňková H (2012) Micro-network creation in industrial surveying. Geodesy Cartography 38(2):70–74 [Taylor & Francis, Co-Published with Vilnius Gediminas Technical University2029-6991 (Print), 2029-7009 (Online)]. doi:10.3846/20296991.2012.692216

Staňková H; Černota P (2013) Application of Ott's mine coordinate system in the past and in present day. Geodesy Cartography 39(2):53–58. doi:10.3846/20296991.2013.806696

Tomek Č, Ibrmajer I, Koráb T, Biely A, Dvořáková L, Lexa J, Zbořil A (1989) Crustal structures of the West Carpathians on deep reflection seismic line 2T. Mineralia Slovaca 21(1):3–26

Vanko S, Wyrzykowski T (1981) Recent vertical crustal movements of the Carpatho—Balkan region. Tectonophysics 71:41–52

Varga C, Lada E (1988) Magnetotelluric measurements on profile. 2T Manuscript Geofyzika s.p., Brno, ELGl, Budapest, 56 pp

Vybíral V, Valušiaková A, Kolár J (1986) Geophysical measurement in Jasenie mine. MS Geofyzika s.p., Brno, Branch Bratislava, 47 pp

Surface Classification Above Gas Pipeline Facilities

Dalibor Bartoněk, Irena Opatřilová and Jiří Bureš

Abstract The article discusses a surface classification above the stored gas facilities in the territory under the administration of the RWE Company in the Czech Republic (CR). The aim of the project was to determine reproductive values of these facilities and a reasonable estimate of the costs that would be spent on a construction of new pipelines. The classification uses a raster dataset of orthophoto with the resolution of 25 cm/1 pixel, a vector dataset of the Fundamental Base of Geographic Data (ZABAGED) and vector datasets of the route of underground utilities. As part of the assignment, it was necessary to classify a surface from two perspectives: from the object standpoint (the surface above gas facilities) and the spatial standpoint (administrative division of the territory of the CR). The authors have succeeded in creating a technological line (workflow) with a high degree of automation. The core of this workflow is a geographic information system (GIS). The technology has been implemented within a special application developed in Python language with a support of the ESRI libraries and ArcGIS 10.0. It is massive task given the territorial coverage of almost the whole of the CR with the area of 64,350 km2 with a data volume of the order of 500 GB. The results demonstrated the effectiveness of proposed procedure to achieve low error rate in the range of 2–3 % across the modeled area.

Keywords Classification · Gas pipeline · GIS

D. Bartoněk (✉) · I. Opatřilová · J. Bureš
Institute of Geodesy Faculty of Civil Engineering, Brno University of Technology, Veveří 331/95, 602 00 Brno, Czech Republic
e-mail: bartonek.d@fce.vutbr.cz

I. Opatřilová
e-mail: opatrilova.i@fce.vutbr.cz

J. Bureš
e-mail: bures.j@fce.vutbr.cz

D. Bartoněk
European Polytechnic Institute, Osvobození 899, 686 04 Kunovice, Czech Republic

K. Růžičková and T. Inspektor (eds.), *Surface Models for Geosciences*,
Lecture Notes in Geoinformation and Cartography,
DOI 10.1007/978-3-319-18407-4_3

1 Introduction

The aim of the project was:

(a) The surface analysis of the RWE gas line in the Czech Republic (CR), i.e. processing of classification data of storage of gas facilities below certain types of terrain surfaces. This analysis was done in order to determine reproductive values of gas facilities (pipelines) and a valuation of costs which would be necessary to spend for building new networks.
(b) The surface analysis in three variations for the route of gas line: a high-pressure gas line, a main line and service pipes. The requirement was in this subject-division to classify a surface in three separate groups:

- 2 groups of a surface type: paved and unpaved
- 5 groups of a surface type: asphalt, main road, local road, unbound, unknown
- 10 groups of a surface type: forest, grassland, bare soil, asphalt, roof-tile, roof-straight, shadow, main road, local road, path.

(c) The output was realized in a graphic form (*.shp format) and in a text form as a structured table in XLSx format with classified data of surface types above the route of gas line.
(d) The extent of the analysis contained 12 regions from standpoint of the administration division in the CR. There are these regions: Karlovarský, Plzeňský, Středočeský, Ústecký, Liberecký, Královéhradecký, Pardubický, Vysočina, Moravskoslezský, Olomoucký, Zlínský and Jihomoravský. So it was processed total 188 municipalities with extended state administration (ORP). The outputs of classification were required in a graphic and in a text form to a territorial detail of the part of municipality within a territorial identification register of the CR.

The solution was based on a pilot project and a developed technology for this purpose called "Classification of data about storage of utilities and facilities below certain types of terrain surfaces" (Bureš et al. 2013).

2 Related Works

Works that deal with the results of image classification can be divided according to the application domain and according to the used methods. The field of remote sensing is represented most frequently in terms of application (Pupin et al. 2014; da Silva et al. 2014; Debes et al. 2014; Iiames et al. 2013; Puertas et al. 2013; Mouelhi et al. 2013; Landgrebe 2003; Liang 2004). Another fields of image classification is used: medicine and biology (Gray and Song 2013), food industry (food quality testing) (Johnson 2013), agriculture (forecast of crop in the given field), ecology

(Makarau et al. 2013) and geodesy (Zhong et al. 2012) (classification of orthophoto).

There are a number of methods for improvement of the results of image classification and these methods are the most common:

- works dealing with the basic usage of image classification (Landgrebe 2003; Liang 2004),
- classical methods of image enhancement (filtering, transformation of brightness, discrete convolution) (Pupin et al. 2014),
- the use method of Support vector machine (SVM) (Iiames et al. 2013),
- a combination of different methods of image capturing of the Earth's surface (da Silva et al. 2014; Makarau et al. 2013),
- processing of time sequence of images (Debes et al. 2014),
- the use of graph theory (factors of graph) (Mouelhi et al. 2013),
- application of correlation analysis with a matrix of errors (Puertas et al. 2013),
- the use of metric spaces (Searcóid 2006) for evaluation of the results of image classification,
- the use of image classification in geodesy and surveying (Černota et al. 2011; Dandoš et al. 2013).

3 Theoretical Foundations

Let a non-empty set of $U \neq \varnothing$ is the universe of discourse and the X is a subset ($X \subseteq U$). The U set represents in our case the entire solved area and the X subset is part of the solved area. The elements of the X set are geographic objects of given locality. Then, let a system of relation of equivalence exists $\{R_i\}$ (i = 1, 2, …, m), where each R_i relation divides the X set into n subsets $X/R_i = \{X_{i1}, X_{i2}, \ldots, X_{in}\}$ such that for every i, j holds:

$$X_{ij} \subseteq U, X_{ij} \neq \emptyset, \mathit{for}\ (i = 1, 2, \ldots, m), \\ (j = 1, 2, \ldots, n) - - - (\text{all subsets are non-empty}) \tag{1}$$

$$X_{ij1} \cap X_{ij2} \neq \emptyset,\ \mathit{for}\ (i = 1, 2, \ldots, m), (j_1, j_2 = 1, 2, \ldots, n), \\ j_1 \neq j_2 - - - (\text{intersection of all subsets for each decomposition,} \\ \text{defined by the } R_i \text{ displays, is empty}) \tag{2}$$

$$\bigcup\nolimits_{i=1,j=1}^{i=m,j=n} X_{ij} = X(\text{union of all subsets within each decomposition,} \\ \text{which is defined by } R_i \text{ displays, is just the whole } U \text{ set}) \tag{3}$$

The R_1 relation of equivalence can be considered as territorial division of the CR according to an administrative arrangement that divides this area into lower administrative units (regions, districts, etc.). If the conditions in (1)–(3) are complied, each sub-area in terms of set theory has a character of the X_{ij} class.

Suppose there is another relation R_2 such that defines decomposition according to the subject (thematic) principle, i.e. according to geographical objects, which are elements of the X set. In our case, geographical objects are gas facilities stored in the ground.

Then the intersection of relations $S = R1 \cap R2,\ S \subseteq X$ defines the modeled area in which we will conduct spatial analysis. Because this territory is defaultly smaller than the original solved territory, this combination of decomposition leads to optimize the entire process of analysis and time saving of solution.

4 Method of Solution

The solution was carried out in the following stages:

1. Data preparation—generate data sets for each ORP for the entire territory of the CR (creation of binding database tables and dials, data sorting, repairing of topological errors in input data, building data sets, debugging of generating scripts)
2. Determination of parameters—measuring of the widths of roads on orthophoto for setting buffers of roads, creation of classification keys, data reclassification according to the classification keys
3. Data classification—classification of pipeline sections above various surface types based on image of orthophoto, classification keys and parameterized communication
4. Editing of errors—checking of completeness of image classification, semi-automated editing of errors of image classification, batch editing of changes to database tables
5. Final inspection—batch sorting outputs and files, checking completeness of the outputs and checksums
6. Completion—completion of data sets for a transfer, creation of XLSs file, creation SHP files for regions, a checking the completeness.

The scheme of technology of data analysis in illustrated in Fig. 1.

The data analysis was solved in the ESRI ArcGIS 10 environment. Total 3 powerful computers of the PC type and 5 notebooks were used for processing, which were connected in computer network (Fig. 2). The accessibility to the data was provided only for researchers so copyrights of data were protected. The processing of automated part was carry on computer machines of the PC type with parameters: CPU Intel Core i5 3.8 GHz, 16 GB RAM, NVIDIA GTX650 2 GB, HD of SDD type and VelociRaptor, OS of Microsoft Windows 7 (64 bit). Processing software was ArcGIS 10.0, Python 2.6 (both 32-bit).

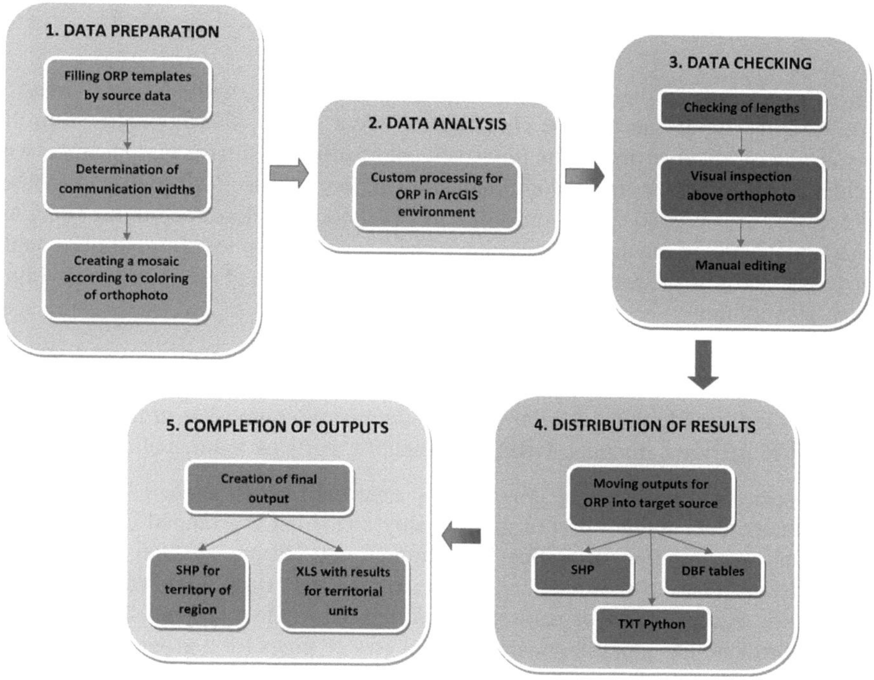

Fig. 1 The scheme of technology of data analysis

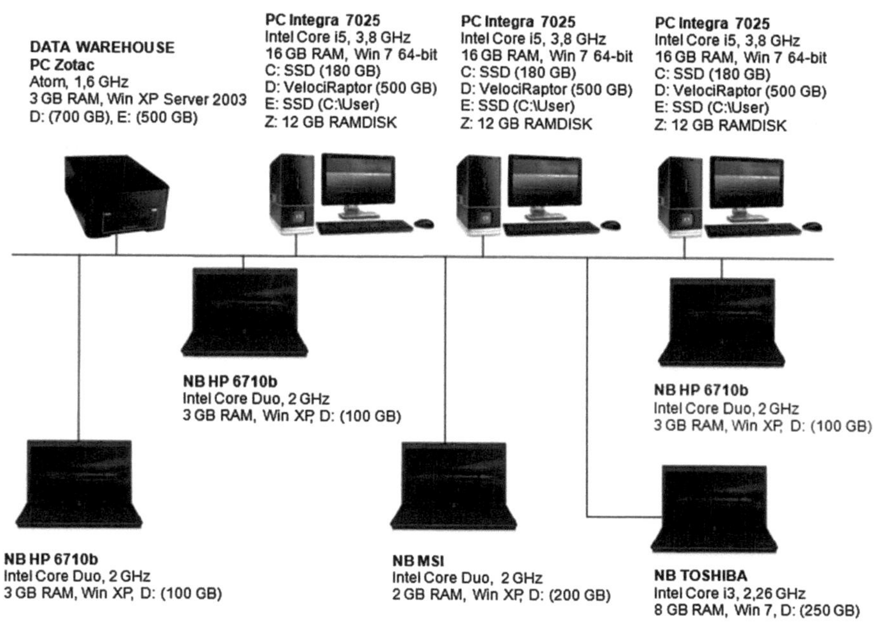

Fig. 2 The scheme of connection and parameters in computer network

5 Experimental Results

The project was implemented on a territory of the CR with the total area of 64,350 km^2. Input data for the classification were sheets of orthophotos with the resolution of 25 cm/1 pixel. The improvement of results of image classification was achieved by filtration from communication layers of the ZABAGED database. Processing was carried out according to the methodology described in the previous chapter. Results of the data analysis have been developed in text form (a pivot table in Excel, see Fig. 3) and graphical form (SHP format, see Fig. 4) in the dividing from standpoint:

1. Object standpoint, i.e. gas lines of types: high-pressure, main line and service pipes
2. Territorial standpoint, i.e. according to administrative division of the territory of the CR (regions, districts, ORP, municipality, parts of municipalities).

The structured table in Fig. 3 is the row structured list of about 13,000 rows (1 row contains data about a part of municipality). This list is arranged in the form of pivot table with options of sorting by common and widespread filter according to categories of a part of municipality, municipality, ORP, district and region (yellow parts in the figure). Surface analysis was carried out in three variants for the course of high-pressure gas pipeline, main lines and service pipes for 2 groups, 5 groups or 10 groups of surface type. The result is cumulative lengths of high pressure gas lines, main lines and service pipes under individual classes of surface for defined

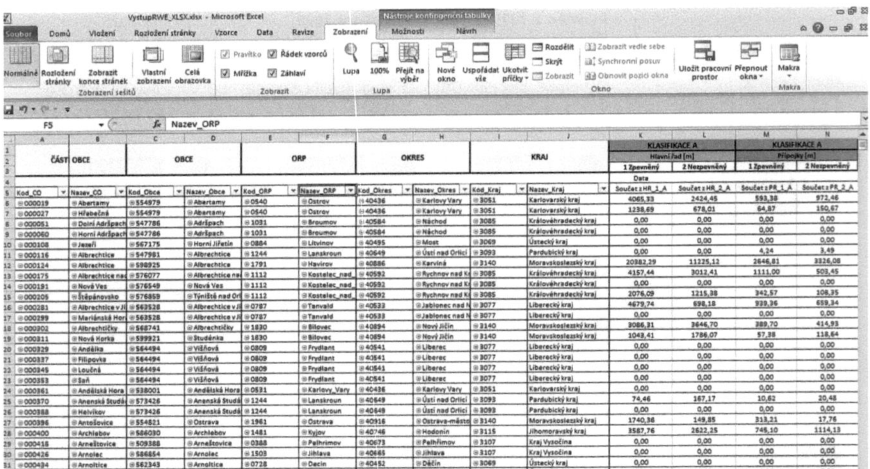

Fig. 3 The example of a pivot table of output in XLSx format

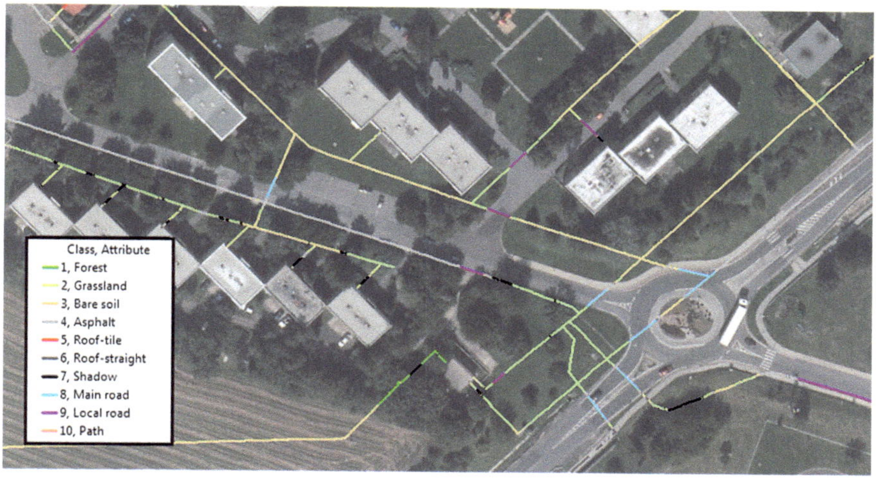

Fig. 4 Graphical output of a structured classified data in SHP format of the route of gas line

selection by a filter. In terms of volume data, it is a very detailed division of extensive territory-bound geo-database. Results include a total of $3 \times (2 + 5 + 10) \times 13{,}000 = 663{,}000$ items.

The graphical output of structured classified data in SHP format for 3 types of gas lines is in Fig. 4. This is a segmented course of gas line with an attribute table for 10 classes, which is georeferenced and viewable over any graphic base (orthophotos, cadastral map, etc.) in GIS of RWE Company.

GIS tools can conduct the function evaluation of the sum of the lengths of types of gas line under individual classes of surface in given categories (according to the definition by alphanumeric attribute) above any area in graphics, which is defined by e.g. fence, closed polygon, etc.

6 Analysis of Error Rate of Processing

The error rate of the F was evaluated as the ratio of the total length erroneously classified sections of the gas line to the total length of the gas line in percentage. The error rate was determined on a sample of 20 % of the territory (40 ORP) using locating and measuring lengths of erroneous sections of the S in a classified image. The orthophoto image was admitted as a reference fact. The error rate was assessed by the human factor by the following equation:

$$F = \frac{A}{C} * 100[\%] \qquad (4)$$

Table 1 Error rate of processing

Error rate	High pressure[%]	Main line [%]	Service pipes [%]
Minimum	0.4	1.4	1.1
Maximum	2.6	5.4	4.6
Average error rate	**0.5**	**1.9**	**1.5**
Standard deviation	0.4	1.4	1.1

Average error rate contains statistical value of errors of surface classification through the whole teritory of the Czech Republic

where
A is the length of incorrectly classified sections of the gas line
C is the total length of the gas line

The overview of the error rate for the high pressure gas line, the main series and service pipes is given in Table 1. The maximum error rate of sample does not exceed 5.4 %. The largest average error rate shows the main line.

7 Conclusions

These following conclusions are from our solution:

(a) The technology enables with high information ability (error rate less than 5.5 %) and relatively rapidly (in the period to 16 weeks) processing data analysis of lengths of gas line sections that lie under the surface of various types:
- 2 groups of a surface type: paved and unpaved
- 5 groups of a surface type: asphalt, main road, local road, unbound, unknown
- 10 groups of a surface type: forest, grassland, bare soil, asphalt, roof-tile, roof-straight, shadow, main road, local road, path.

(b) The results of data analysis for 2 groups (paved and unpaved surface) have the largest information ability.
(c) The technology of data analysis enables repeatability practically without influence of the human factor.
(d) Data analysis shows very high information ability in terms a low error rate and it can be used repeatedly. The error rate is due mainly a limit quality of the default underlying data.
(e) The technology of data analysis has qualitative potential of further refinement with possibilities of future use of new, qualitatively higher, source datasets.
(f) The entire technological process requires about 54 % of skilled manual work, especially in terms of data preparation and qualified decision between processes—see Fig. 5

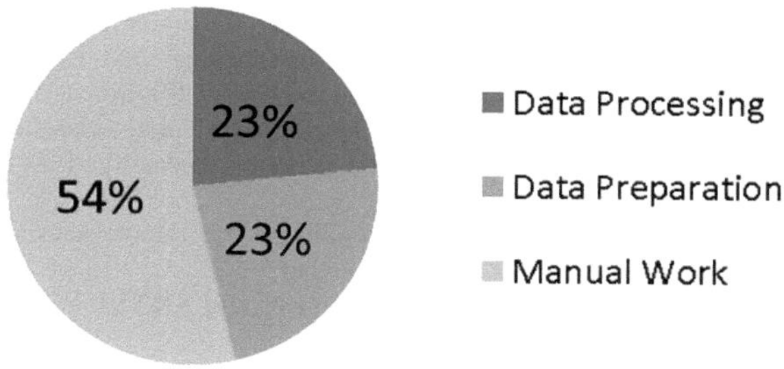

Fig. 5 Portion of time-demand of automated part of data analysis and manual work

The results showed that only one additional dataset for filtering of sub-results suffices for the effective achievement of the required quality classification for the given purpose and this layer refines results of the automated process. Subsequent visual inspection associated with editing then slightly improves the result of classification so that the absolute error rate of the fact, that is represented by the image orthophoto, does not exceed 2–3 %.

The process described in this paper has general character and it can be used for classification of the surface above utilities such as water supply, sewerage, electric power distribution, media distribution, etc.

Acknowledgments The work was solved within the project marked FAST-S-15-2723 and called "Methods development of qualitative appreciation of data of progressive measuring techniques via GIS tools on base of operational analysis" and FAST-S-14-2464 called "Research of the behavior of building structures under operating conditions via geodetic methods" and within research project of MŠMT AdMaS ED2.1.00/03.0097 Nr. HS12357021212200 "Data analysis of surfaces above RWE gas lines in the Czech republic".

References

Bureš J, Bartoněk D, Opatřilová I (2013) Data analysis of surfaces above RWE gas pipelines in the Czech Republic. Final report. Project AdMaS ED2.1.00/03.0097 Nr. HS12357021212200

Černota P, Staňková H, Gašincová S (2011) Indirect distance measuring as applied upon both connecting surveys and orientation one. Acta Montanist Slovaca 16(4):270–275. ISSN 1335-1778

da Silva AQ, Paradella WR, Freitas CC, Oliveira CG (2014) Improving the impervious surface estimation with combined use of optical and SAR remote sensing images. Remote Sens Environ 155–167

Dandoš R, Staňková H, Černota P, Subiková M (2013) Spatial visualization of the infantry blockhouse OP—S 25 "U trigonometru". Adv Mil Technol 8(1):73-84. ISSN 1802-2308

Debes C, Weiss C, Zoubir AM (2014) Development and assessment of a 250 m spatial resolution MODIS annual land cover time series (2000–2011) for the forest region of Canada derived from change-based updating. Remote Sens Environ 140:731–743

Gray J, Song C (2013) A new automatic image analysis method for assessing estrogen receptors' status in breast tissue specimens. Comput Biol Med 43(12):2263–2277

Iiames JS, Congalton RG, Lunetta RS, Brenning A (2013) Balancing misclassification errors of land cover classification maps using support vector machines and Landsat imagery in the Maipo river basin (Central Chile, 1975–2010). Remote Sens Environ 137:112–123

Johnson DM (2013) Comparison between pixel- and object-based image classification of a tropical landscape using Systeme Pourl' Observation de la Terre-5 imagery. J Appl Remote Sens 7 (1):073512, pp 14

Landgrebe DA (2003) Signal theory methods in multispectral remote sensing, vol XI. Wiley, Hoboken, p 508. ISBN 0-471-42028-X

Liang S (2004) Quantitative remote sensing of land surfaces, vol XXXVI. Wiley, Hoboken, p 534. ISBN 0-471-28166-2

Makarau A, Palubinskas G, Reinartz P (2013) Modelling habitat suitability for alpine rock ptarmigan (Lagopus muta helvetica) combining object-based classification of IKONOS imagery and Habitat Suitability Index modeling. Ecol Model 254:22–32

Mouelhi A, Sayadi M, Fnaiech F, Mrad K, Ben R, Khaled P (2013) Alphabet-based multisensory data fusion and classification using factor graphs. IEEE J Sel Top Appl Earth Obs Remote Sens 6(2):969–990

Puertas L, Brenning O, Meza AJ (2013) Estimating area from an accuracy assessment error matrix. Remote Sens Environ 132:202–211

Pupin AM, Alves SD, Friedrich TRB, Ferreira SA (2014) Area-based and location-based validation of classified image objects. Int J Appl Earth Obs Geoinf 28:117–130

Searcóid MO (2006) Metric spaces. Springer undergraduate mathematics series, Berlin. ISBN 1-84628-369-8

Zhong L, Gong P, Biging GS (2012) Segmentation and thematic classification of color orthophotos over non-compressed and JPEG 2000 compressed images. Int J Appl Earth Obs Geoinf 15(SI):92–104

Rockfall Monitoring Based on Surface Models

Snežana Bogdanović, Miloš Marjanović, Biljana Abolmasov, Uroš Đurić and Irena Basarić

Abstract This research addresses surface analysis based on the Terrestrial Laser Scanning data exampled on rockslope site along M-22 highroad near Ljig in Serbia. The slope is about 100 m wide and over 20 m high. The scanning was performed in three epochs 2011, 2013 and 2014, but only the latter two were involved in the analysis due to insufficient quality of the pilot epoch. Gap between the latter two epochs coincided whit local rockfall events detected in the middle section of the slope. Scanning was performed by Leica ScanStation P20 instrument. Two consecutive point clouds were produced and gridded to 3 cm resolution. Comparative assessment of both point clouds and related surface models revealed the volumes and spatial extents of the anticipated rockfalls. Data were analyzed in two software packages: CloudCompare and GeomagicStudio, that have different modeling approaches based on direct point cloud analysis and mashed surface comparison, respectively. We compared advantages and disadvantages of each approach. Finally, we addressed how LiDAR technology can contribute qualitatively and quantitatively for research and monitoring of rockslopes.

S. Bogdanović (✉) · M. Marjanović · B. Abolmasov · I. Basarić
Department of Geotechnics, Faculty of Mining and Geology, University of Belgrade, Djušina 7, 11000 Belgrade, Republic of Serbia
e-mail: snezana.zecevic@rgf.rs

B. Abolmasov
e-mail: biljana@rgf.bg.ac.rs

I. Basarić
e-mail: irenabasaric@yahoo.com

M. Marjanović
Chair of Landslide Research, Faculty of Civil, Geo and Environmental Engineering, Technische Universität München, Arcisstr. 21, 80333 Munich, Germany
e-mail: milos.marjanovic@rgf.bg.ac.rs; milos.marjanovic@tum.de

U. Đurić
Department of Geotechnical Engineering Faculty of Civil Engineering, University of Belgrade, Bulevar Kralja Aleksandra 73, 11000 Belgrade, Republic of Serbia
e-mail: udjuric@grf.bg.ac.rs

K. Růžičková and T. Inspektor (eds.), *Surface Models for Geosciences*,
Lecture Notes in Geoinformation and Cartography,
DOI 10.1007/978-3-319-18407-4_4

Keywords Lidar · Point cloud · Rockfall · Monitoring · Surface analysis

1 Introduction

3D presentation and visualization of terrain are very important in geotechnical investigations of slope stability. One of the established approaches.

Collecting spatial information on millions of points has found its use in modeling of slopes and as supplement/substitute for conventional field data acquisition (Kemeny and Turner 2008; Tonon and Kottenstette 2006). The point clouds allowed high-resolution, computer-based analysis.

There are a number of benefits in using point clouds as discrete models of the rock slopes:

- they enable mapping in inaccessible parts of the slopes,
- high-precision cross-sections and contour lines of the terrain surface,
- more objective and more accurate measurements,
- once scanned, the models are safely stored in digital format and are available for reuse at any time with no additional expenses, unlike the field investigation revisits.

Scanning data can be exported to computer-aided design application for additional modeling such as AutoCad and can also be used as Digital Terrain Model in GIS. The main purpose is to allow the user to interactively view objects or scenes from different angles and directions. Also, parts of the rock face that are difficult to reach can be measured directly. Accurate vertical or oblique cross-sections and contour lines can be extracted from a point cloud or related surface model. This can advance the allocation of potentially loose blocks and help determining their optimal stabilization measures. Another useful application is monitoring of the movement of unstable (rock) slopes (Slob and Hack 2004). In respect to all these and other applications, terrestrial LiDAR scanning was used for several slope and rock cut stability assessments as discussed in (Ablellan et al. 2009; Lato et al. 2009; Sturzenegger and Stead 2009; Sturzenegger et al. 2011).

However, the main disadvantages are:

- costly equipment and processing hardware and software,
- necessity of a multidisciplinary team of engineering geologists and surveying specialists.

In addition, scanning conditions must be appropriate (vegetation, draped mesh or other objects can screen the laser signal, which results in appearance of noise and shadows). Therefore, the scanning must be performed in the season when the vegetation is low and slope is clean. Removal of the signal noise (e.g. vegetation) is very important for a proper interpretation of the laser scanning results because all subsequent measurements based on the point cloud or surface model can be affected (as well as the surface model itself).

In Serbia, the first Terrestrial Laser Scanning (TLS) for determination of its application possibility in geotechnical slope stability studies was carried out at the site "Letnja Pozornica" in Belgrade (Marjanović 2013), and kinematic rockslope analysis near Ljig (Marjanović et al. 2013). In the present paper, we extended the analysis at the latter site in terms of monitoring the rockfall dynamics using the TLS records of different epochs.

2 Study Site

The study site (E20°14′35″, N44°12′02″) is located in the Western Serbia near the town of Ljig, along a very important and very frequent roadway, M-22. One of the motifs for investigating this rockslope arises from the necessity of maintaining the safety along the road. There are no protective wire nets or any other protective measures installed. The slope is over 20 m high and inaccessible for direct measurement in the most part of the slope (Fig. 1). Small to medium rock and block falls as well as raveling and flaking at the location are obvious.

Fig. 1 A photo (by B. Abolmasov) of the highest part of the slope (≈20 m)

The site is composed of the Upper Cretaceous flysch sediments, with layers of degraded sandstone, conglomerate and shale (alternating darker and brighter interlayers in Fig. 1). The area around the slope is not particularly tectonized. The nearest regional structure is "Boljkovački" fault that is deflected in eastward direction, several kilometers away from the site (Filipović et al. 1978). The stratification is monocline and day-lights opposite to the slope. Discontinuities are generally grouped into three characteristic systems: stratification ss 39/44° (strike/dip), joint system JS1 187/79° near-parallel to the slope and joint system JS2 148/79° near-perpendicular to the slope (Marjanović et al. 2013).

3 Methodology

The data were obtained by the Leica ScanStation P20, a terrestrial long-range laser pulse scanner, with relatively narrow beam (range goes to 100 m in optimal conditions, beam diameter at that distance is up to 2.6 mm). This scanner allows the collection of million points per second, providing dense point clouds very quickly. For each point all three coordinates (X, Y, Z) are registered in the coordinate system of the scanner, as well as the intensity of the reflected laser beam, which can give the information about the material fabric of the slope. This particular model of the scanner also contains a high resolution digital camera and allows registering of the light reflection in the visible spectrum, so that each scanned point receives appropriate RGB (red, green, blue) value. Scanning was performed from multiple stations (standpoints) in a similar fashion in each scanning epoch 2011, 2013 and 2014, meaning that 4 stations at approximately same positions, similar range to the object and same number of tie points (Fig. 2) were considered. After decimating and equally distributing the points, the point cloud resolution remained 3 cm in each

Fig. 2 The point cloud of the first scanning epoch 2011. Note that permanent geodetic markers were placed at this stage (1 and 11 and other *white dots*) and used in subsequent epochs as permanent tie points (together with additional mobile points)

epoch, resulting in 5.76 millions of points per point cloud. As indicated before, the present data consider 2013 and 2014 epoch, because of insufficient quality of the pilot epoch, but also because no significant displacement occurred on the slope between 2011 and 2013.

Local rockfalls were visually evident at the site between 2013 and 2014. Scans from late summer 2013 and early spring 2014 were therefore compared to determine, locate and measure the occurrence of detached and accumulated material. Dimensions (2D base) of detachments and accumulations and the 3D surface difference between two epochs were used to estimate the total rockfall volume. Detachments and accumulations were thereby treated as primitives, i.e. prism-like solids, i.e. (base * max. difference)/3.

Analysis were performed in CloudCompare and GeoMagic Studio software packages. CloudCompare is a 3D point cloud (and triangular mesh) processing software. It is an open source software and allows the analysis and work on surface models and point clouds, but it is more focused on the latter. Some limitations are present, particularly in respect to the surface model processing. For our purposes, direct comparison of two point clouds was essential and we discovered sufficient functionality in CloudCompare. GeomagicStudio is a complete toolbox for transforming 3D scanned data into highly accurate surface, polygon and native CAD models. A geometrically simplified (meshed) surface model of the slope face was generated and further analyses have been performed on the mesh.

4 Results and Discussion

Investigated slope has potential for developing block and wedge failures (Marjanović et al. 2013), especially in the upper, weathered sections in shale layers of the flysch sequence. Small to medium block falls in the foot of the slope are obvious (Fig. 1). Slope was scanned in the epochs under similar conditions and with the same scanner. The changes were detected, and the volumes of fallen materials were estimated.

In the Cloud Compare software, analysis was done on the point cloud and the results are visualized in Fig. 3. Comparative assessment of both point cloud shows that there are displacements between scanning epochs that are larger than the error rate of semi-automatic fitting of 2013 and 2014 point clouds (standard deviation = 0.2319 m).

The figure clearly shows three possibly related events (difficult to date exactly, but presumably taking place at early spring of 2014 when substantially higher than average precipitation was introduced in the region). It can be seen that the left part of the rock slope shows a very big difference in range between the two scans, representing the scale of displacement during that period. It is important to mention that the point cloud analysis showed absolute differences (no intermediate surface modeling). The color scale is presented at the right of Fig. 3 (in meters) and it indicates displacements from 0 to 1.5 m, while actual maximum is at 0.5 m. It can

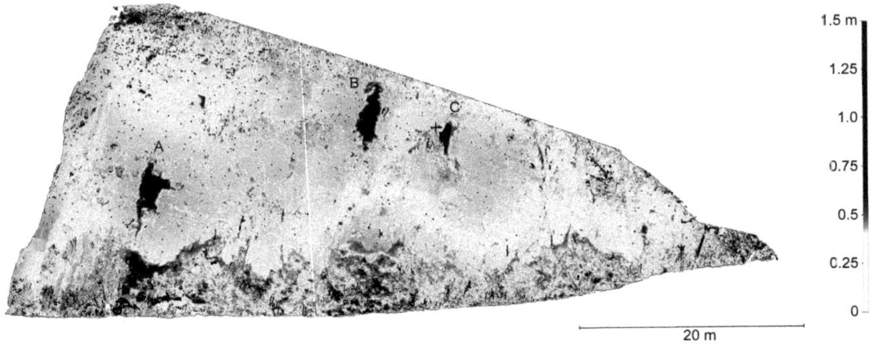

Fig. 3 Comparative analysis of two epoch of scanning-rock cut giving the absolute displacement (CloudCompare software)

be suspected that outliers of over 0.5 m are due to vegetation and other sources of noise (chaotically distributed dark dots). Thus, dark areas represent either detachment or accumulation zones bigger than 0.4 and lesser than 0.5 m, while lighter gray indicates no movement. Three events are evident in the left (A) and upper middle section of the slope (B and C). Except for the A They are not that well indicated by accumulations on the slope foot (also dark patches under A, B and C), presumably due to road maintenance intervention.

The second approach involved software GeomagicStudio. The point cloud was first converted into a surface model. Then, comparison of surface models was made from the two recording epochs, wherein 2013 and 2014 surface models were fitted with sufficient accuracy (standard deviation = 0.15 m). The results are shown in Fig. 4. The color scale is presented at the right side (in meters) and indicates displacements from −2.0 to +2.0 m, while actual max-min range is −0.5 to +0.5 m. The dark patches show the negative changes (detachments) deeper than 0.4 m and

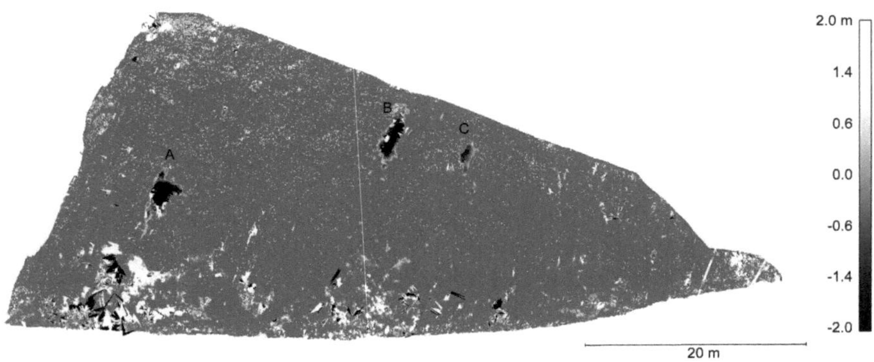

Fig. 4 Comparative analysis of two epoch of scanning-rock cut giving relative displacements (GeoMagic Studio software)

Table 1 Dimensions of registered rockfall

	A	B	C
X (m)	5.67	4.23	2.82
Y (m)	2.40	2.07	1.45
H (m)	0.5	0.4	0.35
V (m^3)	2.27	1.17	0.48

the positive changes (accumulated material) are shown in white. Light gray indicates no movement.

As indicated before, dimension measurements were carried out on the obtained surface models, inferring a prismatic shape (Marjanović et al. 2013). Two-dimensions of prism were measured (X, Y) on the model and the third dimension (H) was read from the color scale (Figs. 3 and 4). Results, presented in Table 1, indicate the largest rockfall volume at position A, equalling 2.27 m^3, followed by 1.17 m^3 at B and 0.48 m^3 at C. The volume of accumulated material at A is estimated to a similar 2 m^3, while the remaining two blocks do not correspond well with their accumulation zone.

5 Conclusions

One of the principal tasks was to compare two different approaches of displacement monitoring, which were performed in two different software packages. Each of these analyzes has its advantages and disadvantages.

CloudCompare is Open Source software so it can be used freely, and the user community is growing strong, offering a lot of resources and support. This software allows for direct analysis on the point cloud, so no conversion to surface model is needed and no intermediate errors and outliers are introduced. Disadvantages are that it shows us only the absolute differences, while direct volume measurements are not possible (it is possible to use primitives to fit the point clouds), but this functionality is currently under development.

GeoMagic Studio is a licensed program, user-friendly but not as supported as its counterpart CloudCompare. It requires conversion of the point cloud into a mashed surface model for any volume-based analysis, while only simple measurements, such as distances and angles can be performed on a point cloud. Relative (positive and negative) surface differences are obtained by analyzing the models, therefore indicating detached and accumulated rockfall material differently. Another advantage of this program is certainly capability of direct volume measuring.

In conclusion the paper presents a successful application of TLS for monitoring of rock slopes and proved applicability of both software alternatives in such scenario, with all the benefits and drawbacks. The future work might be focused on refining the input data, by more accurate noise filtering, perhaps classification-based and direct volume difference computation.

Acknowledgments This research is supported by the Ministry of Education, Science and Technological Development of the Republic of Serbia (project TR36009).

References

Ablellan A, Calvet J, Vilaplana JM, Blanchard J (2009) Detection of millimetric deformation using terrestrial laser scanner: experiment and application to rockfall event. Nat Hazards Earth Syst Sci 9:365–372

Filipović I, Marković B, Pavlović Z, Rodin V, Marković O (1978) Osnovna geološka karta 1:100,000—tumač za list gornji milanovac L 34-I37. Federal Geological Bureau, Belgrade

Kemeny J, Turner K (2008) Ground-based LiDAR rock slope mapping and assessment: rock slope mapping and assessment, Report FHWA-CFL/TD-08-006. Federal Highway Administration Central Federal Lands Highway Division, Lakewood, CO

Lato M, Diederich MS, Ball D, Harrap R (2009) Engineering monitoring of rockfall hazards along transportation corridors: using mobile terrestrial LIDAR. Nat Hazards Earth Syst Sci 9:935–946

Marjanović M (2013) Kinematic analysis of a rock slope using terrestrial 3D laser scanning data. In: Proceedings of the 5th international young geotechnical engineers' conference, vol 2, pp 32–35

Marjanović M, Abolmasov B, Djurić U, Zečević S, Šušić V (2013) Basic kinematic analysis of a rock slope using terrestrial 3D laser scanning on the M-22 highroad pilot site. In: Kwasniewski M and Lydzba D (eds) Rock mechanics for resources, energy and environment. Taylor & Francis Group, London, pp 679–683

Slob S, Hack R (2004) 3D Terrestrial laser scanning as a new field measurement and monitoring technique. Eng Geol Infrastruct Plann Europe Lecture Notes Earth Sci 104:179–189

Sturzenegger M, Stead D (2009) Close range digital photogrammetry and terrestrial laser scanning for discontinuity characterization on rock cuts. Eng Geol 103:17–29

Sturzenegger M, Stead D, Elmo D (2011) Terrestrial remote sensing estimation of mean trace length, trace intensity and block size/shape. Eng Geol 119:96–111

Tonon F, Kottenstette J (2006) Laser and photogrammetric methods for rock face characterization. Alexandria, American Rock Mechanics Association. In: Proceedings of the workshop held in conjunction with US Rock Mechanics Symposium, Golden, 18–19 June http://www.usbr.gov/pmts/geology/goldenrocks.pdf

Coping with Integrating Low-Cost 3D Printing and Surface Models: A Case Study on Prusa i3

Jan Brus and Radek Barvíř

Abstract The article describes one of the possible ways how to create a physical terrain and surface models using a low-cost 3D printer Prusa i3. In the first part, there is a summary of currently used technologies for 3D printing, their descriptions and basic characteristics. Then the open-source 3D printer Prusa i3 based on the RepRap project is shortly described as well as available data used for the testing. The rest part is focused on the whole process of transforming elevation data into a form, which can be printed on the 3D printer. Main attention is given to transferring raster format of digital elevation model into a physical 3D model using Rhinoceros 5, slic3r and pronterface software. The procedure for modelling in Rhinoceros is described using step-by-step instructions and simple formulas. Several freeware alternatives are mentioned as well. Also the user interface of slic3r is described in detail with some recommended options. In the case of the Pronterface software, the user interface and functionality of all important buttons and settings is described. In the final part, the output results, possible problems encountered during the process of printing 3D models and their possible solutions are described in detail.

Keywords 3D printing · Rapid prototyping · Fused deposit modelling · Digital terrain model · Prusa i3 · Rhinoceros 5 · Slic3r · Pronterface

J. Brus (✉) · R. Barvíř
Department of Geoinformatics, Faculty of Science, Palacký University, 17. listopadu 50, 771 46 Olomouc, Czech Republic
e-mail: jan.brus@upol.cz

R. Barvíř
e-mail: barvir.radek@gmail.com

K. Růžičková and T. Inspektor (eds.), *Surface Models for Geosciences*, Lecture Notes in Geoinformation and Cartography,
DOI 10.1007/978-3-319-18407-4_5

1 Introduction

The term 3D printing refers to a technology or process for creating three-dimensional physical models from digital models. In this method, material is gradually added, which makes it different from the process of 3D milling. The digital model is first divided into thin layers using an appropriate software, and then these layers are gradually applied. Currently, there are several 3D printing technologies that differ in the materials used, layer height, material curing and other properties of the resulting model, its accuracy and possible use (Ghawana and Zlatanova 2013; Schwarzbach et al. 2012; Svobodová and Voženílek 2010).

2 3D Printing

2.1 Stereolithography

The first device for the direct production of 3D models was developed in 1986 by 3D Systems. The technology used was called stereolithography, and it still bears the abbreviated designation SLA or SL. A resin solution responsive to light is sequentially irradiated with a laser beam, and it solidifies in irradiated spots, thereby gradually forming a model. In some systems, UV (ultraviolet) light is also used for irradiation in stereolithography. Layer height typically varies from 0.05 to 0.2 mm. After the layer is created, it is levelled using a blade, and the platform descends by one layer height. The entire process is then repeated (Rase 2012; Schwarzbach et al. 2012).

2.2 Selective Laser Sintering

Another technology based on a similar principle is Selective Laser Sintering (SLS), introduced by EOS. Powdered material is applied on the printing area and, using a powerful laser, it is melted and bound. The powder is applied either to the whole area or it is applied directly by the print head only where needed. In unirradiated places, the powder serves as a support structure. Despite high printer cost reaching hundreds of thousands of euros, both the technologies described above are still often used in the industrial production of large-size models, e.g. in the automotive industry.

2.3 Powder-Binder Printing and Inkjet Technology

Powder-Binder Printing differs from SLS in the way of connecting the powdered material, which is done using a liquid binder spurted out similar to standard 2D inkjet printers. Machines of this type usually produce no odour and can print up to 390,000 pieces in various colours. Instead of powder, wax can also be used as a construction material. Z Corporation is one of the companies specializing in the production of 3D printers based on this principle. In some systems, even no powdered material is needed and the model is made of the spurted substance only. In this case, another support material is needed. This building substance can be also coloured. (Butscher et al. 2011). The technology can achieve a very high accuracy of up to 5000 DPI, and therefore it is used to manufacture jewellery models. However, the models are usually brittle comparing to other methods.

2.4 Laminated Object Manufacturing

Other possibilities of 3D printing have been presented by planar lamination of board materials called Laminated Object Manufacturing (LOM). It is a solution on the border between 3D printing and milling. The input material is usually paper coated with polyethylene on one side, but some products work with plastic or metal films. One of these materials is first wound on the printing area, then rolled onto the previous layer and finally cut off using a blade or laser. The advantages of this method include the ability to print large models and production speed. The main disadvantage is the impossibility to reuse the cut-off waste parts after completing the model, but high material consumption is compensated by its low cost. LOM is not a suitable method for fine structures, because they may be damaged or destroyed in the separation process (Ahn et al. 2012).

2.5 Digital Light Projection

One of the latest 3D printing technologies is Digital Light Projection (DLP), developed by EnvisionTEC. Its principle is similar to the SLA, using a liquid photopolymer irradiated by UV light. Due to fixation, a system of supporting elements is created which are mechanically removed after completion of the entire model. The production process is very fast, with very good surface quality. However, at temperatures greater than 70 °C it significantly changes its parameters and becomes brittle (Gauvin et al. 2012; Chua et al. 2010).

2.6 Fused Deposit Modelling

A cheaper solution in terms of cost of the device has been offered by the technology called Fused Deposit Modelling, also known under the acronyms FDM or FFF. It was developed by Stratasys Company. The main principle is melting plastic (sometimes even metal) wires to reach a semi-liquid state and its application, through a nozzle, onto the printing surface, where it solidifies immediately. The print head moves one level up after the completion of each layer. A wide range of materials can be used for model construction. The most well-known are ABS (Acrylonitrile butadiene styrene), PLA (Polylactic acid), polycarbonates, nylon, etc. The support structure is applied using a special nozzle. After completion of the model, the support structure is removed mechanically or it dissolves in the bath. The FDM technology is also used by open-source 3D printers made within the RepRap project, e.g. Prusa i3 printer.

FDM technology was chosen as a good way to produce low-cost maps and physical terrain sourface models. A small disadvantage is that this method is being used for monochromatic models very often, mainly in case of cheap open-source machines. On the other hand, some printers, e.g. DesignJet Color 3D from Hewlett-Packard and Stratasys, enable printing in up to eight colours and CubePro® C full-color printer from botObjects can print even in CMYKW (Cyan + Magneta + Yellow + Black + White) colour palette. On the other hand, the precision is at sufficient level and there are many substances possible to use.

3 Prusa i3

Prusa i3 is a 3D printer composed of open-source components falling under the RepRap project. It was developed by Josef Průša, who has been dedicated to the development of 3D printers since 2009. The device operates on the Fused Deposit Modelling technology and enables printing from ABS, PLA, Petty, Laywood and Laybrick. Other materials are gradually tested. The work area has the dimensions of 200 by 200 by 200 mm, but because of the clips holding the glass platform on a heated board objects cannot be printed to the very edge. In the basic version, the nozzle is 0.4 mm in size and it is made from food materials. Thanks to the open-source design, it is not a problem to exchange it for another compatible type, which also applies to some other parts. The minimum layer height is 0.05 mm, and the step size in the horizontal direction starts at 10 microns. The highest speed is 200 mm/s. The printer uses reliable RAMBo electronics by Ultimachine, with fully integrated stepper motor drivers. The Prusa i3 printer can be purchased for the price of CZK 10,000–25,000 (€350–900), depending on the components used and the willingness to obtain components from different manufacturers and assemble the printer individually.

For the case study, this open-source printer was chosen because of its wide range of applicable materials, modifiability, low cost and worldwide popularity.

4 Datasets

The basic datasets currently available that are suitable for 3D printing of physical terrain models for the territory of the Czech Republic can be divided into two parts —detailed altimetry data supplied by the Czech Office for Surveying, Mapping and Cadastre (ČÚZK) and less detailed global data accessible free of charge.

4.1 The Czech Office for Surveying, Mapping and Cadastre Data

With regard to the ČÚZK data, this article focuses on the new altimetry mapping including the digital terrain model of the 4th generation (DMR 4G) and 5th generation (DMR 5G), and the digital surface model of the 1st generation (DMP 1G). Data collection for these datasets was carried out using airborne laser scanning in 2010 in the middle part of the Czech Republic. The western part of the country was scanned the following year and the eastern region a year later, in 2013 (Uhlířová 2010). All data is distributed in the S-JTSK coordinate system, in text files of the XYZ format. The basic output unit is the state map sheet of 1:5000 of the 2.5 by 2 km size. The datasets including specifications is available through the ČÚZK Geoportal (http://geoportal.cuzk.cz/).

The 4th Generation Digital Terrain Model of the Czech Republic (DMR 4G) represents a view of the Earth's natural surface or the surface modified by human activity in digital form, as the elevations of discrete points in a regular square grid of 5 by 5 m. The points contain coordinates X, Y, H, where H represents the elevation in the Baltic elevation reference system with a total mean error of elevation of 0.3 m in exposed terrain and 1 m in forested terrain.

The 5th Generation Digital Terrain Model of the Czech Republic (DMR 5G) also represents the view of the Earth's natural surface or the surface modified by human activity. However, unlike the less detailed DMR 4G, the points are located in a triangulated irregular network (TIN). The total mean error of the elevation is equal to 0.18 m in exposed terrain and to 0.3 m in forested terrain. Extension of the current partial coverage to the entire territory of the Czech Republic is expected by the end of 2015.

The 1st Generation Digital Surface Model of the Czech Republic (DMP 1G) represents a view of the territory including buildings and vegetation cover in the form of an irregular network of elevation points (TIN) with a total mean error of 0.4 m for precisely defined objects (buildings) and 0.7 m for objects that are not precisely delimited (forests and other vegetation cover). As with DMR 5G, the entire territory of the Czech Republic is expected to be covered by the end of 2015. It will probably be updated using the method of image correlation of aerial survey photographs and aerial laser scanning.

5 Global Digital Elevation Models

Shuttle Radar Topography Mission (SRTM) is the name of a dataset and international research programme of National Aeronautics and Space Administration (NASA) whose aim was to create a digital terrain model and a complex topographic database using radar photography aboard the Endeavour space shuttle in 2000. An area from 60°N to 56°S (approximately 80 % of the Earth's surface) was photographed. The resulting DTM is available in two versions: SRTM 1 in a spatial resolution of 1 arcsecond (approximately 30 m at the equator) and SRTM 3 with resolution 3 arcseconds (approximately 90 m at the equator) (Werner 2001). The spatial resolution in the territory of the Czech Republic is around 77 m per pixel (SRTM 3) or 26 m per pixel (SRTM 1). Elevation accuracy is provided up to 20 m.

ASTER Global Digital Elevation Model (GDEM) is the digital terrain model created by NASA in collaboration with Japan's Ministry of Economy, Trade and Industry (METI). The data was obtained by the eponymous multi-spectral scanner Advanced Spaceborne Thermal Emission and Reflection Radiometer (ASTER), located on the Terra satellite, with the possibility of scanning stereoscopic pairs. These spectral pairs were used to create a global DSM for the area between 83°N and 83°S, i.e. with three times higher spatial resolution compared with SRTM 3. In the territory of the Czech Republic, the spatial resolution is around 26 m. However, the quality of the data in different areas is not the same. The area of Central Moravia displays a relatively significant distortion in dozens of metres due to the small amount of usable stereopairs. Variations in the rest of the country are typically up to 10 m (Tachikawa et al. 2011).

Both of the global DTMs mentioned above can be downloaded free of charge on the website of the USGS (United States Geological Survey) for the territory defined by the user (http://gdex.cr.usgs.gov/gdex/). Raster images are downloaded in the geoTIF format.

To summarize, it can be generally said that to create a model at a scale of 1:200,000 and smaller with 3D printing, the NASA SRTM 3 dataset can easily be used. For a scale of 1:100,000, it is preferable to use more detailed SRTM 1 or ASTER GDEM. Conversely, DMR 4G can be used in larger scales up to 1:10,000, DMR 5G and DMP 1G possibly up to a scale of 1:2500. However, these values are only indicative and are based on experience with model production using the Prusa i3 printer with the 0.4 mm nozzle. In order to increase the detail of data, various interpolation algorithms can also be used. In the case of this study no interpolation algorithms were used because of adequately detailed data. It always depends on the specific application, quality requirements, hardware, etc. When selecting data or scales, the following formula may be helpful:

$$m = \frac{1000 \cdot rr}{ns} \quad (1)$$

m = lowest possible scaling model number
rr = raster spatial resolution [m]
ns = nozzle size [mm]

In the case of DMR 4G with a spatial resolution of 5 m and printer with the nozzle diameter of 0.4 mm, the maximum possible scale of the model using the highest possible detail of the printer is therefore 1:12,500.

6 Creating 3D Terrain Model

6.1 Rhinoceros 5

Rhinoceros 5 is s software for creating 3D models and performing simple analyses that is used mainly in design, architecture and modelling. However, its tools can be used in a wide range of other fields, including 3D cartography. The program is available with a licence for commercial use, as well as with a student licence for individual academics and students, or as a laboratory licence for up to 30 school computers. A three-month fully functional trial version can be downloaded on the website of the software manufacturer for testing. In this article, we will focus on the use of the program for creating a 3D model based on the digital terrain model, which will then be converted into a physical model using 3D printing.

The only input data for the production of a 3D terrain or surface model in the Rhinoceros environment is an integer raster whose pixel values only take non-negative values (unsigned). TIF and JPG formats with a bit depth of 8 and 16 bits were successfully imported. Raster images with higher bit depths, as well as raster images with negative pixel values (signed) could not be imported into Rhinoceros 5 3D at all. Raster files with decimal pixel values were always imported in the form of a flat grid of zero height in its entirety. If there are some NoData pixels used in the DTM raster, it is also very important to set the right NoData value, because Rhinoceros detects and process them as the rest of regular pixels.

File processing is carried out using the “Heightfield” tool. In the first step, the path to the file is set. Consequently, the starting corner and the second corner of the model on the x axis or dimension of the model in the direction of the x axis must be selected. The start can thus be chosen, for example, in point 0,0,5 (the lowest point on the modelled surface will be located 5 mm above the beginning of the z axis, by which the resulting model obtains a 5 mm base) with the model length of 180 mm.

The second horizontal dimension in the direction of the y axis is automatically calculated according to the ratio of the input raster sides. In our example, the raster has a square shape for simplification, and the size of the model in the direction of the y axis will therefore be again 180 mm.

In the following window, several additional parameters are set. The first is the number of patterned points in the x and y axes, which will be set, for example, as 450 by 450, so that the model contains one elevation point for every 0.4 mm. The second parameter is the height between the highest and lowest point of the surface of the model specified in millimetres. To complete this information it is necessary to know the real difference of the minimum and maximum elevation in the area, planned horizontal model scale and the multiple of the z coordinate (Z-factor). The values are substituted in the formula:

$$h = 1000 \cdot rd \cdot hs \cdot z \tag{2}$$

h = height [mm]
rd = real difference in elevations [m]
hs = horizontal scale
z = Z-factor

In the exemplary situation of creating the model of an area at a scale of 1:15,000 with elevation difference of 600 m in reality with highlighted shapes of the terrain using the Z-factor 1.5 (i.e. the vertical scale of the model will be 1:10,000), the formula would looks as follows:

$$h = 1000 \cdot 600 \cdot \left(\frac{1}{15000}\right) \cdot 1.5 \tag{3}$$

Of the three options offered below in the window, select “Interpolate surface through samples” and confirm it by clicking OK. The “Set image as texture” option does not have any effect in this case. After running the tool, the surface is created.

Another part of the process is creation of a box around the area that has just been created using the “Box” tool. The beginning in the example will be at point 0,0,0. Subsequently, the opposite corner of the base is entered, or both horizontal dimensions are gradually entered. Therefore, the dimension of 180 should be entered twice successively. Next, it is only necessary to select the height of the box by making sure that it exceeds the surface representing the Earth’s terrain at its highest point.

After that, it is necessary to divide the created area into two parts using the “Split” tool. Then it is possible to remove the upper part of the box and create a solid from the remaining areas using the “Create solid from intersecting surfaces” tool. The resulting solid can be considered as a model ready for export to STL format.

However, the solid can be further modified, e.g. by cropping, rotating, adjusting individual vertices, adding 3D text describing the model, etc. The 3D inscription

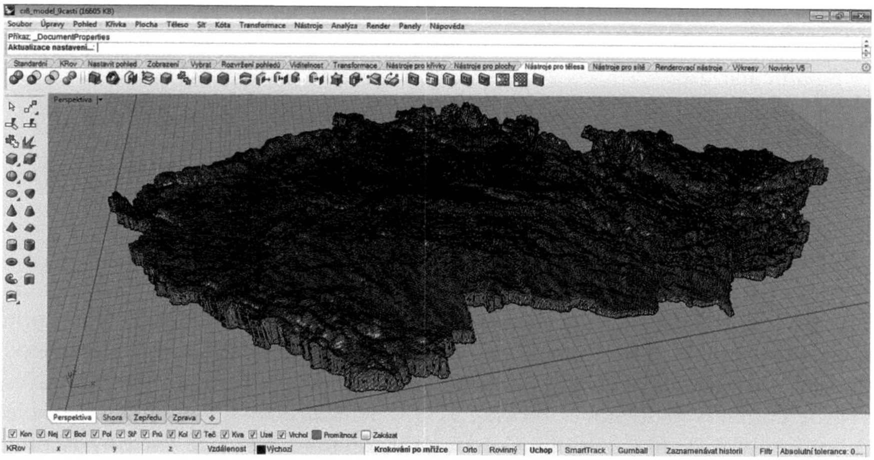

Fig. 1 Final 3D terrain model created in Rhinoceros 5 software

can be created using the "Text objects" tool. After turning and moving it to the desired position to intersect with or at least touch the main solid, it is possible to combine all the letters and the model itself into a single solid using the "Boolean union" tool. The position of the text, however, should be selected considering the printability of such a model. The letters protruding no more than 2 mm from a side wall were not printed very well when testing, but the inscription was readable. On the contrary, 3D text located on the top wall of the model would cause a loss of information about the terrain form at the given point. The best quality of the inscription was reached by sinking the text into the bottom wall of the model. The font size should also be considered to ensure that the text is well modelled.

Before exporting, it is necessary to select the entire solid and then, using the File > Export Selected commands, enter a name and select STL as a file type. Other export settings can remain unchanged. The export process is quite demanding of the amount of available RAM memory. To create a relatively simple test model, RAM of at least 8 GB is needed. The difficulty increases with the number of fracture edges (Fig. 1).

6.2 Alternatives

The simple application BMP2IGES or online converter available at http://www.embossify.com can be considered cheaper and, with limited use, even free alternatives allowing the transfer of JPG files into STL model form with several basic parameters. On the contrary, DEM format files can be converted into STL form and edited by the AccuTrans 3D software. However, the output quality of each application is different and it is necessary to consider whether each software is sufficient for a particular use.

7 Printing the Model

7.1 Slic3r

Other settings of model parameters, such as the infill density and structure, print speed, temperatures, layer height etc., are conducted outside the Rhinoceros environment. An example of an application for dividing the model into individual layers is slic3r, currently in the stable release 1.1.7 (Ranellucci 2014). This program enables converting of STL files into the G-code programming language used by RepRap printers. At the same time, slic3r makes it possible to set important parameters affecting the strength of the model, print speed and material consumption.

In the first tab, PLATER, STL files are loaded, models can be rotated, scales can be changed, and it is also possible to view and crop models (Fig. 2). On the right side, different settings for the model itself, filament and printer can be modified. Finally, the G-code file is exported here.

The second tab, PRINT SETTINGS, is crucial for print settings. In the "General" section, it is possible to specify the layer height and the number of layers from each side of the model. It is also necessary to pay attention to the infill structure and density, which has a significant impact on material consumption, especially in large models. The following sections are devoted to the support material, movement and print speed, and other less important parameters. When printing the physical models

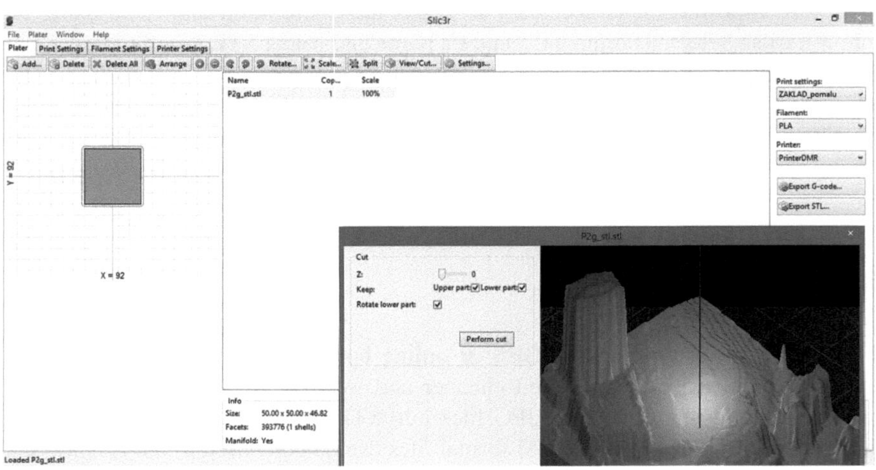

Fig. 2 Slic3r 1.1.7 environment with imported surface model of an urban area

of the terrain of smaller sizes (up to 10 cm) from ABS, it has proved to be efficient and fast to set a 10 % infill, two layers on the bottom wall of the model, one layer at the edges and three layers on the top model part with the visualisation of the earth's surface. In larger models, it is useful to increase the number of layers on each wall, as well as the infill density. Comparison of different infill structures in terms of material and time consumption is clearly described in the slic3r manual (http://manual.slic3r.org/expert-mode/infill). Layer heights range from 0.15 to 0.4 mm were tested. The smaller is the layer height, the more accurate and smoother the model surface is. Production of small models (up to 10 cm) takes a few hours. Printing larger models (up to 20 cm wide) may take up to one day or even more, with the consumption reaching dozens of metres of the 3 mm-thick filament.

The FILAMENT SETTINGS tab focuses on material used for 3D printing. It is used for setting the roll diameter in millimetres and both, extruder and bed temperatures, which are different for each substance. In the case of small models, ABS plastic can be used. For larger models (more than 15 cm in diameter) ABS plastic tends to twist and crack, so it is better to use PLA material.

The last tab, PRINTER SETTINGS, describes the printer parameters. It is used, for example, for entering the bed size and coordinates of its centre.

7.2 *Pronterface*

The last program within the entire manufacturing process of 3D models is an application to control the 3D printer. In the case study, the pronterface program was used to control open-source RepRap printers (Yanev 2014). After opening the application, you the printer can be connect using USB port and a G-code file loaded. Immediately after loading, the estimated material consumption and printing duration is displayed on the right side of the screen. The left lower part of the environment is used for setting extruder and bed temperatures to the desired values, depending on the material used. There are also buttons for manual shifting of the print head in space and for moving it to the home position. It is also possible to view the appearance of individual layers (Fig. 3). After preparing the printer, loading the file and reaching the desired temperatures, printing can begin. After that in the event of any problems, it can be paused or restarted.

8 Results

Over the testing process, several terrain models of the Moravian attractive terrain shapes and famous sights were made successfully (Fig. 4). These models were scaled 1:20,000–1:30,000, each 9.5 × 9.5 cm wide and 2–3 cm high with 3-mm-high base. Z-factor was set 1–3 depending on terrain conditions of individual areas. The models were made of 3 mm-thick green, blue or white ABS filament. Printing

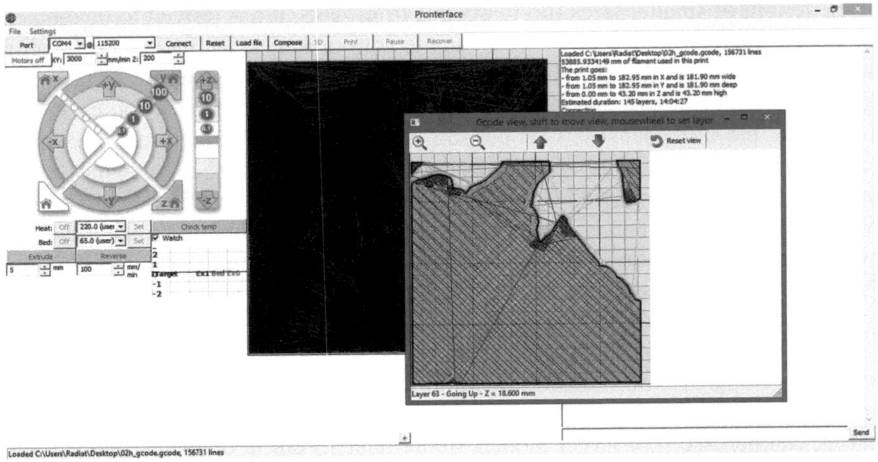

Fig. 3 Pronterface environment with imported model and single layer scheme

Fig. 4 Small (9.5 × 9.5 cm) terrain models from ČÚZK data made of ABS plastic

of each model took a few hours and the price of the material used fluctuated CZK 30–50 (€1–2) per model.

Besides the terrain models, a small surface model was printed as well (Fig. 2). The model captures a part of a high-rise hotel building with several trees and nearby houses at scale 1:2000. The model is 5 × 5 × 4.7 cm large. The Z-factor was left 1 to prevent shape distortion of the buildings. The same blue ABS filament was used as in the case of other models. During printing one copy, the printer head moved 5 mm sideways and continued printing in that position, so there was a step on the model. Other copies were faultless.

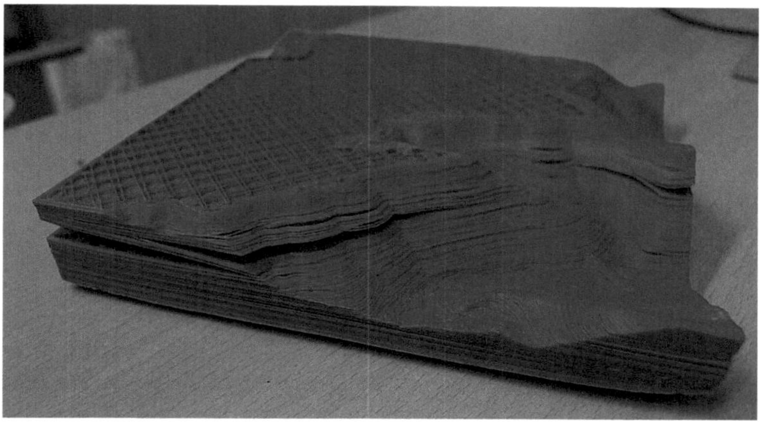

Fig. 5 Larger (18 × 18 cm) ABS model unfinished due to twisting and cracking

Conversely, there were no large models made of ABS substance printed successfully. As it is shown in Fig. 5, during the printing process the model started to twist, e.g. the right bottom corner, and crack so the process had to be stopped. The large models also tend to detach from the printer bed, even if the bed is heated.

Because of this ABS substance characteristic, PLA material, which does not expand so much when heated, was also used. Unfortunately, there has been no successful model printed of PLA on our Prusa i3 printer yet. The PLA filament always melted in the central part of the extruder, crystallized there and completely plugged the extruder, so the substance could not run through the printer head. This issue usually happened several minutes after the print initiation. Because of this, the new advanced extruders compatible with the RepRap printers were ordered. Theoretically, the problem could be also caused by a bad quality of the PLA filament.

9 Discussion

Although, based on the procedure described above, the model production process using 3D printing can look relatively trouble-free at first glance, it is not always so, starting with raster conversion into a form to be importable into Rhinoceros 5 and ending with the actual printing. During the printing process using relatively inexpensive open-source printers a number of other problems may arise. The reason is that every such printer is quite unique, and although in general the settings described above apply, it is sometimes necessary to adapt these values in a specific device based on your own experience. An example of this is extruder temperature settings, which should be 230–260 °C according to the specifications of the manufacturer of the ABS material used for testing, but in the actual printing a temperature of 285 °C proved to be ideal. This is probably also the reason why different sources recommend different temperature ranges for the same type of material.

Another characteristic feature of such printers is bad printing of a model or another problem without a clear cause. It is therefore often necessary to try several solutions with an uncertain outcome. Therefore the experience of the person operating the printer also plays an important role.

When using the ABS plastic, we can encounter problems such as gradual twisting, cracking and subsequent detaching of the model from the bed during creation, which prevents its completion, especially with larger models. There are several causes, as well as possible solutions. Twisting as a result of different temperatures in different parts of the model is caused by the character of the ABS plastic, but it can be exacerbated by insufficient heating of the bed, wrong choice of infill density combined with the number of layers on the individual walls, print speed, etc. Such problems may be solved by adjustment of the relevant parameters, use of a more adhesive bed or application of a different filament.

One example is the natural material PLA, produced from corn and potato starch, which is substantially less prone to twisting than ABS. Under normal conditions, the resulting model is more flexible compared to ABS, but at temperatures above 60 °C it begins to lose its solid structure and becomes supple. In some cases, however, this material sticks to the feeder walls during printing due to the heat rising from the extruder. It crystallizes here and may cause interrupted printing or even print head clogging. The solution recommended to users is, for example, obtaining a better quality extruder or a system of fans for cooling the feeder.

Fortunately, a great advantage of open-source 3D printers is a constantly growing community of users who are willing to help with their experience at several forums.

10 Conclusion

The entire production process of physical terrain and surface models is quite complicated in the beginning, but after proper settings of individual parameters and automation of data transfers it may become an interesting way of making plastic models of any area. In production of unique pieces or smaller quantities of identical models, the 3D printing technology is also significantly cheaper than the previously used methods of producing 3D maps.

As with most other fields, the higher is the purchase cost, the better is the product quality and the easier is whole the process, whether we are talking about hardware or software used. However, it is necessary to consider the goals we want to achieve, because the various methods of 3D printing, materials used, etc. always excel in a specific field of use.

The key parts of the process are obtaining suitable elevation data, choice of the ideal horizontal and vertical scale for each area, effective settings of the model printing parameters and finally successful printing. Resulted 3D models can be then used in many other application where 3rd dimension is appropriate like in spatial planning (Pechanec et al. 2011).

Acknowledgments The authors gratefully acknowledge the support by the project IGA_PrF_2014007 of the Palacký University, Olomouc.

References

Ahn D, Kweon J-H, Choi J, Lee S (2012) Quantification of surface roughness of parts processed by laminated object manufacturing. J Mater Process Technol 212(2):339–346

Butscher A, Bohner M, Hofmann S, Gauckler L et al (2011) Structural and material approaches to bone tissue engineering in powder-based three-dimensional printing. Acta Biomater 7(3): 907–920

Chua CK, Leong KF, Lim CS (2010) Rapid prototyping: principles and applications. World Scientific, Singapore, ISBN 9812778977

Gauvin R, Chen Y-C, Lee JW, Soman P et al (2012) Microfabrication of complex porous tissue engineering scaffolds using 3D projection stereolithography. Biomaterials 33(15):3824–3834

Ghawana T, Zlatanova S (2013) 3D printing for urban planning: a physical enhancement of spatial perspective. Urban Reg Data Manage UDMS Ann 2013:211

Pechanec V, Burian J, Kilianová H, Němcová Z (2011) Geospatial analysis of the spatial conflicts of flood hazard. Moravian Geogr Rep 19:41–49

Ranellucci A (2014) Slic3r—G-code generator for 3D printers. [Version for 1.1.7 2014]

Rase W-D (2012) Creating physical 3D maps using rapid prototyping techniques. True-3D in Cartography (pp 119–134)

Schwarzbach F, Sarjakoski T, Oksanen J, Sarjakoski LT et al. (2012) Physical 3D models from LIDAR data as tactile maps for visually impaired persons. True-3D in cartography (pp 169–183)

Svobodová J, Voženílek V (2010) Relief for models of natural phenomena. In: Landscape modelling. Springer, Netherlands, pp 183–196

Tachikawa T, Hato M, Kaku M, Iwasaki A (2011) Characteristics of ASTER GDEM version 2. In: IEEE international geoscience and remote sensing symposium (IGARSS), IEEE, pp 3657–3660

Uhlířová K (2010) Možnosti využití leteckého laserového skenování pro vodohospodářské účely. Sborník-Sympozium GIS Ostrava

Werner M (2001) Shuttle radar topography mission (SRTM) mission overview. Frequenz 55(3–4):75–79

Yanev K (2014) Printrun. https://github.com/kliment/Printrun

Line-of-Sight Derived Indices: Viewing Angle Difference to a Local Horizon and the Difference of Viewing Angle and the Slope of Line of Sight

Jan Caha and Alexandra Rášová

Abstract The determination of visibility within GIS is often treated as a Boolean phenomenon. The surface is classified into two classes—visible and invisible parts of surface. This classification is result of the algorithm that compares whether a point of interest rises above all points on a line of sight that can be constructed between a viewing point and a point of interest. This comparison reduces all the information originally contained within the line of sight into simple true or false statement. However, the issue visibility is in fact more complicated than simple true/false statement. One of the other elements affecting visibility that has been already identified is the distance from the viewpoint or the relation of a point of interest and a local horizon. The distance affects the suitability of visibility: closer areas have obviously better visibility than distant places. Besides distance there are also other factors that can significantly affect visibility or more precisely the suitability of visibility. In this paper we would like to introduce two indices that can be calculated from the line of sight and that have potential to help user with better interpretation and also reasoning about visibility. The first index is based on the difference of the vertical viewing angle of a local horizon and a point of interest. The second index is the difference of a viewing angle and slope of line-of-sight (LoS) at the point of interest. Both these factors influence how well a point of interest is visible from a viewing point. The paper presents the methods and algorithms for calculation of these indices as well as case study to present these indices in a practical example.

Keywords Visibility analyses · Viewshed operation · Line of sight · Viewing angle

J. Caha (✉)
Institute of Geoinformatics, VSB-Technical University of Ostrava, 17. listopadu 15, 70800 Ostrava, Czech Republic
e-mail: jan.caha1@vsb.cz

A. Rášová
Slovak University of Technology in Bratislava, Radlinskeho 11, Bratislava, Slovakia
e-mail: alexandra.rasova@stuba.sk

K. Růžičková and T. Inspektor (eds.), *Surface Models for Geosciences*, Lecture Notes in Geoinformation and Cartography,
DOI 10.1007/978-3-319-18407-4_6

1 Introduction

Visibility analyses (often referred to as viewshed operation (Fisher 1993, 1996) in GIS usually provide Boolean result classifying the surface into visible and invisible parts of the surface. The operation is known to be rather sensitive on quality of the input data—mainly accuracy of the surface. There has been a significant amount of research dedicated to this topic e.g. Fisher (1992, 1994) and (Nackaerts et al. 1999). While precision and accuracy of the surface is very important for the calculation of visibility it is only one of the aspects affecting the calculation. Another issue is that the nature of visibility in the real world is not Boolean. Indeed, the area can be classified into categories "visible" and "invisible" with respect to the observer. However, when describing the situation, the observer will probably describe what was visible well, what was barely visible and what he/she could not see at all. This is what Ogburn (2006) describes as "levels of visibility".

In this article the line-of-sight (LoS) refers to a line consisting of points with known x, y and z coordinates. LoS can be visualized as a profile of the surface between the viewpoint and the point of interest (Fig. 2). The process of calculating visibility can be divided into two main steps. First, the points of LoS on the grid are determined and their elevations are calculated. There are several variants of this part of the algorithm that are described by Fisher (1993). This step leads to the creation of the actual LoS, which is later used in the second part of the algorithm that is relatively simpler and consists of calculations on the LoS. While classic viewshed operation determines only if the LoS intersects the surface between the viewpoint and the point of interest, there are also other types of visibility that can be calculated. For example Fisher (1996) mentions the horizons viewsheds. The horizon viewsheds provide information about the position of a point of interest and local/global horizons, which enriches the information of classic Boolean viewshed. As mentioned by Fisher (1996) it is substantial for GIS users to understand the operations correctly and use them to answer queries that the operations are designed for, otherwise serious misunderstandings may occur.

In this paper we would like to present two indices that significantly enrich the information provided by the viewshed analyses. These indices are calculated from the LoS and provides user with better information about the visibility of points on the LoS.

2 Line of Sight Determination and Visibility Calculation

2.1 Inferring LoS from Grid

The visibility analyses are most often calculated on raster data model even though the variants of the algorithm for TIN do exist as well (Llobera 2003) (Fig. 1). Fisher (1993) identifies four possible approaches to the inferring of elevations on LoS:

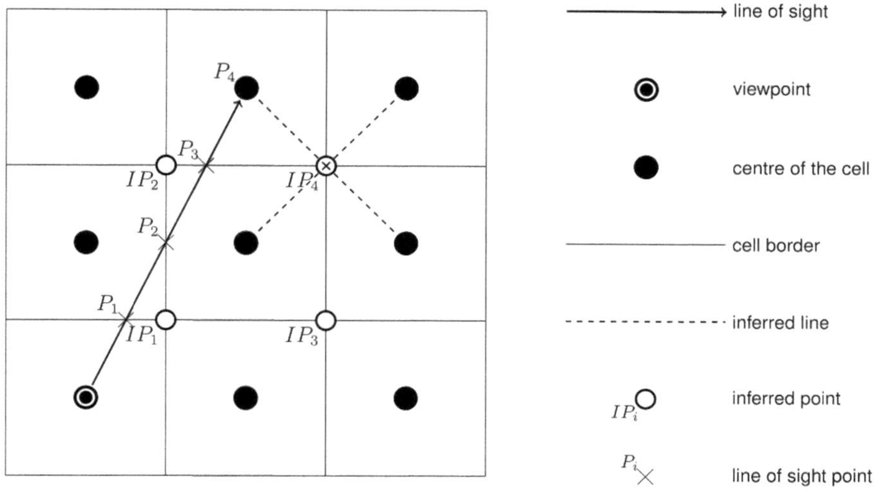

Fig. 1 The stepped model for inferring elevations from the grid

linear interpolation between grid neighbours, triangulation of the grid, grid constraint of the mesh and the stepped model. The choice of the inferring method does affect the result of the viewshed operation, because it determines how much and what types of elevations (besides the centres of cells of the grid) are added to the LoS. However, it does not affect the calculation of the indices from the LoS because the algorithms are not dependant on the specific structure of LoS. In this research the stepped model for the inferring of points and their elevation on the LoS is used. Figure 1 shows how the points on the LoS are obtained from the surface. The inferred points are calculated as mean value of four neighbouring cell values (as show on IP4) and the points on LoS are calculated as weighted mean of two inferred points on the cell border that the point lies on (e.g. P3 is calculated as weighted mean of IP2 and IP4).

2.2 Calculation on LoS

The LoS consists of viewpoint denoted V and a set of points P_i (Fig. 2). The viewpoint has the elevation Ve that consists of elevation of the surface at the location of the viewpoint plus an offset that describes height of the observer above the surface. Each point P_i has the distance from the viewpoint denoted as P_id and the elevation P_ie. Points are sorted according to their distance from the viewpoint from the closest to the farthest. The point with the highest distance is the point of interest. However, that does not apply in all cases; the calculation of the global offset viewshed (Fisher 1996) requires LoS to extend farther behind the point of interest. The points between

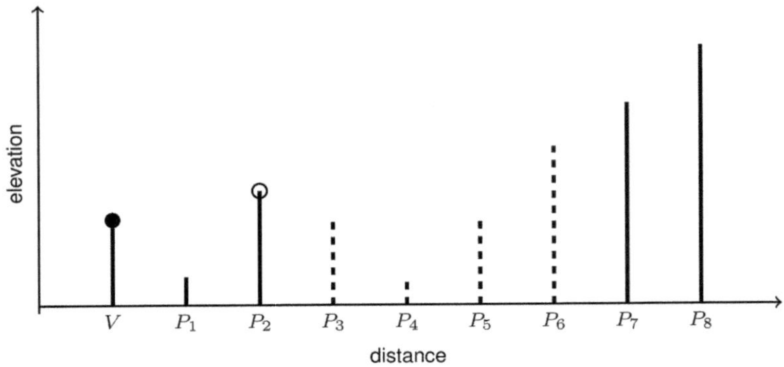

Fig. 2 The LoS with visible points shown as *full lines*, invisible points shown as *dashed lines*. The point P_2 is a local horizon

the viewpoint and the point of interest are used to determine whether the point of interest is visible. For each point P_i a viewing angle can calculated as:

$$P_i\alpha = \frac{180}{\pi}\arctan\left(\frac{P_ie - Ve}{P_id}\right) \quad (1)$$

A point P_n is visible if $P_m\alpha < P_n\alpha$ for all $m < n$. The determination of classic Boolean visibility is relatively simple calculation. Besides visibility, there are other indices that can be calculated from LoS, e.g. difference of the vertical angle of the point of interest from horizontal plane at height of the observer (Neteler and Mitasova 2004) or vertical difference of a point of interest from local or global horizons (Fisher 1996). For the purpose of this research, the local horizons on LoS are of interest as they will be further used in calculations. A local horizon is a visible point on LoS that is immediately followed by at least one point that is not visible. LoS may have several local horizons. A global horizon is a horizon where the terrain meets the sky; there are no further visible points behind it (Fisher 1996).

3 Proposed Indices

As mentioned previously, the LoS contains a lot of information about visibility, which, as Fisher (1996) showed, can be used for various applications, e.g. landscape planning or fire detection. In this section we would like to present two indices that should help users to reason about visibility and provide them with information not only about what is visible but also how well it could be visible.

Both proposed indices are based on viewing angle of the point of interest and its relation to other characteristics of LoS. Through the rest of the text the highest local horizon between the viewpoint and the point of interest will be denoted LH. In the same way as other points on LoS LH have the elevation LHe, the distance LHd and the vertical viewing angle LHα from the viewpoint.

3.1 Viewing Angle Difference to a Local Horizon

Fisher (1996) provided an index named local offset viewshed that is calculated as the difference of LHe and P_ie. Such index could be used to identify whether an object (e.g. building) of height x will be visible from the viewpoint. Indeed, this is a useful characteristic for the landscape planning or any other application where the difference of heights matters. With slight modification, this index can help the user to identify areas that are visible, but the difference of their elevation to the local horizon is so small that these points barely rise over the horizon. To identify such areas the actual difference of elevations is not important because it is not possible to compare elevations at different distances from the observer. To allow comparison, these elevation differences must be converted to differences in angles (according to Eq. (1)). For every point of interest the difference angle to a local horizon (Fig. 3) can be calculated as:

$$LH\Delta\alpha = LH\alpha - P_i\alpha \tag{2}$$

Two special cases may occur: if the P_i is not visible, then the value of LH$\Delta\alpha$ = −1 denotes invisibility, and if there is no local horizon between the viewpoint and P_i then the value of LHα = −90 is used for calculation to emphasize that the user is not limited by a local horizon thus the difference is calculated from the downward direction.

The utilization of viewing angle difference instead of elevations, as proposed by Fisher (1996), allows comparison of the differences at various distances. Elevations,

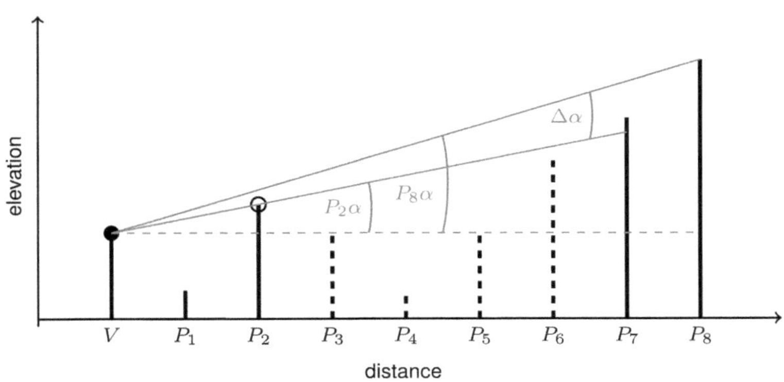

Fig. 3 Visualization of the angle difference to a local horizon (P_2)

on the other hand, can only be compared if they occur at approximately similar distance from the observer. This index provides information about how significantly the point of interest rises above the horizon; obviously points with small values will be barely visible. Also small value LHΔα can be used to identify points that can be easily affected by uncertainty of the surface. If the angle difference is small, then a small rise of the elevation of local horizon will render the point invisible, so this index can be used to point out areas that may be influenced by uncertainty.

3.2 Difference of Viewing Angle and the Slope of LoS

The slope of a surface is quite an important factor that indicates how well are specific parts of the surface visible. Imagine an observer standing in a relatively flat area with some hills farther away. This observer can see relatively well in the area close around him, but as the distance grows, the flat land is no longer well visible, because the slope of the surface aligns with the viewing angle. The hills on the other hand are likely to be well visible to him, because the slope there is higher. This is caused by the slope of surface intersecting with LoS with higher angle. An example is show in Fig. 4 where the viewing angle is denoted as $P_8\alpha$ and the slope of LoS is denoted as $P_8S\alpha$. The slope is defined by the point of interest (P_8) and the point that is just before the point of interest (P_7). If the point P_7 has higher elevation the angle difference would be smaller making the visibility of the point P_8 worse. On the other hand, if the elevation of P_7 is smaller than the angle difference would be higher and the visibility would be better. The reason why there is a need to calculate the angle difference and not only the slope of LoS is the situation when observer at elevated terrain looks on lower flat terrain. Such terrain is well visible to him, because the angle difference is higher.

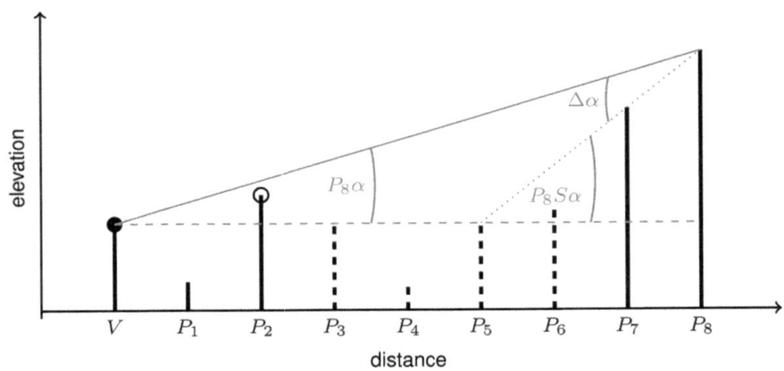

Fig. 4 Visualization of the angle difference to a local horizon

Formally the slope of LoS at point P_i is defined based on height of this and the previous point $P_{(i-1)}$ of LoS:

$$P_i S\alpha = \frac{180}{\pi} \arctan\left(\frac{P_i e - P_{i-1} e}{P_i d - P_{i-1} d}\right) \tag{3}$$

If the point of interest is P_1 then elevation of the viewpoint V is used as $P_{(i-1)}e$ (obviously the surface value is used—that means Ve without the offset of the observer) and $P_{(i-1)}$ d = 0.

The angle difference of the slope is than calculated as:

$$S\Delta\alpha = P_i S\alpha - P_i \alpha \tag{4}$$

The result is always a positive number. For the invisible points on LoS the value $S\Delta\alpha = -1$ is assigned.

4 Case Study

Visibility analysis is often used in archaeological studies to inquire about spatial relationships of monuments and their placement in the landscape, such as the analysis of Bronze Age cairns of Mull (Fisher et al. 1997), stone circles in Ireland (Flanagan 2006), Nasca geogplyphs in Peru (Lambers and Sauerbier 2008).

Proposed indices were used in the analysis of roundels located in the Western Slovakia, Piešťany district. Roundels are circular ditched structures from the Middle Neolithic era (first half of the 5th millennium BCE) that can be found in Central Europe (Germany, Austria, Czech Republic, Slovakia). Although they occur in various configurations and differ in sizes and forms, they share some common features: round form, four gates, one or more ditches, and a palisade. The function of roundels is still unknown, there are assumptions about their social, cultural, and ritual usage, astronomic calendar, solar temple, or more pragmatic function as a fortification, marketplace, or a shelter for people and animals (Melichar and Neubauer 2010).

Considering the nature of these monuments, it is relevant to analyse their placement in the landscape in relation to other roundels, significant landscape features, or astronomic phenomena. In this study we calculated the viewing angle difference and the difference of viewing angle and the slope of LoS for 3 roundels: Prašník, Šterusy, and Borovce. This dataset was provided by the Institute of Archaeology of the Slovak Academy of Sciences in Nitra. The location of these objects is visible from aerial photographs and was confirmed by geophysical measurement to verify their origin. The digital elevation model used in the analysis was the digital terrain model "DMR-3.5" with 10 m resolution provided by the Geodetic and Cartographic Institute Bratislava (GKU) (Fig. 5).

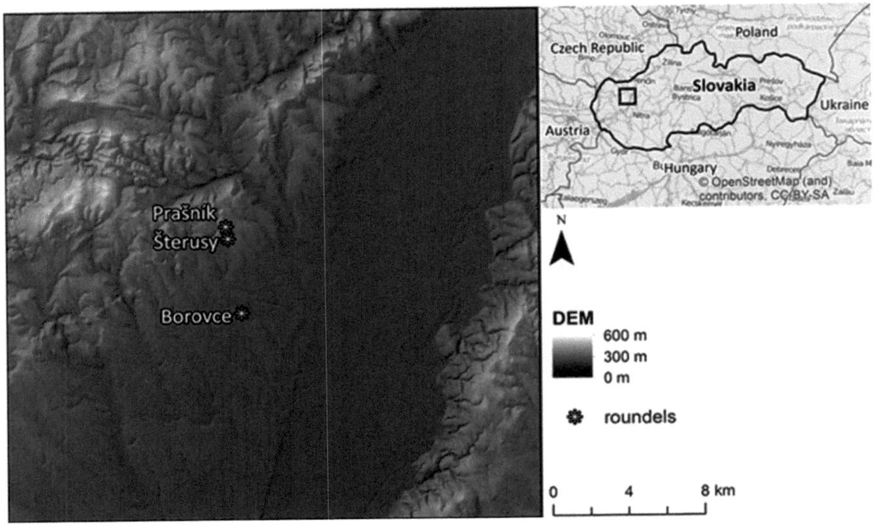

Fig. 5 The digital elevation model DMR 3.5 with observer points (roundels)

Computed indices—viewing angle difference to a local horizon and the difference of the viewing angle and the slope of the LoS—are valuable mostly in the analysis of the areas visible from the roundel in the landscape, but they can be used also in the analysis of mutual visibility of the roundels. In Fig. 6 there is the line of sight between two roundels, Borovce and Prašník, along with the illustration of the viewing angle difference and the difference of the viewing angle and the slope of the LoS. These indices confirm the mutual visibility of these points: viewing angle difference of 1.6° is higher than the human eye acuity (about 1 arc minute), and the difference of viewing angle and the LoS slope of 37°. This is also the case of the

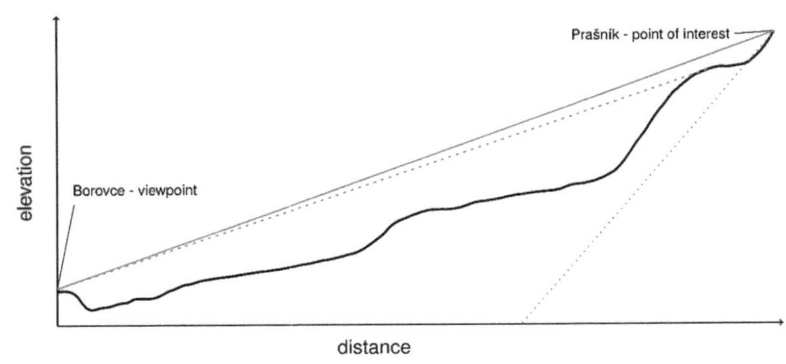

Fig. 6 LoS between two roundels. *Full gray line* shows viewing angle, *dashed gray line* shows viewing angle to a local horizon and the *dotted gray line* shows slope of LoS at the point of interest

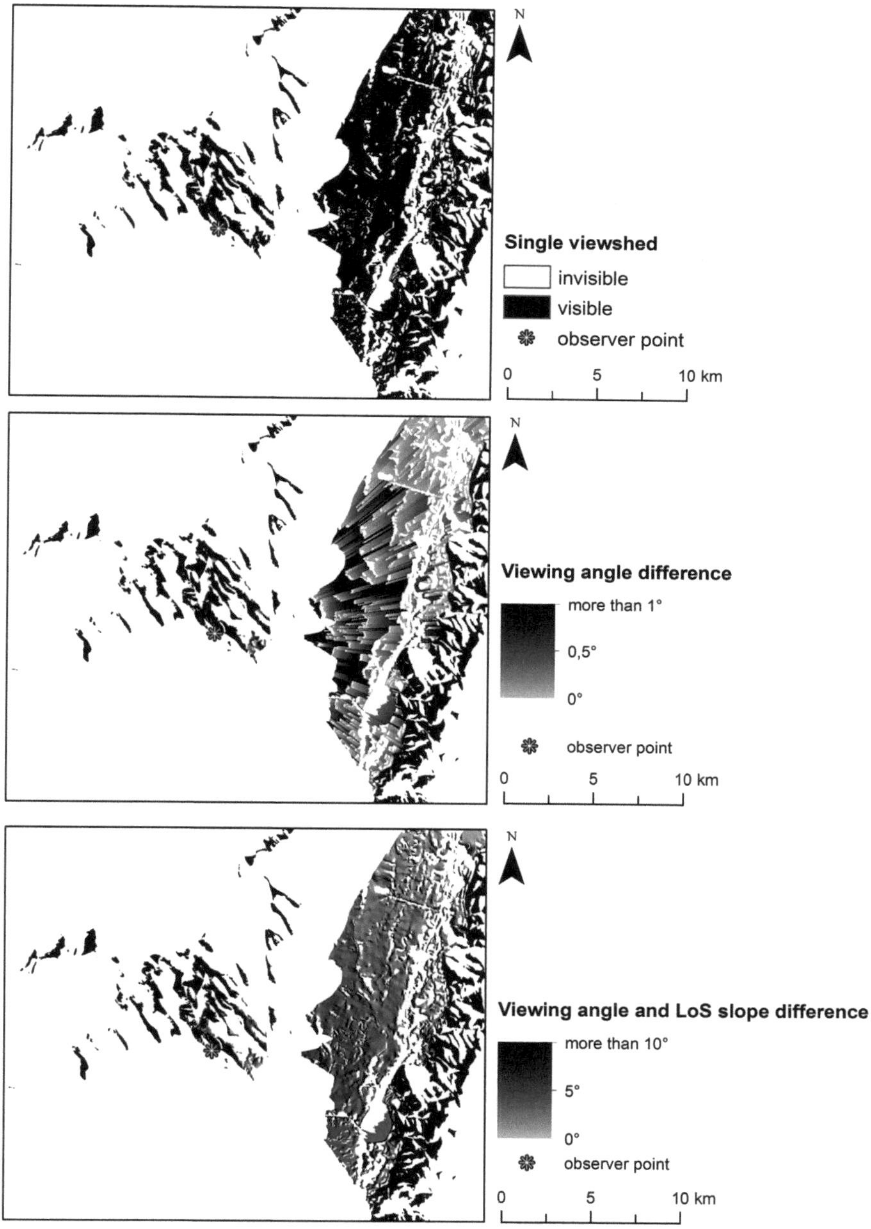

Fig. 7 Single viewshed (*top*), viewing angle difference (*middle*) to a local horizon, and the difference of the viewing angle and the slope of the LoS (*bottom*) computed for the roundel "Borovce"

other pair of visible roundels, Borovce and Šterusy (Roundels Prašník and Šterusy are not mutually visible despite the fact that they are very close to each other), which has the difference of viewing angle about 7° and the difference of viewing angle and the LoS slope about 35°.These values—and also the values of reciprocal indices computed for a situation with exchanged target and observer point—speaks for relatively good visibility between these points, especially considering the roundel construction, which could have been few meters high (offset above the ground), that would increase the value of the difference of the viewing angle.

More interesting is to examine the areas visible from the roundels (not specific points) to examine what portion of the landscape could have been observed and how well could this visibility been. An example is in Fig. 7, where the single viewshed, the viewing angle difference to a local horizon and the difference of the viewing angle and the slope of LoS is displayed for the roundel Borovce. While the single viewsheds provides only the information about visibility or invisibility of cells in the computing area, the two indices provide a piece of information with higher interpretation value about the suitability of visibility: the hills and hillsides have much higher values of both indices than the flatter areas in the valley. In the viewing angle difference we can observe that the areas with higher value of the index closer to the observer act radially as barriers for areas right behind them.

Using these indices it is possible to analyze the visibility more qualitatively and the results represents the reality better than the binary viewshed operation. They can be used to compare the visibility patterns of structures such as roundels in archaeological research to find whether the location of one of them could have been more preferable than the other, in future research maybe also with consideration of significant landscape features.

5 Conclusions and Future Work

As mentioned previously, visibility is more complex phenomenon than its representation in classic Boolean viewshed suggests. Fisher (1996) suggested several indexes that were designated for the usage in landscape planning. In this paper we propose two indices with more general utilization. The main purpose is to provide the user with more information that can be obtained from the line of sight. As mentioned by Fisher (1996), there are many possible queries regarding the visibility that the user may ask. The binary viewshed does not sufficiently cover all these queries, so new indices calculated from LoS are both necessary and useful. It is interesting that new variants of visibility algorithm (the boolean variant) are being proposed (e.g. Tabik et al. 2013), yet information that could be obtained from LoS is studied very rarely.

The proposed indices—viewing angle difference to a local horizon and the viewing angle difference to the slope of LoS should provide user with more information about the fitness of visibility and allow to reason about the visibility in a qualitative way. The binary information provided by classic Boolean viewshed is significantly enriched, as described on the case study. For both indices a limit values, denoting what is distinguishable in real world, could be identified. That would allow specific parts of visible areas, e.g. viewing angle difference to a local horizon smaller than 1 arc minute, to be proclaimed as invisible or problematically visible. However, specific values of these limits are a question to be answered. This is an issue for a further research. Another interesting topic for further research is the utilization of the proposed indices on visibility calculated from a line (Růžičková et al. 2012). In such cases the indices can provide information about how well are specific areas visible from a line, such line can represent for example a hiking trail or a road.

Both these indices could be also used in their reversed variants. The reverse variant of the viewshed algorithm according to Fisher (1996) is the calculation of pixels from which a specific point can be seen. Due to a problem with intervisibility, this is query differs from the one about the area visible from this point (Fisher et al. 1997). In the same way both proposed indices have a reversed variant. The reversed variant is describing the angle difference of a viewing angle to a specific point from a local horizon for each pixel of the surface. The second index then describes the difference of a viewing angle to a specific point from the slope of the LoS for each pixel of the surface. Both these variants would provide interesting information about how well a specific point could be seen from the area of interest. This issue will be studied in further research as it is of importance for archaeological studies.

References

Fisher PF (1992) 1st experiments in viewshed uncertainty-simulating fuzzy viewsheds. Photogram Eng Remote Sens 58(3):345–352

Fisher PF (1993) Algorithm and implementation uncertainty in viewshed analysis. Int J Geogr Inf Syst 7(4):331–347

Fisher PF (1994) Probable and fuzzy models of the viewshed operation. In: Worboys M (ed) Innovations in GIS 1. Taylor & Francis, London, pp 161–175

Fisher PF (1996) Extending the applicability of viewsheds in landscape planning. Photogram Eng Remote Sens 62(11):1297–1302

Fisher P, Farrelly C, Maddocks A, Ruggles C (1997) Spatial analysis of visible areas from the bronze age Cairns of Mull. J Archaeol Sci 24(7):581–592

Flanagan DL (2006) A spatial analysis of stone circles in Southwestern Ireland: master's thesis. The University of Texas at Arlington, Arlington, p 160

Lambers K, Sauerbier M (2008) Context matters: GIS-based spatial analysis of the Nasca Geoglyphs of Palpa. In: Wagner GA, Reindel M (eds) New technologies for archaeology multidisciplinary investigations in Palpa and Nasca, Peru. Springer, Berlin, pp 321–338

Llobera M (2003) Extending GIS-based visual analysis: the concept of visualscapes. Int J Geogr Inf Sci 17(1):25–48

Melichar P, Neubauer W (2010) Mittelneolitische Kreisgrabenanlagen in Niederösterreich. ÖsterreichischeAkademie der Wissenschaften, Wien

Nackaerts K, Govers G, Van Orshoven J (1999) Accuracy assessment of probabilistic visibilities. Int J Geogr Inf Sci 13(7):709–721

Neteler M, Mitasova H (2004) Open source GIS : a GRASS GIS approach, 2nd edn. Kluwer Academic Publishers, Boston

Ogburn DE (2006) Assessing the level of visibility of cultural objects in past landscapes. J Archaeol Sci 33(3):405–413

Růžičková K, Růžička J, Cirbus J, Matuszczyk M (2012) The influence of observer line geometry on results of visibility analysis. In: 12th international multidisciplinary scientific GeoConference. Conference proceedings, 17–23 June 2012, vol 2. Albena, Bulgaria, p 1177–1184

Tabik S, Zapata E, Romero L (2013) Simultaneous computation of total viewshed on large high resolution grids. Int J Geogr Inf Sci 27(4):804–814

Initial Results of a Surface Deformation by Using InSAR Techniques: Case Study of Babadağ (Denizli), Turkey

Fatma Canaslan Çomut, Şule Gürboğa and Aydın Üstün

Abstract InSAR (Interferometric synthetic aperture radar) techniques and applications are very useful for monitoring and mapping different types of deformations such as landslides, subsidence and volcanic activities on the Earth surface. In this research, we investigated Babadağ landslide (Denizli in western Turkey) by using the InSAR techniques and compare our previous research from Karaman city, central Turkey. An active landslide in Babadağ county has been issued in the study by using DInSAR to monitor the vertical deformation amount. According to initial results, the application of DInSAR has not have sensitive like PSInSAR in Karaman for the recognition of subsidence. In the further step, PSInSAR technique will be applied for the Babadağ landslide.

Keywords InSAR · PSInSAR · Babadağ landslide · Denizli · Karaman · Turkey

1 Introduction

Deformation rates of the Earth's surface has been investigated with various methods by different disciplines, especially earth scientists. Because, deformation rates are the results of natural phenomena and give many information about the processes that affect the Earth. One of the most effective methods of remote sensing to understand the precise meaning of the earth deformation is Interferometric

F.C. Çomut (✉)
Disaster and Emergency Directorate of Denizli (AFAD), Denizli, Turkey
e-mail: fatma.c.comut@afad.gov.tr

Ş. Gürboğa
Department of Geological Engineering, Akdeniz University, Antalya, Turkey
e-mail: sule.gurboga@gmail.com

A. Üstün
Department of Geomatics Engineering, Kocaeli University, Izmit, Turkey
e-mail: aydin.ustun@kocaeli.edu.tr

K. Růžičková and T. Inspektor (eds.), *Surface Models for Geosciences*,
Lecture Notes in Geoinformation and Cartography,
DOI 10.1007/978-3-319-18407-4_7

Synthetic Aperture Radar (InSAR) technique. In this research, we tried to understand the potential of DInSAR techniques to realize the vertical displacement, and then we compare the result with our previous PSInSAR application for subsidence in Karaman (central Turkey). For the recognition of surface movements (landslide and subsidence), we investigated the capability of Synthetic Aperture Radar Interferometry for Babadağ landslide. Remote sensed data allow examining previous movement evidence and also these data can be used to predict the recent and future events for the detection and monitoring of earth surface processes.

2 Mathematical Approach

In the processing of SAR data, after the the creation of complex master and slave images, corresponding pixels are overlapped with each other (co-registration). In this way, the phase difference between each conjugate pixel values gives a deformation rate after eliminating the error on the data. InSAR measurements are especially sensitive for topography, mass movement, atmospheric conditions, spatial distinctions between satellites and the electrical properties of the earth. All these factors have been given in formula 1 (Ferretti et al. 2001; Colesanti et al. 2003). InSAR phase signals are mean changes in all these factors and depending on the application, some factors can be considered as noise. For example, if the main goal is to create a Digital Elevation Model (DEM), deformation can be considered then as noise. Similarly, while achieving change where the time series analysis methods is difficult to separate the stable phase topographic factors.

$$\Delta\phi = \varphi_{flat} + \varphi_{topo} + \varphi_{orbit} + \varphi_{defo} + \varphi_{tropo} + \varphi_{iono} + \varphi_{scat} + \varphi_{noise} \tag{1}$$

$\Delta\varphi$	Interferometric phase difference
φ_{flat}	flat Earth phase
φ_{topo}	topographic phase
φ_{orbit}	phase effect of orbit error
φ_{defo}	phase effect related to surface deformation
φ_{tropo}	tropospheric phase effect
φ_{iono}	ionospheric phase effect
φ_{scat}	phase effect related to electrical characteristics of scattering objects
φ_{noise}	combined noise

Flattening of Earth (φ_{flat}), topographic phase (φ_{topo}) and surface deformation (φ_{defo}) are three important effects that are the equation of the distance between the two transitions (array) and they are the function of satellite orbit and topography.

Changing in distance is related to the phase in Eq. (2)

$$\Delta\varphi_{flat,\ topo,\ defo} = \left(\frac{4\pi}{\lambda}\right) \times \Delta R \tag{2}$$

where λ is radar wavelength ΔR is change in distance between the first and second passes of satellite. ΔR and flat Earth phase are related to shape of Earth and calculated by using the orbit of satellite and geodetic datum (WGS84). SAR images are generated assuming that Earth is flat. Because of this assumption, phase difference appears and the Eq. (3) is related to parallel phase:

$$\varphi_{flat} = \left(\frac{4\pi}{\lambda}\right) \times B_{\amalg} \tag{3}$$

$B_{\amalg}$ = parallel phase for each pixel.

Topographic phase is the combination of interferograms related to a part of topography up to reference ellipsoid of topography and it is portional to the vertical base (Eq. 4).

$$\varphi_{topo} = \left(\frac{4\pi}{\lambda}\right) \times \left(\frac{B_{\perp}}{R \times \sin\theta}\right) \times \Delta z \tag{4}$$

φ_{topo} topographic phase effect
$B_{\perp}$ perpendicular (vertical) base length
R the distance between target and satellite
θ antenna look angle
Δz topography from reference ellipsoid

In Eqs. (3) and (4), values of flat of Earth and topographic phase is provided in relation to the $B_{\amalg}$ and $B_{\perp}$ length of base. Therefore, any error in the orbital data is evaluated in the residual and shown as φ_{orbit} at the end.

Phase effect with respect to surface deformation (5);

$$\varphi_{defo} = \left(\frac{4\pi}{\lambda}\right) \times \Delta R_{defo} \tag{5}$$

Deformation is also a part of interferometric phase and measured in the line-of-sight of radar (Osmanoğlu, 2011).

3 Target Areas

Babadağ landslide in Denizli city has been defined as case study for this research and its initial results have been compared with our previous research (Fig. 1). Whereby we apply DInSAR method for the mass movement in vertical direction and measure the capabilities of InSAR techniques for different deformation types.

In Babadağ landslide, 4 images of ALOS PALSAR satellite that traded on L-band have been preferred for the analyses. They have FBS (Fine Mode Single Polarisation), 28 MHz, 10 m resolution, 70 × 70 km in size and HH polarization images. These are the most suitable images that are equal to SLC (Single Look Complex) of Envisat ASAR to get best results in DInSAR applications. In our previous research (Canaslan et al. 2014), Karaman surface subsidence, 15 Envisat ASAR images were used to perform PSInSAR application. In Table 1, some characteristics of the satellites are given for the analyses of Babadağ landslide.

Fig. 1 Location map of the Denizli and Karaman cities

Table 1 General characteristics of satellites used in this study

	ALOS PALSAR
Observation mode	Fine beam FBS
Date	January 2006
Orbit height	692 km
Return period	46 days
Frequency	1.26 GHz
Band	L
Wavelength	23.6 cm
Polarization	HH/HH + HV
Angle	34.3 der.
Area	70 km × 70 km

3.1 Babadağ Landslide

Babadağ is a county 30 km far from the Denizli city (western Turkey) with a population of about 4500 people. Gündoğdu town in the border of Babadağ that has the highest risk for landslide was issued for many scientists that they have attracted attention about high risk for the population living in Gündoğdu (Fig. 2). Approximately 2000 people were living in Gündoğdu settlement and after the recognition of structural disasters the area has been evacuated. For the reason, the area was investigated by InSAR technique in this research to understand the ability of InSAR in such situation. To apply the analyses, the satellite images of European Space Agency have been used. The required images have been selected ALOS PALSAR and Envisat ASAR that are the most appropriate for use in urban areas for the Babadag and the surrounding areas. ALOS PALSAR images are extremely effective for the detection of landslides so they have been deemed appropriate to be processed.

Some scientific researchers measured these breakdowns on the buildings and this gave a chance to compare the results of InSAR technique with them. So, the capability of InSAR has been appeared. This area has been recognized as Disaster Zone by Repealed by General Directorate of Disaster Affairs and closed to any construction.

As a result, a large part of the Gündoğdu settlement sitting on an active landslide mass has been moved to the center of Denizli. 406 houses and 101 shops in this context were constructed (Fig. 3).

All these previous work have been done by using other scientific methods without a study covering larger areas in the region as InSAR. Scope of InSAR

Fig. 2 Damages resulted from landslide at station AE4: **a** location of station AE4 at January 2005, **b** view from station AE4 at 31 August 2007 with construction team, **c–d** damages at 27 June 2009, **e–h** damages at 22 March 2012 (Kumsar et al. 2012)

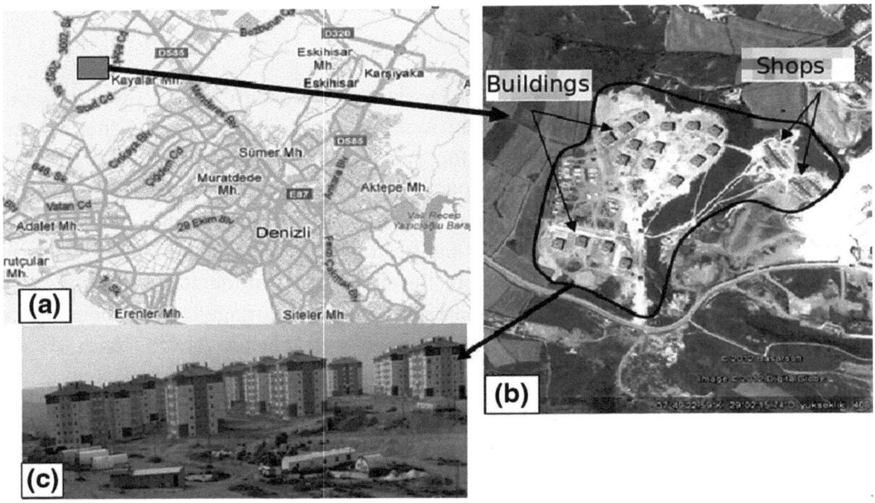

Fig. 3 New constructions of houses and shops in Denizli

applications in this area is to cover more broad area and compare the results of InSAR with previous works. As it is mentioned before, 4 images have been selected to perform the work and their track/frame is shown in Fig. 4.

The two important points in the image selection are acquisition time of images (temporal baseline), and the relative position of satellite at the time of acquisition of

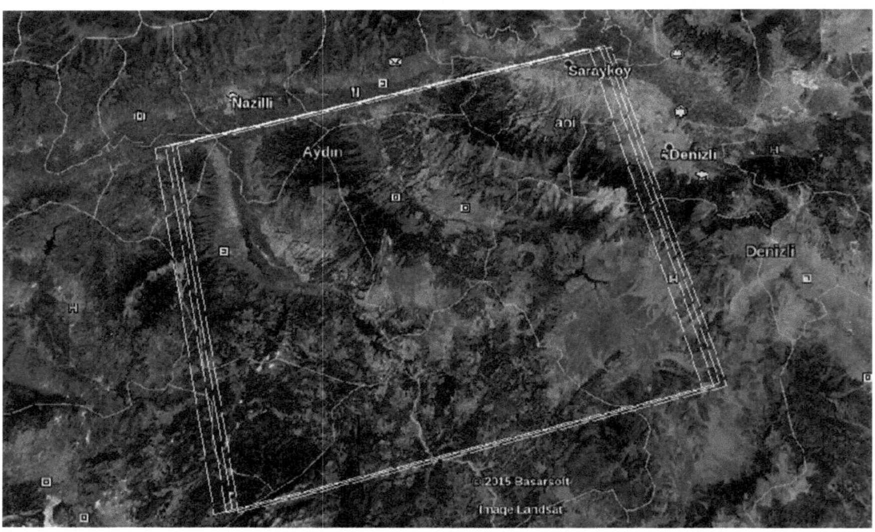

Fig. 4 ALOS covering the study area

the images. The intense rainfall period in March and December in 2009 and 2010 when the landslide has been expected to the most active time were chosen to perform the analysis. In addition, approximately 9 months difference between the acquisition time of images have been selected during that time difference the change in the working area may occur more than the length of radar wave, so that it was made possible to reveal more clearly the amount of deformation. The acquisition time of images and relative baseline length have been listed and shown at Fig. 5. In Fig. 5a, the legend on the right side indicates the coherence values between images. Four graphs in Fig. 5b show the normal baseline, temporal baseline and positions of seasonal characteristics of images with each other.

The acquisition times of images in terms of seasonal conditions are very important for the final product. For example, especially for the landslide zones, acquisition time must be chosen in the same season for differential InSAR

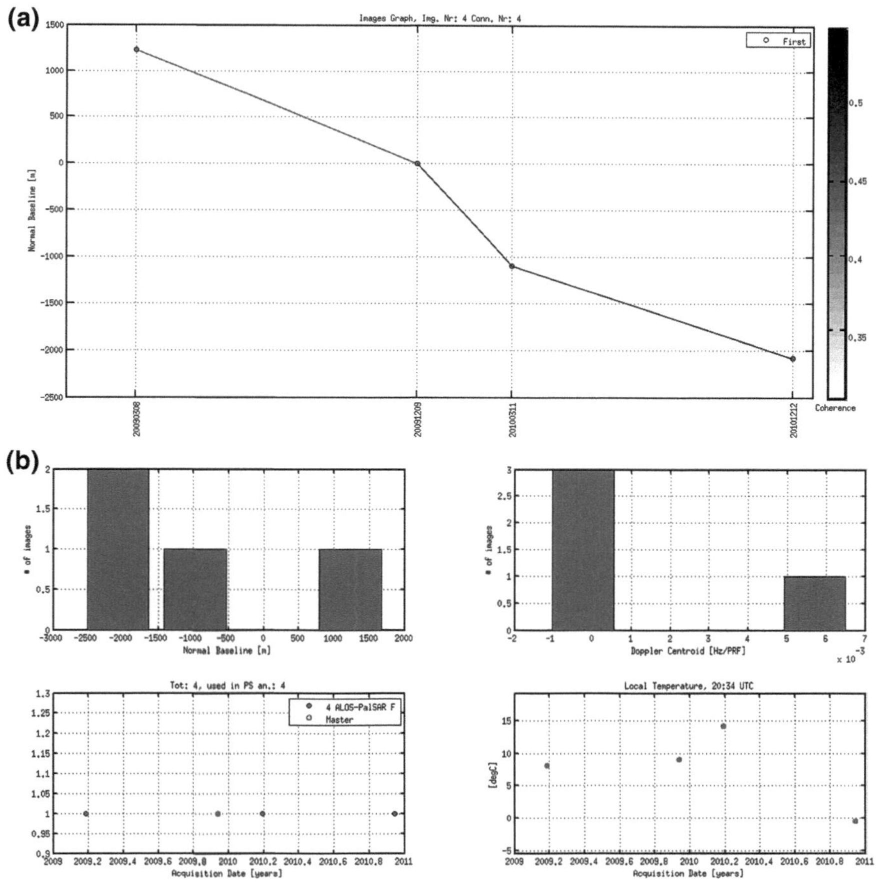

Fig. 5 The date of selected images and spatial distribution of baseline shown by tables

application like winter months in which heavy rainfall occurs. In addition, seasonal values for each image must be necessarily included in the image processing. For this purpose, graphics including image acquisition time related seasonal conditions is shown in Fig. 5b. All of these graphics are used to understand the proportion of correlation between deformation and seasonal values.

3.2 InSAR Application

To analyze the image in this study, Sarproz software developed in 2009 by Perissin is used (www.sarproz.com). 4 images of HH polarization SAR data have been examined by means of baselines and based on the optimal length of image, 20091209 dated image has been selected as the master image. Figure 6 shows a generated image of the master SAR data.

After this process, other slave images have been created for the overlapping process (co-registration). The proper coregistration is extremely important for the further stages of work (creation Georeferencing and interferograms). For the purpose, suitable Digital Elevation Model of the study area should be created. For this study SRTM data is used and Digital Elevation Model was produces shown in Fig. 7.

Babadağ county and near vicinities have been chosen as target area (Area of Interest, AOI) not all areas of Fig. 8. AOI Spot is specified for this frame as latitude 37.7988 and longitude 26.8684 on the geographical coordinates.

The next step is Georeference process for which ground control point should be determined. The verification data obtained from orbital correction or external digital elevation model is used. Because of the sensitivity of both methods depend on the accuracy of its interior features; a more precise Georeferencing could be done by means of a point with strong reflection property from reflectivity map and its coordinate (image and geographic) can be determined manually. This application would

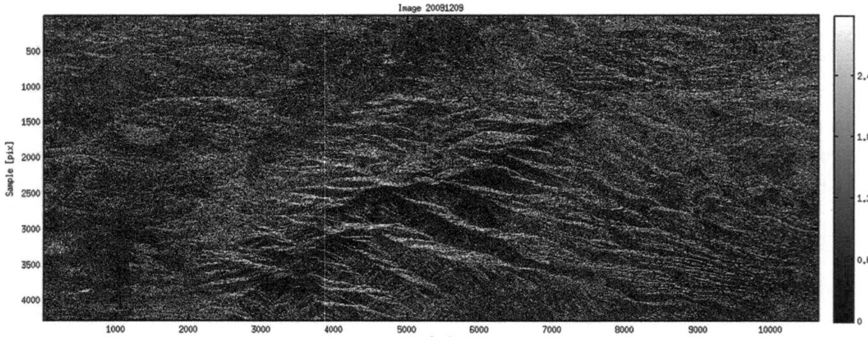

Fig. 6 Master image with 20091209 acquisition time

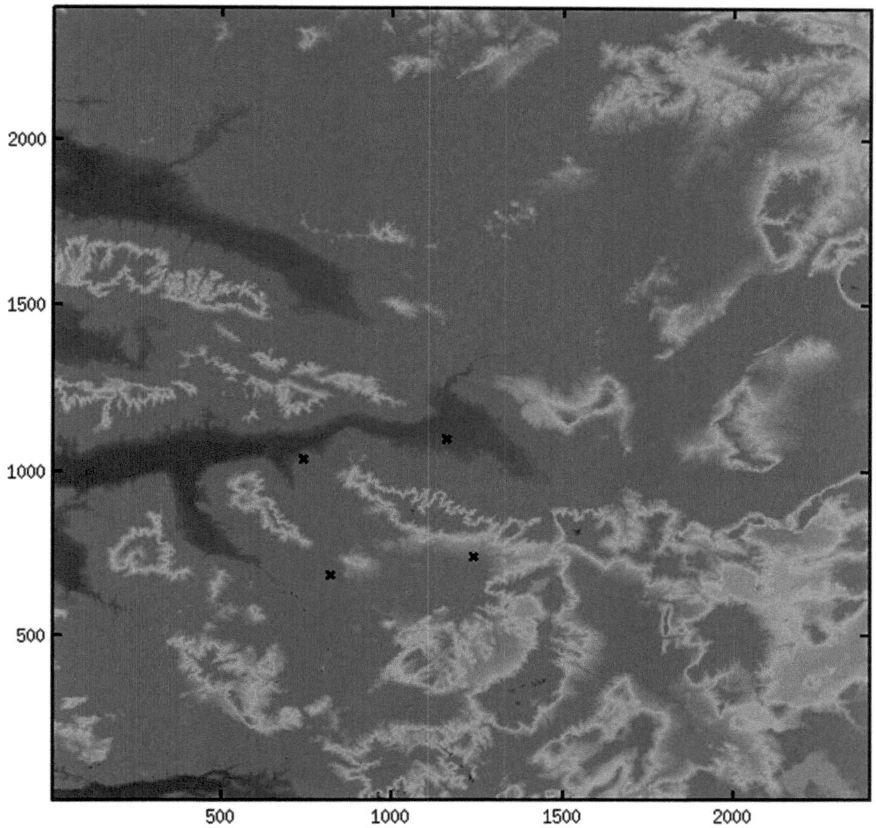

Fig. 7 Digital elevation model of study area derived from SRTM data. *Dark gray* (lower elevation) and *light gray* (higher elevation) colors are relative to each other. *Black crosses* define the selected area to create interferograms

have improved the accuracy of image. The ground control point at Fig. 9 which will be the basis for further data processing steps has been selected as ground control point.

Point with the best reflection properties in the frame covering our working area is center of Geyran county border, 37.713253° latitude and 28.710206° longitude on geographic coordinates and 4870 range (sample = y), 656 azimuth (line = x) on the image coordinates. After the main processing steps of InSAR (master extraction, slave extraction, co-registration, external DEM selection, georeferecing, flattening, removing DEM errors, filtering and unwrapping), 3 interferograms have been produced from 4 images. The highest correlation between the interferograms produced from 8 March 2009 to 9 December 2009 dates is given in Fig. 10.

The area indicated by the light blue is coincided with the area (Fig. 11) that is previously declared dangerous for landslide. So, all the houses and shops were evacuated and were carried out demolitions of old houses and work places.

Fig. 8 Babadağ and near vicinities that are focused to create interferograms

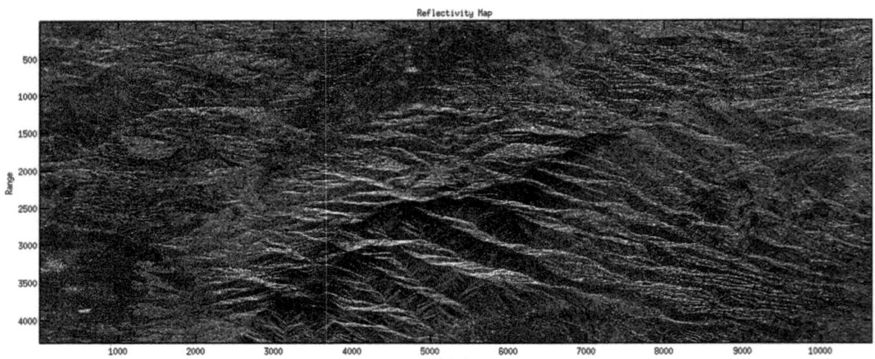

Fig. 9 Location of ground control point based on its high reflective feature on the reflective map

Even the results of landslide research is a preliminary work in this research, differential InSAR technique is very effective for the detection of landslide areas. After compilation of initial results obtained by InSAR in Babadağ, we would like to compare PSInSAR results of our previous research that have been applied in Karaman city (Canaslan et al. 2014). Karaman city is located in southern part of the Konya Closed Basin (Fig. 1). This area is faced with an important risk which is uncontrolled underground water wells. So, these wells create surface subsidence dangerously. Canaslan et al. (2014) has applied PSInSAR techniques around Karaman city and got very good results (Fig. 12).

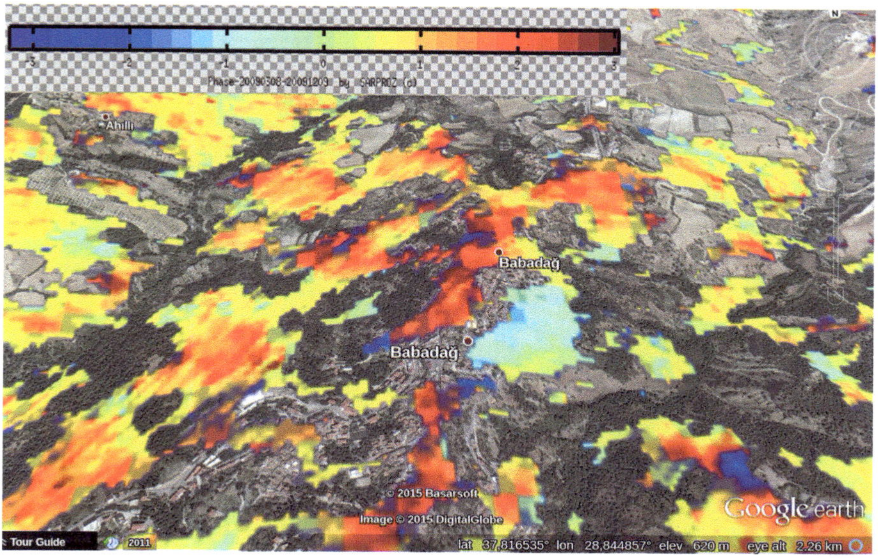

Fig. 10 Landslide area (*light blue colored area*) between two dates 8 March 2009 and 9 December 2009 (276 days)

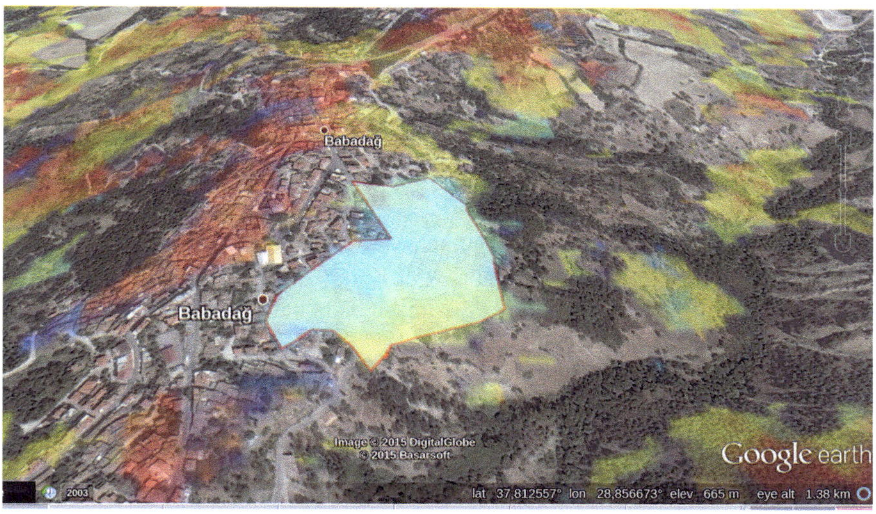

Fig. 11 Overlapping of *light blue colored* and border of previously evacuated area by the decision of Council of Ministers

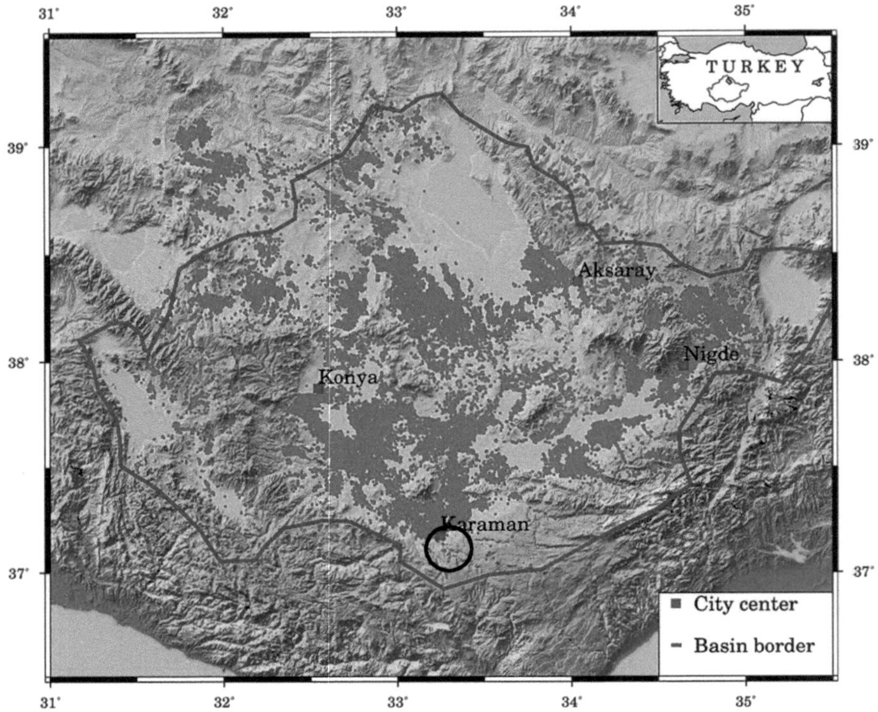

Fig. 12 Location map of Karaman city. *Black circle* is showing the area of interest

Total of 15 Envisat ASAR images taken between 2003 and 2010 (Fig. 13) years in descending mode for the application of PSInSAR have been selected for the study (Canaslan et al. 2014). In this research, PSInSAR (Ferretti et al. 2000) technique that has been developed as an alternative to the conventional InSAR (DInSAR) was applied over the urban area of Karaman city. The dataset shows fair overall coherence with five long-baseline image pairs having coherence below 0.15 (Fig. 13).

As a results of this previous research, point based deformation amounts have been obtained from the Karaman city center (Fig. 14).

As it is seen from the point base deformation map, in some points the amount of motion is more than 9 mm/year that are located on the southern parts of the Karaman city. These points should be investigated by further research. Light blue colored points represent almost stable areas. PS method presents a linear trend of subsidence in Karaman more than 4 cm within observed period. This is very good resolution such urban areas where a number of constructions are present.

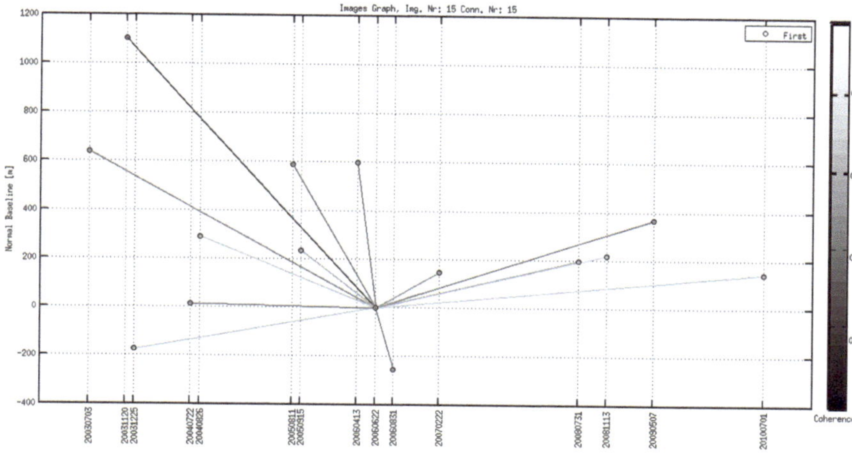

Fig. 13 Location of Envisat ASAR data to each other for PSInSAR analyses (Canaslan et al. 2014)

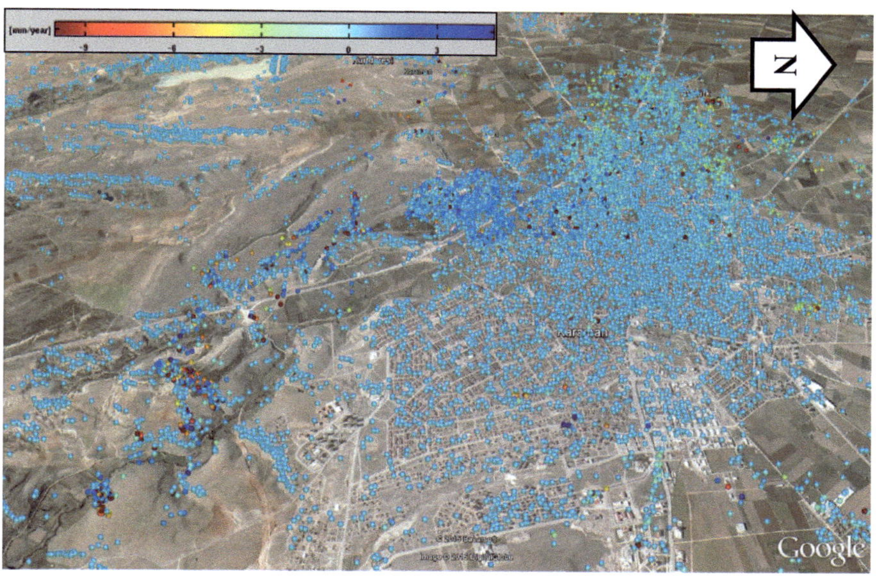

Fig. 14 PS processing result of Karaman area (15 × Envisat ASAR 2003–2010) using SarProz Software (Canaslan et. al. 2014)

4 Conclusions and Further Works

In this research, we focused on the analyses of Babadağ landslide by using the InSAR technique and compared to our earlier research in which PSInSAR technique had been used. Even Babadağ landslide analyses have representative good results for the amount of vertical deformation; PSInSAR technique has given good results more than DInSAR. It is clearly understood from the results that the determination of surface deformation has been obtained in cm scale in PSInSAR techniques in Karaman. PSInSAR has shown valuable additional information from DInSAR processing to monitor vertical deformations of the area of interest. To compare the results of InSAR techniques for the Babadağ landslide, the further work should also include processing of other multitemporal techniques such as Quasi-PS InSAR over Denizli Babadağ landslide.

Acknowledgments The authors are very thankful to ESA for the Envisat data (provided under ESA Category-1 project, contract No.C1P5863).

References

Canaslan FÇ, Lazecky M, Perissin D, Üstün A (2014) Monitoring of land subsidence in Konya closed basin between 2003 and 2010 by means of PSInSAR processing: first results in Karaman District. In: Centeris 2014—conference on enterprise information systems

Colesanti C, Ferretti A, Locatelli R, Novali F, Savio G (2003) Permanent scatterers: precision assessment and multi-platform analysis. IGARSS 2003, Tolosa, pp 1–3

Ferretti A, Prati C, Rocca F (2000) Nonlinear subsidence rate estimation using permanent scatterers in differential SAR interferometry. IEEE Trans Geosci Remote Sens 38:2202–2212

Ferretti A, Prati C, Rocca F (2001) Permanent scatterers in SAR interferometry. IEEE Trans Geosci Remote Sens 391:10–20

Kumsar H, Çelik S, Aydan Ö, Tano H, Ulusay R (2012) Babadağ Gündoğdu Heyelanının Çok Parametreli İzleme Teknikleri İle İncelenmesi ve Doğal Afet Kapsamında Değerlendirilmesi

Osmanoğlu B (2011) Applications and development of new algorithms for displacement analysis using InSAR time series, Ph.D. thesis. University of Miami

Results Comparison of the Flow Direction and Accumulation Algorithms Together with Distributed Rainfall-Runoff Models in Czech Switzerland National Park

Vladimír Fárek and Jan Unucka

Abstract Algorithms for calculation direction (FD) and accumulation of runoff (FA) are one of the most widely used software of GIS modules. A both commercial platforms ESRI ArcGIS and IDRISI used in this case study and also the used open source tools such as GRASS GIS or SAGA GIS dispose of such modules. It is expected that the currently established algorithms generate good enough results, so crucial factors for such analyses become quality and resolution of the digital terrain model. Considering HPPS ČR and POVIS systems, which are both provided by Czech Hydrometeorological Institute, these analysis serves only as the basic mechanisms to obtain the complementary informations about the watershed parameters. However, similar analyses serves also as the fundamental basis for more the sophisticated analytical apparatus, which represent currently distributed rainfall-runoff models. This paper focuses on comparing the results of different algorithms FD and FA with the fully distributed rainfall-runoff models MIKE SHE and SIMWE. The pilot area is a Czech Switzerland National Park and various small anthropogenically least affected catchments, such as Jetřichovický potok, Dlouhá and Suchá Běla, Červený potok, Koutský potok. An important factor here is extremely complicated relief NP Czech Switzerland, so it was necessary to use LIDAR DTM, specifically from the project GeNeSIS.

Keywords Flow accumulation · Rainfall-runoff modelling · SIMWE · MIKE SHE · Czech Switzerland national park

V. Fárek
Czech Hydrometeorological Institute, Kočkovská 18/2699, 400 11 Ústí nad Labem, Czech Republic
e-mail: vladimir.farek@chmi.cz

J. Unucka (✉)
Czech Hydrometeorological Institute, K Myslivně 3/2182, 708 00 Ostrava-Poruba, Czech Republic
e-mail: jan.unucka@chmi.cz; jan.unucka@vsb.cz

J. Unucka
Faculty of Mining and Geology, VSB-Technical University of Ostrava, 17. listopadu 15/2172, 708 33 Ostrava-Poruba, Czech Republic

K. Růžičková and T. Inspektor (eds.), *Surface Models for Geosciences*, Lecture Notes in Geoinformation and Cartography,
DOI 10.1007/978-3-319-18407-4_8

1 Preface

Today, we can remark a higher utilization of the fully distributed rainfall-runoff models (hereinafter RR) together with the more frequent use of open source tools in the context of GIS and mathematical modelling in hydrology. The reasons are quite simple—increasing the number and quality of thematic geodata and consequently the increasing potential of open source tools not only for "basic" hydrological analysis of digital terrain model (hereinafter referred to as DTM). So, the spatial data with higher quality and open source tools play more important role even in the operative hydrology including the operative hydrologic forecasts. One of the traditional problematic moments of the hydrological forecasts are often small basins with a complete absence of operational monitoring network or a representative number of hydrometric measurements. The issue of determining the runoff characteristics and prediction of discharge not only for flood episodes in these small basins is still important not only in the context of Operational Flood Forecasting Service of Czech Republic (HPPS ČR) but also abroad. Although this basins are often less important from a water management perspective, they are important source areas of flood discharges quite frequently and important tributaries of the basins of higher order during such extreme events. The impact of basin's geomorphologic and hydrologic parameters on rapid increase of discharge volume during convective precipitation events is quite problematic in Czech republic in last years. These factors represent somewhat negative effects on the natural conditions of basins and socio-economic activities of man. From this point of view, it is desirable to analyze basin parameters and key hydrological characteristics as efficiently and accurately as possible. These analyses and determination of selected attributes of the hydrological basin then necessarily take place with the support of GIS and hydrological models that allow not only the spatial analysis of morphometric (slope orientation, vertical and horizontal curvature etc.) and basin's hydrological parameters (e.g. the spatial distribution of runoff coefficient, values of CN curves,etc.), but also in the context of open source tools also various indices of water and runoff concentration. Consequently, the natural step is then preprocessing of RR or hydraulic models (hereinafter HD) within the GIS preprocessing tools and performing the basin schematization using particular RR and HD models. The phenomenon of the creation and calibration of the RR models on unmeasured (or so-called ungauged) catchments is described systematically mainly with the team Wagener (2014), respectively Beven (2009, 2012). Beven's most recent monographs or even Haan et al. (1994) created the significant contribution for these issues too. Using a fully distributed models not only in these terms and conditions intensively studied Vieux (2004) or Bedient et al. (2013). Because these basins are located especially in border areas with significant part of forested areas, we can also suggest in such context works of Chang (2006), De La Cretaz and Barten (2007) or Stednick (2008). This article presents the results of the previous solution phases of authors team. Field survey of selected watersheds, spatial analysis in GIS of the pilot catchments and RR modelling using the fully distributed models both within the open source tools and commercial software packages. Another necessary step is verification of the modelling results, which in selected watersheds already partly taken place.

2 Used Tools and Methods

ESRI ArcGIS, GRASS GIS, SAGA GIS were used for DTM analyses and RR & HD models preprocessing at the level of software packages for GIS. A fully distributed models SIMWE, MIKE SHE were used at the level of hydrological (RR) models together with the semi-distributed model HEC-HMS. Finally, the hydraulic model MIKE 11 was used for selected streams to achieve better production of hydrographs in DHI MIKE platform and also the more complex communication with unsaturated and saturated zone. Detailed descriptions of these particular software packages can be found on the websites of manufacturers and developer communities in the case of open source software GRASS GIS and SAGA GIS. Partial descriptions of these tools in relation to this type of analysis also discusses by the team of authors, for example Fárek and Unucka (2010), Fárek et al. (2014), Unucka (2014) or Unucka et al. (2011). An important factor in the use of fully distributed RR models such as SIMWE or MIKE SHE is to achieve the spatial distribution of the geomorphic and hydrologic parameters more precisely within the catchment scope instead of aggregation of these parameters on schematized subbasins using the semi-distributed approach. The availability of high quality DTM is the next of the major factors for successful accomplishing of the hydrological analyses and modelling. Pilot basins have such values of several geomorphic parameters, that we can call them as attributes of the so-called extreme relief. Interpolation of DTM is quite problematic task even in locations with rocks and canyon valleys, which can be clearly demonstrated by derivation of such DTM and confrontation of these results with field observations and measurements. DTM GeNeSIS is based on the laser scanning (LiDAR with the cell resolution of 1 × 1 m. For the morphometric analysis and derivation of the basic hydrological parameters were chosen ESRI ArcHydro tools and Spatial Analyst, morphometric analyses were performed in GRASS GIS modules including *r.flow*, *r.terraflow, r.watershed, r.topidx and r.slope.aspect*. The SAGA GIS made further morphometric analyses as the depth and width of valleys, moisture index and the index of topographic convergence. More detailed documentation of the abovementioned analyses and description of the specific modules can be found in websites of each GIS platforms and also in Maidment and Djokic (2000), Wilson and Gallant (2000), Maidment (2002), Neteler and Mitasova (2008) or Hengl and Reuter (2009). Abovementioned morphometric analyses and schematization of the distributed RR models SIMWE and MIKE SHE were performed on all pilot watersheds. Comprehensive schematization of MIKE SHE and consequent schematization of HD MIKE 11 was then carried out on Jetřichovický potok (JP) and Suchá Bělá (SB) basins at this stage of research (see Figs. 2 and 3). Model MIKE SHE was schematized with the cell resolution of 5 × 5 m—however used DTM enables more detailed schematization. Other data sets haven't such accurate spatial resolution, and another important factor was the increasing of the simulation time (MIKE SHE together with MIKE 11 and MODFLOW up to 59 h).

3 Pilot Area Description

Researched Jetřichovický creek basin together with the adjacent basins is located in the central part of the National Park Czech Switzerland near the border with Germany. Pilot area is the part of the province Czech Highlands and sub-province Krušnohorský system, Krušnohorská hilly area and Děčínská highland in terms of geomorphological classification of Czech republic. From the groundwater scope, pilot area is part of the hydrogeological zone 4660—Cretaceous structures of Lower Kamenice and Křinice rivers. Number of hydrological sequence of Jetřichovický potok subbasin is 1-15-01-0120 and of Suchá Bělá is 1-14-05-0260. The highest elevation value of JP basin (derived from LiDAR DEM) is 487 m above sea level, mean basin elevation is 335 m above sea level, the lowest elevation of basin (outlet and contribution into Křinice river is located in the area of Zadní Jetřichovice abandoned village) is 243.7 m above sea level. Basin area derived from DEM using r.watershed is 13.75 km^2. The height ratios of the pilot basin in relation to other basins of the National Park Czech Switzerland are shown in Fig. 1 (hypsographic curve of the main basins of the broader pilot area shows Graph 1). Watershed is composed of Cretaceous sandstone sediments, established on the old crystalline basement situated on the eastern edge of the Krušnohorská unit. The landscape relief is then formed by Turonian sandstone structure, to whose formation volcanic activity during the Tertiary has contributed. In addition the gradually denuded, initially underground volcanic infill is formed by volcanic relief elevations, contributing particularly to its vertical zonation. For large structures dominate the structure of the platforms and canyons (especially Elbe, Kamenice, Křinice). The area is characterized by the occurrence of vertical rock walls, abris, rock towers and top plateaus. Gorges and ravines are usually wet, with reduced insolation and with the significant occurrence of frost hollows. The investigated basins are essentially completely covered predominantly by coniferous forests with reduced herb layer, in which targeted forestry interventions are gradually converting to mixed forests. Given the small slope of the ravines thalwegs and large permeability bedrock surface flows are often ephemeral, flowing nature. According to the Czech national basin typology can be examined segments grouped 1.1 "Source area of the forest", 1.6 "Source area of the wetland, fen and bog ", 2.1 "Natural foothill torrent—forest track " and 2.4 " Natural foothill torrent—rock track, gorge" (Šlezingr 2006). Spatial and morphometric properties of inundation areas (especially around the closing profiles of particular catchments) play an essential role in the runoff

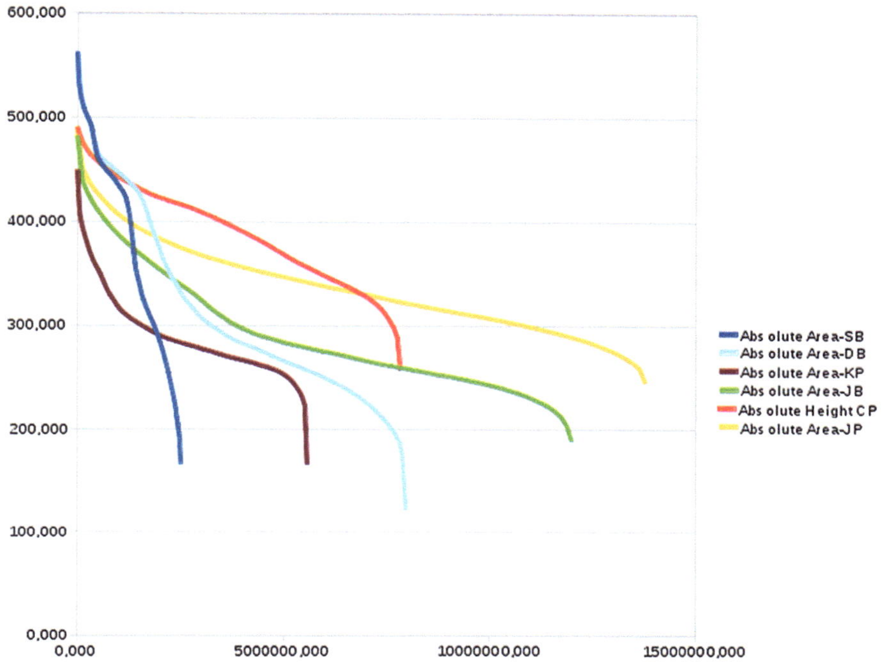

Fig. 1 Hypsographic curves of the pilot subbasins. *DB* Dlouhá Bělá, *CP* Červený potok, *JB* Jetřichovická Bělá, *JP* Jetřichovický potok, *KP* Koutský potok, *SB* Suchá Bělá

hydrograph parameters and culmination values beside the parameters of thalwegs, cross-sections and fluvial network spatial organization. The team of authors tried to take into account these aspects using the combination of fully distributed RR model MIKE SHE and hydraulic model MIKE 11 beside the second fully distributed model SIMWE together with "more conventional approaches" based on the semi-distributed models represented by HEC-HMS in this case study (optionally, together with fully distributed representation of several methods such is gridded SCS-CN) (Table 1).

4 Selected Results

Given the extent of this contribution, the authors will focus on the pilot basin Jetřichovický potok and Suchá Bělá. The main reasons are, at first, the complexity of the current field research (including ADCP measurements) within those basins and second, the finalization of complex distributed model MIKE SHE (together with MIKE 11 and MODFLOW schematizations). Any further stages of the authors research will then focus on the next pilot areas. Figures 2 and 3 show the

Table 1 Basic parameters of the pilot subbasins

Subbasin name	Hydrological order number	Hypsometric integral	Melton's index	Area (ha)	Mean height (m n.m.)	Forested land (%)
Červený potok	1-15-01-0100-0-00	0.566	0.084	783.4	388.3	99.5
Suchá Bělá	1-14-05-0260-0-00	0.535	0.249	252.5	378.7	99.2
Jetřichovická Belá	1-14-05-0200-0-00	0.349	0.085	12.0	291.0	99.0
Jetřichovický potok	1-15-01-0120-0-00	0.377	0.066	1377.0	335.5	99.5
Koutský potok	1-14-05-0220-0-00	0.435	0.120	558.0	289.7	97.0
Dlouhá Bělá	1-14-05-0260-0-00	0.432	0.156	797.5	313.2	98.1
Grosser Zschandbach	–	0.423	0.118	978.4	350.4	95.2
Bílý potok	1-14-05-0080-0-00	0.578	0.109	435.7	419.5	79.0
Kachní potok	1-14-05-0240-0-00	0.343	0.209	484.4	316.7	92.3

comparison of the surface runoff grids simulated in tools SIMWE/GRASS GIS and DHI MIKE SHE. Table 2 summarizes the simulation results of the semi-distributed RR model HEC-HMS and distributed RR models MIKE SHE and SIMWE in the reference points shown in Figs. 2 and 3. There is a fairly close good correspondence of the particular models results according to the demands of operational use of the RR models in the operational hydrologic forecasting service (HPPS ČR). Figure 4 shows a detail of the results of analyses FD (flow direction) and FA (flow accumulation) in GRASS GIS using different algorithms and modules (namely *r.flow, r.terraflow, r.watershed*). An interesting tool in the context of the aforementioned analyses is the GRASS module *r.clump* which appeared in the last versions of this GIS platform. Its use was beneficial especially when we needed to compare results of MIKE SHE and SIMWE. Spatial distribution of particularly low values of surface runoff in the basin near the catchments boundaries declined correlations of grids generated by MIKE SHE and SIMWE were quite complicated. It was obvious that there was a considerable understatement within the whole subbasin scale especially with regard to the place of thalwegs, where already the result correlations were higher. For a more detailed description of the module *r.clump* we can refer to the man pages at GRASS GIS website. We can state, that this tool combines adjacent cells with nearly identical values within a narrow interval and then the resulting continuous field indicates the ordinal values. Using this module helps identify sites in pilot watersheds with significant coincidence of the cell values and

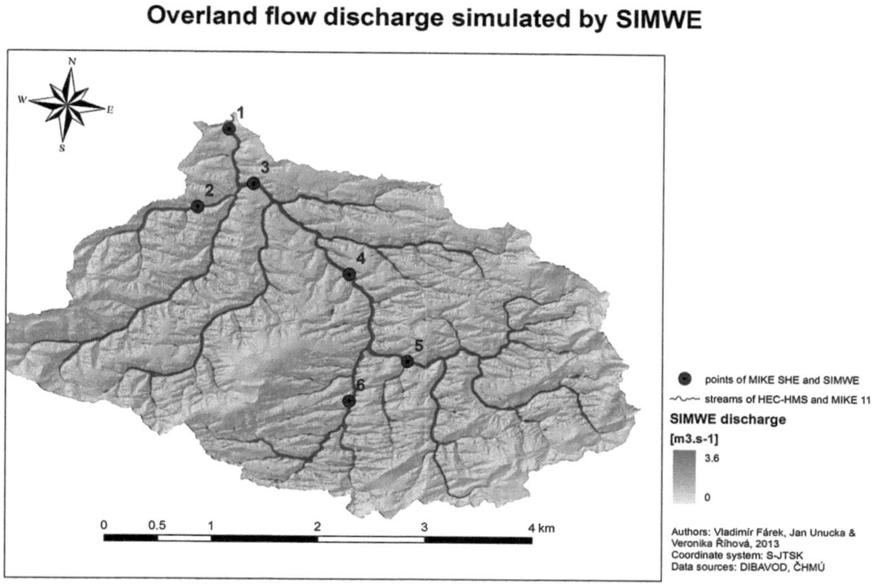

Fig. 2 Overland flow simulated by SIMWE model with reference comparison points

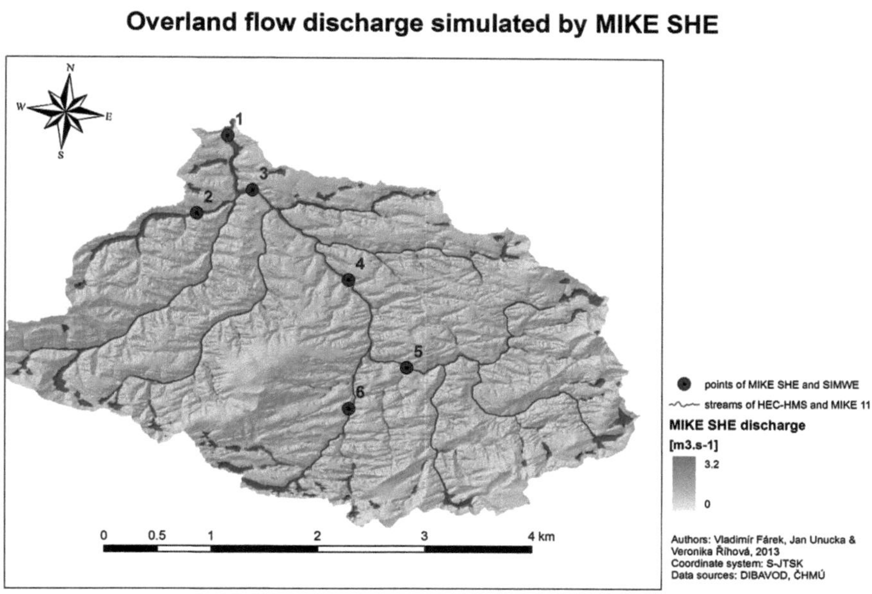

Fig. 3 Overland flow simulated by MIKE SHE model with reference comparison points

Table 2 Overview of culmination discharge values on reference comparison points

Reference point	S-JTSK X	S-JTSK Y	MIKE SHE/11	HEC-HMS	SIMWE
1	−734,992	−953,184	3.35	3.1	**3.6**
2	−735,281	−953,934	0.14	0.12	**0.16**
3	−734,761	−953,709	3.11	3.0	**3.3**
4	−733,858	−954,571	2.45	**2.68**	2.6
5	−733,309	−955,410	1.17	1.1	**1.35**
6	−733,852	−955,794	0.95	0.92	**1.02**

Bold text indicates the maximum values for each reference point

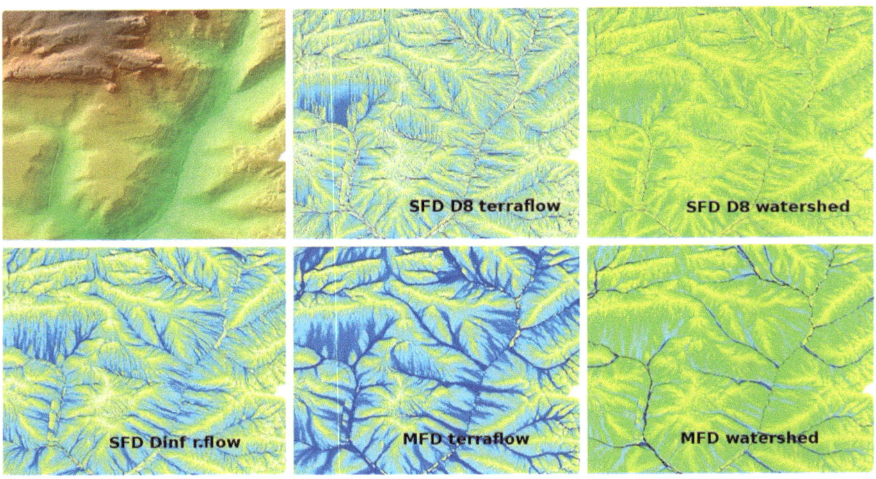

Fig. 4 Comparison of particular results of the FD and FA algorithms on the selected segment of the Suchá Bělá subbasin

eliminates "dead" space in which within one model promotes minimal surface runoff, while the other not, and from that reason the standard regression analysis reduce the value of the correlation coefficient. Figure 5 shows the results of the comparison module *r.clump* for the rasters of depth of overland runoff water simulated by distributed RR models SIMWE and MIKE SHE.

5 Discussion

It is quite apparent that the input hypothesis of the dependence of both real and calculated surface runoff on the particular morphometric parameters of selected pilot basins was confirmed, while selected results are illustrated in particular figures and tables. The effectiveness of using a fully distributed SO models for small

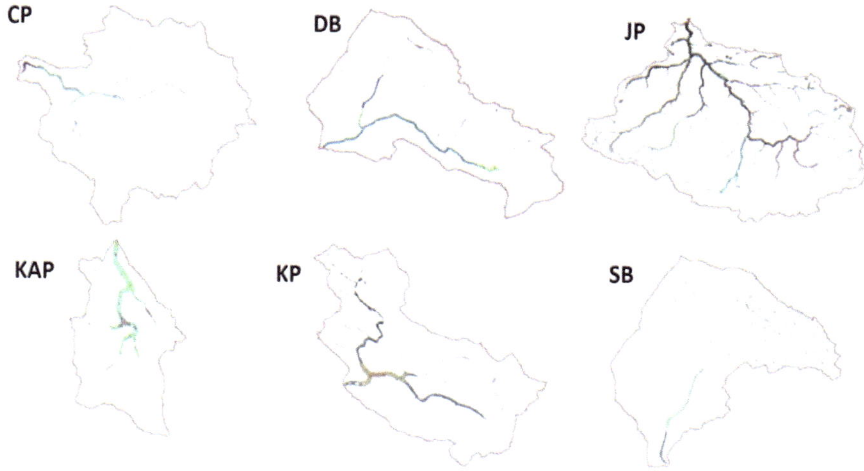

Fig. 5 Comparison of the fully distributed models SIMWE and MIKE SHE with the support of module r.clump for particular basins. (*KAP* Kachní potok)

ungauged catchments, inter alia, is accented by authors like Ruiz-Villeneuve et al. (2013), Callow and Boggs (2013), Grimaldi et al. (2013) or Ilorme and Griffis (2013). Work Ruiz-Villeneuve et al. (2013), inter alia, emphasized the importance of similar scenarios in terms of simulation and prediction of the impact of flash floods, which ultimately point to the experience of "flood years" 2009–2013 in the pilot river basins mentioned in this contribution. A collective work Wagener (2014) and Beven (2009, 2012) remain the fundamental monographs focused on this topic so far. The availability and quality of the input data belong to the most important and frequently mentioned factors in this regard, together with software packages and methods used for RR and HD modelling. MIKE SHE represents undoubtedly one of the most comprehensive tools, especially while its potential for linking with the HD model MIKE 11 (and thus simulating hydraulic transformation of flood waves using the real geometry and not an approximation by simplified geometric shape) and groundwater flow models MODFLOW & FEFLOW is employed. Another advantage is the possibility of using the continuous methods rather than event methods, which is the fact that accented by Grimaldi et al. (2013) among others. MIKE SHE importance as a powerful tool for the comprehensive and fully distributed analysis, simulation and prediction is discussed by Singh and Frevert (2006) and Beven (2012). General aspects of the use of fully distributed models are primarily engaged in Vieux (2004). The disadvantage of using MIKE SHE compared to SIMWE is much higher demands on input data and hardware performance. Consequently, the effective parameterization of the MIKE SHE schematization can't be described as trivial especially within the building complex model in combination with MIKE 11 and MODFLOW/FEFLOW. Beven (2012) also draws attention to the fact that simple robust models can produce more accurate results (or reasonable in ungauged catchments) before model calibration in some cases, as represent particular simulations of selected flood events during 2010–2013 period. This also discuss Singh and Frevert (2006) and we can say that this partial conclusion is confirmed by the works of authors in the moment, when sufficient analysis of surface runoff is needed neglecting hypodermic and groundwater runoff within the RR process. Unless we accept the fact that surface runoff during flash floods situation is dominant, then simple open source models, which are represented by SIMWE tool in this case study, have an attractive potential for such tasks. Notable finding in this regard is the fact that FA values are closer depending on the length of the stream than on the basin area. Another interesting conclusion is that the transition to 3D modelling in this type of relief generates the difference of particular values up to 27 %. Definitely seems more appropriate using of the MFD algorithms (though within the module r.flow this algorithm is not able to continue with the "flow" calculation after some hiatus in terrain singularities), while the SFD algorithms are less useful especially in the flatter upper catchment area (upland sandstone plateaus). The most frequent symptoms of the SFD algorithms are inability to diverge flow (based on naturally from the principle of the algorithm) and excessive production of unrealistic parallel flow paths. Correlation coefficients between the values of FA and results of the RR models have relatively low values (0.15–0.21), but these values are influenced by the size of the grids and the consequently by the principle of calculation (without r.clump). In contrast, the correlation coefficient between models SIMWE and MIKE SHE highlights the very good correlation of their results (0.77–0.92). It can therefore be concluded that the solutions for surface runoff analysis SIMWE RR model represents a suitable tool in pilot catchments. In case, when the complex analysis of water dynamics within the catchment is required, we have to shift to the more complex and sophisticated numerical solutions, which can be achieved by using of the fully distributed RR model MIKE SHE. Interestingly we found that open source software products GRASS GIS and SAGA GIS are much more stable in OS Unix/Linux when working with large raster datasets (up to 20 GB). However, regardless of the OS, it does not make sense to perform this analyses using the machine with capacity of less than CPU i7 as less than 16 GB of RAM. Finally, existence of two physical hard drives (ideally one of them as SSD) is the big advantage.

6 Conclusion

The above preliminary results lead to some interesting facts. First of all, we can say that the importance of input hypothesis of the verification of the GIS analysis results using a fully distributed rainfall-runoff models were largely confirmed yet were particularly confirmed by the field survey on the pilot basins JP and SB. The industrial standards and validated SW represented by RR models HEC-HMS, MIKE SHE and MIKE 11 HD together with fully distributed open source RR model SIMWE were applied as the verification apparatus for FD and FA algorithms. Especially in the case of HEC-HMS and MIKE SHE/MIKE 11 models have been applied in a number of case studies worldwide, so we can say that both models are verified and validated (even within the abovementioned FEMA/NFIP industrial standards evaluation). There will be a need to increase the number of case studies using the model SIMWE, to which goal this authors of this text hope to contribute. At the same time we can say that open source GRASS GIS and SAGA GIS are becoming better and better and even more complex platforms for morphometric and hydrological analysis of the DTM. If the basic requirements about the input spatial data is fulfilled (currently using the best available LiDAR DTM among others), the results of RR modelling are quite “realistic” and they are confirmed particularly by the field verification survey. These basic hydrological analyses are mostly based on variants determining FD and FA (for pilot watershed rather complicated extreme relief pilot river basins), or it can be used more sophisticated analysis indebted to offer platforms GRASS GIS and SAGA GIS as the depth of the valley, the valley width, curvature and relief of thalwegs, time of concentration indexes or topographic wetness index. The logical next step in the research pilot river basins will be the measurement of selected hydrological characteristics and other parameterization models MIKE SHE and MIKE 11. The spatial resolution of DTM seems to be a crucial factor both for more simple hydrological analyses in GIS (FD and FA) and more complex analyses based on the RR modelling. Finally this article attempted to enlarge the knowledge base of the “forgotten” parts of our country and the its Sudetenland locations, that represents abandoned village Zadní Jetřichovice for example, which are much interesting and definitely worth a comment from many different perspectives.

Acknowledgments The authors would like to thank to the project TAČR č.TA04021123 “New methods of the surface water bodies morphology measurements and their utilization in landscape and emergency planning” without its support realization of this article would not have been possible. We thanks also to Povodí Moravy, s.p., Czech Hydrometeorological Institute and thank you also, Morava River Basin, and the Czech Hydrometeorological Institute and State Administration of Land Surveying and Cadastre for the support not only at the level of the data base for the model building and simulations.

References

Bedient PB, Huber WC, et Vieux BC (2013) Hydrology and floodplain analysis, 5th edn. Prentice Hall, London, p 815. ISBN: 978-0-273-77427-3

Beven KJ (2009) Environmental modelling: an uncertain future? Routledge, London, p 310. ISBN: 978-0-415-46302-7. Blackwell, p 451. ISBN: 978-0-470-71459-1

Beven KJ (2012) Rainfall-runoff modelling. The primer, 2nd edn. Wiley, Chichester, p 457. ISBN: 978-0-470-71459-1

Callow JN, Boggs GS (2013) Studying reach-scale spatial hydrology in ungauged catchments. In Journal of Hydrology, vol 496, pp 31–46. ISSN: 0022-1694

Chang M (2006) Forest hydrology, 2nd edn. Taylor & Francis, London, p 474. ISBN: 978-0849353321

De le Crétaz A, Barten PK (2007) Land use effects on streamflow and water quality in the northeastern United States. CRC Press, Boca Raton, p 342. ISBN: 978-0849391873

Fárek V, Unucka J (2010) Modelování povrchového odtoku v extrémním reliéfu. In sborník symposia GIS Ostrava, Ostrava, VŠB-TUO, p 9. ISBN: 978-80-248-2171-9

Fárek V, Unucka J et al (2014) Využití GIS a distribuovaných srážko odtokových modelů v odhadu hydrologických parametrů malých povodí. Případová studie z NP České Švýcarsko a Jetřichovického potoka. In Vodní hospodářství 1/2014, p 55. ISSN: 1211-0760

Grimaldi S, Petroselli A, Arcangeletti E, Nardi F (2013) Flood mapping in ungauged basins using fully continuous hydrologic–hydraulic modeling. In Journal of Hydrology, vol 487, pp 39–47. ISSN: 0022-1694

Haan CT, Barfield BJ, Hayes JC (1994) Design hydrology an sedimentology for small catchments. Academic Press Inc., London, p 588. ISBN: 978-0123123404

Hengl T, Reuter HI (2009) Geomorphometry. Concepts, software, applications. Elsevier, Amsterdam, p 775. ISBN: 978-0-12-374345-9

Ilorme F, Griffis VW (2013) A novel procedure for delineation of hydrologically homogeneous regions and the classification of ungauged sites for design flood estimation. In Journal of Hydrology, vol 492, pp 151–162. ISSN: 0022-1694

Maidment DR (ed) (2002) ArcHydro. GIS for water resources. ESRI Press, Redlands, p 220. ISBN: 978-1589480346

Maidment D, Djokic D (eds) (2000) Hydrologic and hydraulic modelling support with geographic information systems. ESRI Press, Redlands, p 232. ISBN: 978-879102804

Neteler M, Mitasova H (2008) Open source GIS. A grass GIS approach, 3rd edn. Springer, New York, p 417. ISBN: 978-0-387-35767-6

Ruiz-Villaneuva R, Díez-Herrero A, Bodoque JM, Ballasteros Canóvas JA, Stoffel M (2013) Characterisation of flash floods in small ungauged mountain basins of Central Spain using an integrated approach. In Catena 110, pp 32–43. ISSN: 0341-8162

Singh VP, Frevert DK (2006) Watershed models. CRC Press, Boca Raton, p 653. ISBN: 978-08493-3609-6

Šlezingr M (2006) Říční typy. Úvod do problematiky úprav vodních toků. FAST VUT Brno, p 299. ISBN: 80-7204-481-8

Stednick JD (ed) (2008) Hydrological and biological responses to forest practices: the alsea watershed study. Springer, New York, p 322. ISBN: 978-1441928436

Unucka J (2014) Environmentální modelování 1. Skriptum PřF OU. p 209. SVZZ CZ.1.07/2.3.00/35.0053 & PřF OU

Unucka J, Jařabáč M et al (2011) Srovnání možností využití semidistribuovaných a distribuovaných srážkoodtokových modelů v lesnické hydrologii na příkladu povodí Ostravice. In Zprávy lesnického výzkumu 1/2011. ISSN: 0322-9688

Vieux BE (2004) Distributed hydrologic modeling using GIS. Kluwer Academic Publishers, Dordrecht, p 289. ISBN: 978-1402024597

Wagener T, Wheater HS, et Gupta HV (2004) Rainfall-runoff modelling in gauged and ungauged catchments. Imperial College Press, London, p 306. ISBN: 978-1860944666

Wilson JP, Gallant JC (eds) (2000) Terrain analysis. Principles and applications. Wiley, London, p 479. ISBN: 978-0471321880

Surveying of Small Water Reservoirs for Water Management Purposes

Jakub Fuska and Viliam Bárek

Abstract Small water reservoirs are important part of the landscape that is changing under various anthropogenic and natural processes. Knowledge of the actual parameters, such as actual water reservoir storage capacity or the amount of sediment accumulated in the water reservoir is crucial for the further operation and for the fulfilling of the water reservoir purposes. Data collection and processing for obtaining of the actual information is executable as non-contact mapping with the use of sonar and GNSS receiver. This paper describes the experience of the Department of Landscape Engineering with the use of this method for the data collection and data processing to the format of raster and vector digital elevation models of water reservoir bottom. Achieved average accuracy in the creation of these models is in the interval of 13–29 mm. These parameters make the use of non-contact mapping the available, accurate and quick option for the surveying in the water management.

Keywords Echosounding · Water reservoir · Siltation · Digital elevation model

1 Introduction

Water reservoir bottom and its shape are changing in time due to the various processes within the body of the water reservoir, but mostly under the influence of processes in its watershed. The most important factor that affects the reservoir is the process of sedimentation, transport and accumulation of eroded material in the body of water reservoir. This process leads to siltation, which results in decrement of

J. Fuska (✉) · V. Bárek
Faculty of Horticulture and Landscape Engineering, Department of Landscape Enginering, Slovak University of Agriculture, Hospodárska 7, 949 76 Nitra, Slovakia
e-mail: jakub.fuska@uniag.sk

K. Růžičková and T. Inspektor (eds.), *Surface Models for Geosciences*, Lecture Notes in Geoinformation and Cartography,
DOI 10.1007/978-3-319-18407-4_9

water storage capacity. Knowledge of the actual status of siltation is necessary for the decision-making in the reservoir maintenance and operation, but it can also offer the view to the processes of erosion and sediment transport in the watershed. Smaller input of the finer soil particles occurs when they travel for a longer distance due to the wind erosion effect (Urban et al. 2013).

Mapping of the water reservoir bottom can be executed with various approaches, methods and equipment. Common applications are based on the geodetic surveying of the identical cross-sections in different time stage (Holubová 1998; US GS 97-4138 1997) or as the creation of the cross-sections of the digital elevation model (DEM) of the water reservoir bottom (Ceylan et al. 2011). Another method is using the DEM of reservoir in various time stages and its comparison on overall scale—as the creation of the map of sediment thickness (US GS 03-383 2003; US GS 2005-5040 2005; Weis and Kubinský 2014; Fuska et al. 2014). These tasks are easily executable in GIS environment, which is advanced and effective tool for the creation and usage of the models with the goal of the research, predict and know the spatial chronological and functional aspects of geographical sphere (Varga et al. 2013).

For the creation of water reservoir bottom DEM can be used various input data, such as historical contour maps or technical documentation or more accurate and actual data of the direct surveying (GNSS, total station, photogrammetry, LIDAR). Surveying with the methods of geodesy are usually precise and accurate, but the necessity of the emptied reservoir may complicate the surveying. On the other hand, the application of the non-contact surveying (sonar, radar) offers the availability of the data collection during the proper operation of the reservoirs or ponds, when the water level is at its common water level.

2 Materials and Methods

Application of the DEM's varies in accordance to the research purposes, therefore the used model creation and interpolation, but also the input data requires different approach. In general it is available to use both raster and vector models, if it is necessary, there are created more models of the reservoir bottom in different time stages.

2.1 Surveyed Reservoirs

The use of non-contact surveying and its applications for the purposes of this paper are demonstrated at three water reservoirs in Nitra district: Golianovo, Lukáčovce and Kolíňany (Fig. 1; Table 1).

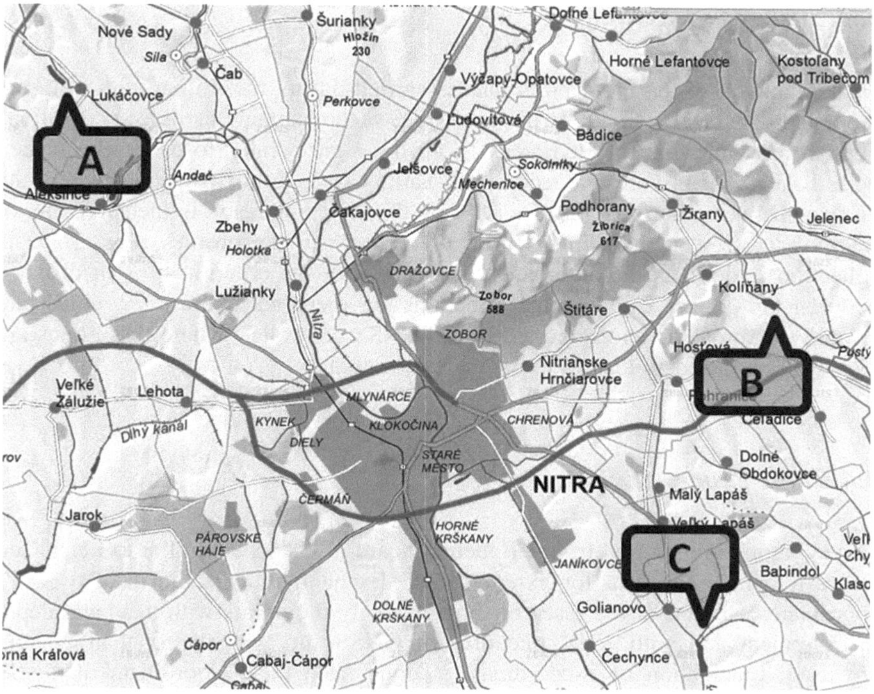

Fig. 1 Surveyed water reservoirs: *A* Lukáčovce; *B* Kolíňany; *C* Golianovo

Table 1 Parameters of surveyed water reservoirs

Name	Operation elevation of water level (m ASL)	Water volume at operation water level (m^3)	Water surface area at operation water level (ha)
Golianovo	179.5	336,000	27.2
Lukáčovce	163.3	81,200	5.8
Kolíňany	149	106,000	13.0

2.2 Surveying Equipment

Data collection is executed with the setup of GNSS receiver and fisher echosounder. Position is obtained with Leica GNSS 1200+ receiver that allows the RTK measurement in high accuracy (usually less than 50 mm in 2D). Depth measurement is done with the fisher echosounder Garmin GPSmap 421 s with 200 kHz dual beam (14°/45°) probe. Depth measurement data as the NMEA string is sent by NMEA cable to the GNSS receiver, where it is stored as the annotation for each surveyed point. Echosounder probe is fixed in given distance from the GNSS pole tip. Whole equipment is mounted to the boat and fixed with the set of ropes to obtain the normal direction of echosounder and GNSS pole to the water level.

2.3 Surveying Method

Data collection of the input point data is carried during the sailing with raft boat propelled with electric engine. This vessel allows the mapping in various conditions (also in reservoirs in protected areas). Sail routes are parallel to the reservoir dam with distance 20 m between each route. Sailing is done in two routes, where the distance between the route profiles is 10 m, this mapping result in the surveying of reservoir cross-sections with distance 10 m between the profiles (Fig. 2). Data collection is done as the automatic surveying with the criterion of 2 m distance between the points. Sailing route directions are prepared as the lines, which are created in AutoCAD and imported to the GNSS controller, where these lines are displayed and used as the aid to navigate the sailing.

2.4 Data Processing

Surveying job is imported from CF memory card of GNSS controller to Leica Geo Office to gather the data for export to MS Excel to process the data to XYZ-coordinate of points at the reservoir bottom. Only the points with available depth measurement and with XYZ Accuracy of GNSS surveying better than 50 mm are used. Calculation of Z-coordinate is done from the Z-coordinate of GNSS surveying and depth measurement of echosounder (Fig. 3).

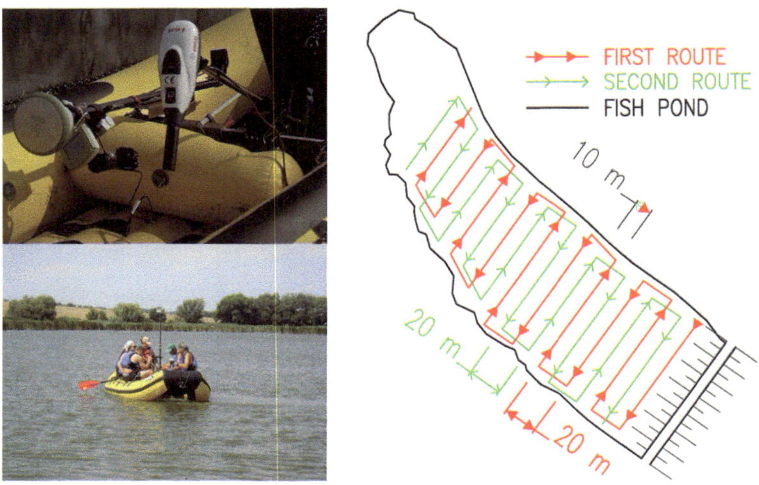

Fig. 2 Survey equipment (*top left* and *bottom left*) and example of sailing routes orientation (*right*) Surveying of fish pond Lukáčovce

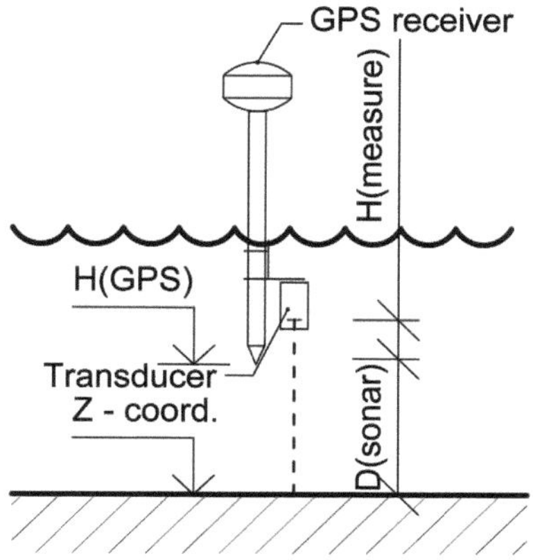

Fig. 3 Calculation of the elevation of the point at the reservoir bottom

All point data are loaded in ArcMAP and exported to shapefile. Afterwards the amount of points is reduced, only the points distributed in approximately 10 × 10 m grid are kept.

To describe the shape of the reservoir shoreline it is usually difficult to obtain the data by direct surveying as the shoreline and reservoir banks are inaccessible in most parts due to strong vegetation growth (reed, cattail, brushes and trees). In accordance to this situation the shoreline is gathered as the boundary from ortophotomap, additional surveying is done to obtain the elevation—usually near the dam or at the open parts of shoreline (if available).

$$Z_{coordinate} = H_{GPS} + H_{mount} - H_{measure} - D_{sonar} \quad (1)$$

2.5 Digital Elevation Model of Reservoir Bottom

Data of the non-contact surveying (points at the reservoir bottom), direct surveying (elevation of water level at the shoreline) and from ortophotomap (shoreline shape) are used to create the DEM's in ArcMAP, for the purposes of small water reservoirs are used following model creation methods:

- Topo to raster—cell size 2 × 2 m, suppression of sink filling
- Spline with tension—cell size 2 × 2 m, weight of tension 0.1
- TIN model—using the rules of Delaunay triangulation

2.6 Historical Data of the Water Reservoirs

For the purposes of the reservoir bottom development assessment it is necessary to know the elevation and topography in different time stages. For the creation of the reservoir bottom DEM from the past can be usually used the available project documentation data, such as contour plans or geodetic surveying data. These data are in fact unavailable for most of the small water reservoirs, because the technical drawings, design plans and geodetic surveying data are often missing. This situation is result of the transfer of the competences and responsibilities between the state authorities and companies (Slovak water management enterprise, Hydro—ameliorations state enterprise, etc.), where also the archives were moved and large part of the documents is missing. Usually there are available documents of the Operation manuals or Programs of the technical and safety maintenance of the reservoirs, where is usually only numerical data about stored volumes at certain water levels.

If there is available contour plan or geodetic field survey in the point-class representation, there can be DEM in raster or vector format created with same settings as DEM of actual reservoir bottom.

2.7 Analysis and Calculations with the Use of Water Reservoir Bottom DEM

Created elevation models of the water reservoir bottom can be used for various purposes, which can be divided into following groups:

- Volume characteristics: actual and original water storage capacity (water volume), actual volume of sediments
- Reservoir bottom profiles creation

3 Results and Discussion

3.1 Analysis of Water Storage Capacity

This analysis is shown at the example of fish pond in Lukáčovce. This fish pond is second in the system of the three fish ponds. First fish pond does not exist anymore as it was buried with the excavated sediment from second and third pond and there is no existing data or design drawing of this fish pond. Actual function of the surveyed second fish pond is fish production. Data collection was done in two steps:

- Direct GNSS surveying in the January 2013, when the fish pond was emptied; there was 48 points surveyed at the reservoir bottom
- Non-contact mapping in August 2013 duriing the common operation of fish pond, there was 2337 points surveyed

For the creation of the actual fish pond bottom DEM was used “Topo to raster” and “Spline with tension” raster models and vector TIN model.

Water storage capacity describes the amount (volume) of water that is stored in the water reservoir usually in the conditions of water level at the operation level. This data is crucial for the management of the water reservoir to fulfil its purposes —for example the amount of water that is available for the irrigation or water supply purpose. This capacity is developing in accordance to the siltation of water reservoir (reduction of capacity) or excavation of sediments (increment of capacity). Volume of stored water can be calculated from both raster and TIN model of reservoir bottom. In case of raster models the actual water reservoir bottom DEM and elevation of the water level (usually the common operation water level) are used for the creation of the water depth as the subtraction of bottom elevation from the water level (Fig. 4).

Afterwards the sum of the water depths at each cell is multiplied by the area of one cell to obtain the total volume of water reservoir storage capacity.

$$V_{water} = \sum_{n=1}^{i} Depth_{water,i} \times Area_{cell} \tag{2}$$

Calculation of storage capacity with the use of TIN model is done as the calculation of volume between the surface of reservoir bottom (TIN model) and surface of water level (polygon of water reservoir shoreline with the elevation of water level) (Table 2).

Results of the mapping and storage capacity have shown that the current storage capacity of fish pond in Lukáčovce has been decreased to the approximately 58 % of the presumed water storage capacity mentioned in operation manual of this fish pond.

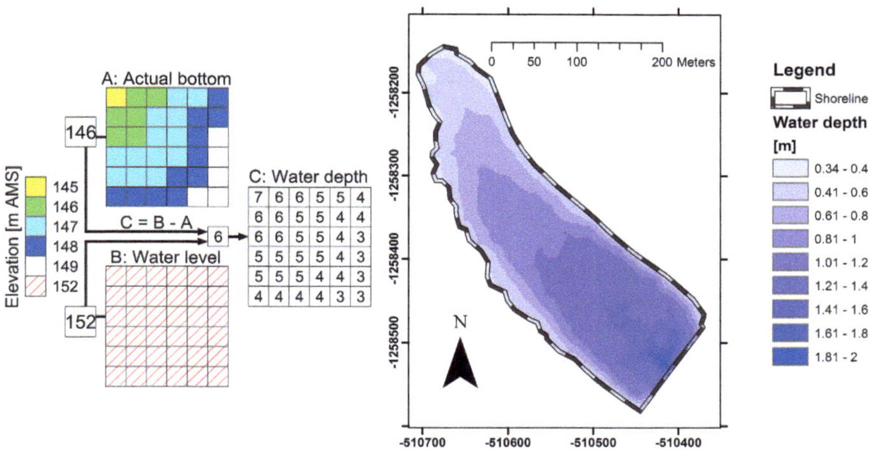

Fig. 4 Map of water depth creation scheme (*left*), Map of water depth in Lukáčovce fish pond (*right*)

Table 2 Actual storage capacity of fish pond in Lukáčovce fish pond

Model interpolation method	Topo to raster	Spline with tension	TIN	Presumed volume
Water volume (m^3)	47,479	46,803	46,376	81,200
% of presumed volume	58.5	57.6	57.1	100

3.2 Analysis of Actual Volume of Sediments in Water Reservoirs

This analysis is shown at the example of water reservoir in Golianovo, which is partially silted in the water inflow part. The strong growth of reed occurs in this part that made it inaccessible—only the visual checking was available that shows the inflow part is fully silted and growth with reed and cattail plants.

Surveying of this water reservoir was done in three days:

- March 2011, there was 717 points surveyed
- April 2011, there was 318 points surveyed during the first day and 1214 points during the second day

For the original and actual bottom DEM creation was used the "Topo to raster" interpolation method. For the creation of original bottom DEM was used the original design ground plan with display of contours. Dem of actual water reservoir bottom was created from the set of points from non-contact surveying.

Volume of the sediment with the use of raster models is calculated similarly to the calculation of the water storage capacity. First step is the creation of the map of the sediment layer thickness created as the subtraction of the raster DEM of original bottom from the raster DEM of actual bottom. This calculation requires the DEM of original and actual bottom have the same cell size and boundary.

Volume of the sediment in the reservoir is calculated as the sum of sediment thickness at each cell multiplied by the cell size of the sediment thickness map (Table 3).

$$V_{sediment} = \sum_{n=1}^{i} Thickness_{sediment,i} \times Area_{cell} \quad (3)$$

Calculation has shown that there is approximately 82,000 m^3 accumulated in this reservoir, but the accuracy of this calculation is scattered by the missing and precise

Table 3 Amount of sediment accumulated in water reservoir in Golianovo

Part of reservoir	Sediment amount (m^3)
Silted part	43,414
Functional part	38,578

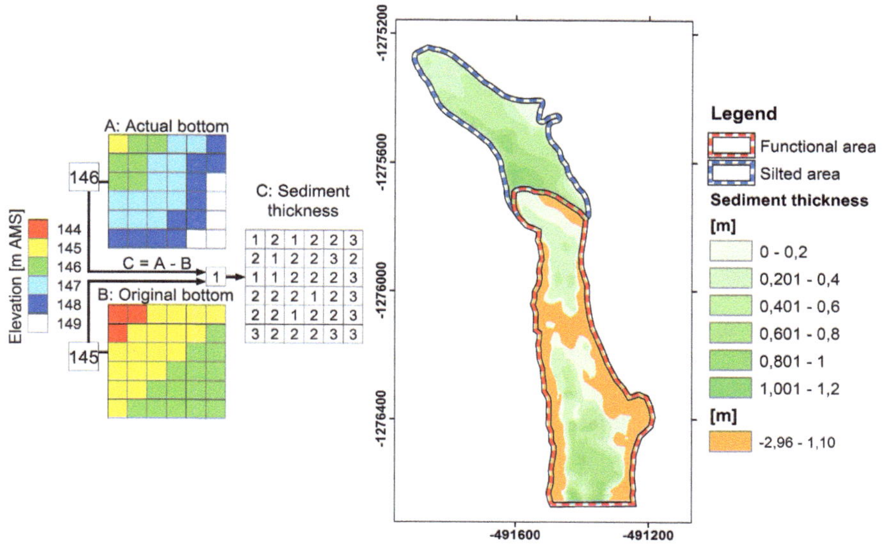

Fig. 5 Map of sediment thickness creation scheme (*left*), Map of sediment thickness in Golianovo water reservoir (*right*)

mapping of the silted part. There is also missing large part of the reservoir documentation, maps and plans from the period of life span of this reservoir that documents the reconstruction and maintenance processes in this reservoir. We presume that the large part of reservoir near the shoreline was deepened during the operation of the reservoir that results in displaying of the negative sediment amount (Fig. 5). This is the result of the process of subtraction of the original elevations from the actual elevations.

3.3 Reservoir Bottom Profiles Creation

This method was shown at the water reservoir in Kolíňany. This reservoir was built for the purposes of the irrigation water supply; nowadays it functions as the fish production pond and as the recipient of treated wastewater from the local water treatment facility.

Non-contact surveying was executed during two days in July 2013:

- During the first surveying was collected 4411 points
- During the second surveying was collected 4404 points

Design and proposal of the various tasks and works, such as excavations, reconstruction of the entrance to the reservoirs and fish ponds for the trucks,

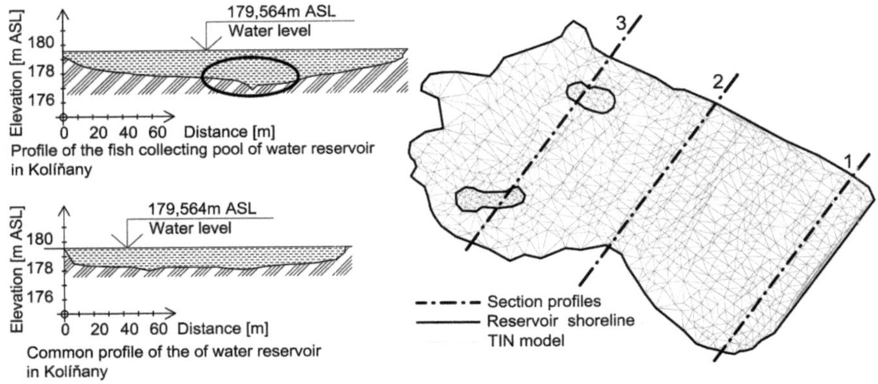

Fig. 6 TIN model of Kolíňany water reservoir bottom (*left*), results of the bottom profiles construction in AutoCAD (*right*)

requires the input data in the format of bottom cross-sections and profiles. These tasks are easily available with the use of TIN model, which is exported from the ArcMAP to AutoCAD for further processing consisting of creation of the section planes, rotation of the cross-sections and modification of scale to non-uniform value to emphasize the elevation proportions. This process is easily executable with the extraction of the cross-section vertices, modification of elevation values and automatized redrawing with the use of AutoCAD scripts (Fig. 6).

3.4 Accuracy of the Created Water Reservoir Bottom Models

Overall performance of data collection, processing and DEM creation was tested at the fish pond in Lukáčovce as the comparison of elevation of the points surveyed directly with the GNSS receiver Leica with the elevation of these points derived from both raster models and from TIN model.

General accuracy of the mapping is given by the partial accuracy of the used equipment. GNSS mapping with the use of Leica 1200+ with the corrections from SmartNet network offers the accuracy better than 50 mm in XYZ coordinates under the common conditions of RTK mapping. Sonar depth measurements with the use Garmin GPSmap 421 s usually offers the accuracy about 20 mm.

Results have shown that the best accuracy is achieved with the use of TIN model, where the average accuracy is 13 mm. Average accuracy of the raster models is slightly worse (29 mm accuracy for Topo to raster and 22 mm for Spline with tension model) (Fig. 7; Table 4).

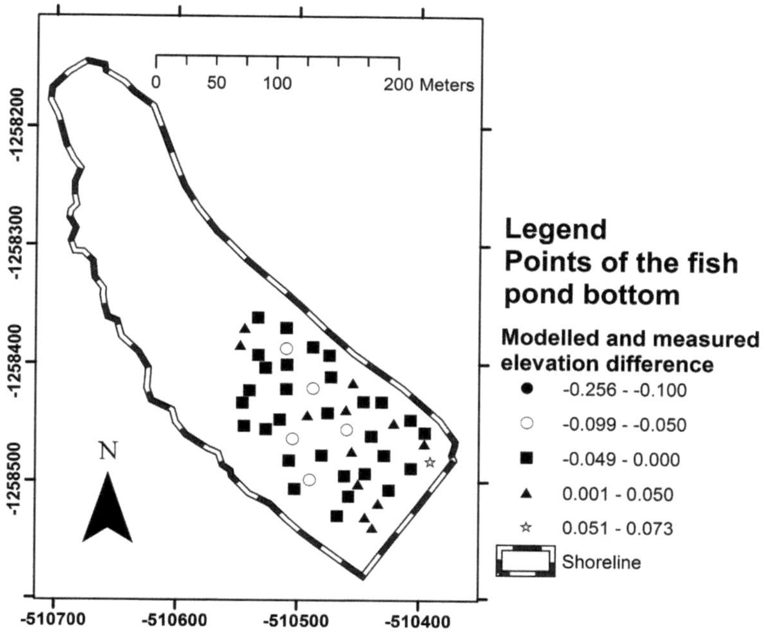

Fig. 7 Difference of the elevations of the surveyed points and models created with various methods at the Lukáčovce fish pond bottom: Topo to raster (*left*), Spline with tension (*middle*); TIN (*right*)

Table 4 Accuracy of reservoir bottom models

Statistical measure	ΔZ: TIN	ΔZ: Topo to raster	ΔZ: Spline
Average	−0.013	−0.029	−0.022
Standard deviation	0.028	0.048	0.031
Minimum	−0.068	−0.256	−0.114
Maximum	0.057	0.072	0.044
Range	0.125	0.328	0.158
Median	−0.016	−0.028	−0.016
Absolute minimum	0	0.001	0.001
Variance	0	0.002	0.001
Root mean square	0.031	0.057	0.038

4 Conclusion

Water reservoir bottom is the entity that is changing over time due to the activities and processes in the watershed and in the own body of the water reservoir. Mapping of the reservoir bottom topography and its changes is in the focus of various studies with the use of various tools, from simple mapping with the leading lines or sounding poles to more sophisticated non-contact methods of data collection. GNSS and sonar mapping offers the possibility of the mapping during the common operation without the necessity of the emptying the water reservoir. Overall accuracy of the created DEM is comparable to the accuracy that is generally achievable by the GNSS surveying (TIN model: 13 mm; Topo to raster: 29 mm; Spline with tension model: 22 mm). Application of the mapping and model creation is useful in the assessment of the water reservoir for the evaluation of the actual water storage capacity for the purposes of the usage of the water reservoir or for the proposal of the maintenance processes and works. Usage of the TIN model for the creation of the cross-sections can also offer a strong support and input data for the design and reconstruction of water reservoirs. Use of the equipment for the non-contact surveying requires the preparation of the vessel direction navigation lines to obtain the input data in the proper distribution, which is processed to the pattern of square network of points for the creation of the digital elevation models in raster and vector models.

Acknowledgments This study was supported by the projects VEGA nr. 1/0027/12 "Development and validation of indirect assessment of anthropogenic immission load" and VEGA 1/0456/14 "Management of the Soil Moisture Regime as a Tool for Climate Change Adaptation".

References

Ceylan A, Karabork H, Ekozoglu I (2011) An analysis of bathymetric changes in Altinapa reservoir. Carpathian J Earth Environ Sci 6(2):15–24. North University Center of Baia Mare, Romania

Fuska J, Bárek V, Pokrývková J, Halaj P (2014) Comparison of actual and presumed water capacity of fish pond in Lukáčovce. Journal of International Scientific Publications: Ecology & Safety. Bulgaria, Burgas

Holubová K (1998) Problémy systematického sledovania erózno-sedimentačných procesov v oblasti vodných diel. Výskumný ústav vodného hospodárstva, Bratislava, Slovakia

Urban T, Lackóová L, Halászová K, Streďanský J (2013) Wind erosion in the agricultural landscape. The wind erosion equation used in GIS. Publishing house of the University of Agriculture in Krakow. Krakow

U.S. Geological Survey Water-Resources Investigations Report 97-4138 (1997) Changes in bottom surface-elevations in three reservoirs on the Lower Susquehanna River, Pennsylvania and Maryland, following the January 1996 flood—implications for nutrient and sediment loads to Chesapeake Bay. U.S. Geological Survey, United States of America. http://pa.water.usgs.gov/reports/wrir97-4138.pdf. Online at 20 Sept 2014

U.S. Geological Survey Open-File Report 03-383 (2003) Bathymetric and geophysical surveys of Englebright Lake, Yuba-Nevada Counties, California. US Geological Survey, California, United States of America. http://pubs.usgs.gov/of/2003/0383/pdf/of03-383.pdf. Online at 15 Sept 2014

U.S. Geological Survey Scientific Investigations Report 2005–5040 (2005) Comparison of Preconstruction and 2003 Bathymetric and Topographic Surveys of Lake McConaughy. U.S. Department of the Interior and U.S. Geological Survey, Reston, Virginia, United States of America, http://pubs.usgs.gov/sir/2005/5040/sir2005-5040.pdf. Online at 13 Sept 2014

Varga V, Lackoóvá L, Streďanský J, Urban T (2013) Comparison of volumetric and deflametric method with wind erosion equation (WEQ) to determinate soil erosion by wind events on selected soil unit. Acta horticulturae et regiotecturae, 16(1). Slovak University of Agriculture in Nitra, Slovak Republic, pp 18–23

Weis K, Kubinský D (2014) Analysis of changes in the volume of water in the Halčianske reservoir caused by erosion as a basis for watershed management. Geografie 119(2):126–144. Czech Geographic Society, Praha, Czech Republic

Detailed Digital Terrain Models in the Research of Deserted Settlement: Hydrological Modelling and Environment of Settlement Areas

Lukáš Holata and Radek Světlík

Abstract This paper presents the use of detailed terrain models in archaeological research. We apply a LiDAR data (DTM of the 5th generation provided by the Czech Office for Surveying, Mapping and Cadastre) in the areas of five villages abandoned during the 15th and the 17th century. Surveyed sites are preserved in the woodlands in the form of relief formations and there have been carried out many surface surveys and geodetic-topographic surveys. Except the detection of previously unknown remains of human activities in LiDAR data (especially field boundaries, tracks or mining areas) we use also an analytical potential of detailed DEM to affect parameters of physical landscape. In particular, we focused on the attachment of deserted villages to water sources and slope gradient of their ploughlands. This allows us to reopen the old-fashioned questions regarding the role of the natural environment in shaping cultural landscape and try to answer them in the new perspective. Although three of considered deserted villages are situated outside of water sources, hydrological modelling of DTM 5G generates higher values of topographic wetness index within close proximity to each of these villages. The ploughlands of deserted villages were situated also in the areas with high slope gradient. These outcomes extend or revised the 'traditional statements' in older Czech literature.

Keywords LiDAR data · Deserted villages · Reconstruction of settlement areas · Hydrological modelling · Late middle ages and early modern period

L. Holata (✉)
Department of Interdisciplinary Activities, New Technologies Research Centre, University of West Bohemia, Univerzitní 8, 306 14 Pilsen, Czech Republic
e-mail: luk.holata@gmail.com

R. Světlík
Department of Archaeology, Faculty of Philosophy and Arts, University of West Bohemia, Univerzitní 8, 306 14 Pilsen, Czech Republic
e-mail: radek.svetlik@gmail.com

K. Růžičková and T. Inspektor (eds.), *Surface Models for Geosciences*, Lecture Notes in Geoinformation and Cartography,
DOI 10.1007/978-3-319-18407-4_10

1 Introduction to the Archaeological Research of Woodlands and Deserted Medieval and Post-medieval Settlement

Except excavations, archaeology is also based on non-intrusive methods of field survey and tries to collect data from the surface. Particular remains of past human activities are preserved on the surface in the form of relief formations up to the present time. In the central Europe, these anthropogenic micro-topographic features can be discovered especially in the current woodlands (e.g. Kuna and Tomášek 2004). One of the most numerous types of sites preserved in this environment are deserted medieval (or post-medieval) villages, in many cases including their hinterlands (field boundaries, tracks, mining areas etc.; for the Czech Republic e.g. Klápště 1978; Smetánka and Klápště 1981; Černý 1992; Vařeka 2006, 2009). These monuments testify for the extensive reduction of settlement patterns particularly during the 14th and 17th century which occurred across most of Europe. Therefore, deserted settlements provide an extraordinary opportunity to study medieval or post-medieval rural environment and the research with quantity of themes or subtopics has a very long tradition within several disciplines (cf. several titles from different European regions: Beresford 1954; Beresford and Hurst 1971; Chapelot and Fossier 1985; Gissel et al. 1981; Nekuda 1975; Pohlendt 1950; Smetánka 1988).

New element in the research concerning deserted settlement is represented by airborne laser scanning (ALS). In the Czech archaeology, LiDAR data has been applied from 2011; currently Digital Terrain Model of the Czech Republic—5th generation (DTM 5G) is used in the largest extent. This technique in the research of deserted settlement has been usually applied to detect still unknown relief formations and thus for localisation and identification of new or hitherto unrecorded traces of past human activities which should be subsequently verified by surface survey—ground-truthing (cf. Lasaponara et al. 2010; Čapek 2013; Čapek et al. 2013). There is necessary to note that LiDAR data is particularly suitable for recording of the line features (field boundaries, tracks) which are on the contrary very hardly documented by surface survey (e.g. Klápště 1978; Smetánka and Klápště 1981)—some parts of these features could be only up to 0.5 m above/under level of terrain and thus visible only within specific conditions (to the general difficulties with the archaeological researches in the woodlands cf. Crow et al. 2007; Doneus and Neubauer 2008; Lennon and Crow 2009). Therefore, LiDAR data has changed the current state of the art because the previous reconstruction of medieval field patterns has been only sporadic (considered as an effect of woodland management; remarkable exception is constituted by numerous traces of deserted medieval fields in Drahany uplands—Černý 1992).

Nevertheless, the application of DTM 5G is much wider. This data has an extraordinary potential in many surface analyses and to study (micro)topography of deserted settlement compared to DEMs generated from different sources (e.g. The geographic base data of the Czech Republic—ZABAGED). This allows us to reopen the old-fashioned questions (often still unanswered) regarding the role of natural environment in the life of past communities. Czech archaeology and historical geography was concerned with the environment of medieval villages, role of natural parameters in selection of settlement areas or their abandonment especially in 1970s and 1980s. Although these interpretations were considered later as 'environmentally deterministic', recent works in historic landscape research have started to re-explore the role of the natural environment in shaping cultural landscapes (especially Williamson 2004; recently Rippon et al. 2014).

On the basis of LiDAR data, we have focused on the two parameters of physical landscape—hydrological character (drainage or water flows reconstruction) and topography (slope gradients of ploughlands). Both parameters have been considered (together with other factors—geological, pedological, geomorphological, climatic etc.) as significant in the location of settlement as well as in the existence of village communities, especially with regard to rural economy (cf. Klápště 1973, 1978; Pokorný 1973; Smetánka 1973, 1978; Snášil 1976; Kalina 1978; Žemlička 1978; Boháč 1983). Despite the fact that direct attachment of settlements to the water sources (streams, spring areas) is usually mentioned, three of considered deserted villages (Bukov, Kamenice and Sloupek) are situated outside of these positions (on the basis of surface surveys and ZABAGED). Therefore, through the hydrological modelling we try to determine hypothetical flows and other water sources in past landscape and explore the limits of water supply. Lay-out of ploughlands has been only estimated to favourable areas with only vague statements about agricultural exploitation of slight slopes. Despite many attempts, there still lacks a particular idea in which environment of the ploughlands could be extended. The aim of the paper is to assess the validity of previous statements using new techniques (ALS) in research.

2 Source Data and Methodology

DTM 5G provided by the Czech Office for Surveying, Mapping and Cadastre were taken for the study areas in the Pilsner region (Fig. 1). These data is the part of west zone of the scanning in the Czech Republic (with using of Riegl LMS-Q680) which was realised from the 11th October 2010 to the 27th June 2011. The altitude was determined 1200 m over ground level for woodlands areas. Data were delivered in ASCII format. Final data are represented as the irregular network of elevation points with density of ca. 0.3/m^2. These data has been evaluated in ArcMap 10.2 and Surfer 12. At first, it has been interpolated to DEMs with using algorithm Natural Neighbor (cell size is 0.5 m). Micro-topographic features including anthropogenic relief formations have been revealed by the method of anisotropic Sky-view factor (Zakšek et al. 2012).

Final DTMs have been confronted with the previous results of the surface surveys and geodetic-topographic surveys in the areas of five deserted villages (cf. Veselá 2006, 2008; Holata 2009; Vařeka et al. 2011). Abandonment of

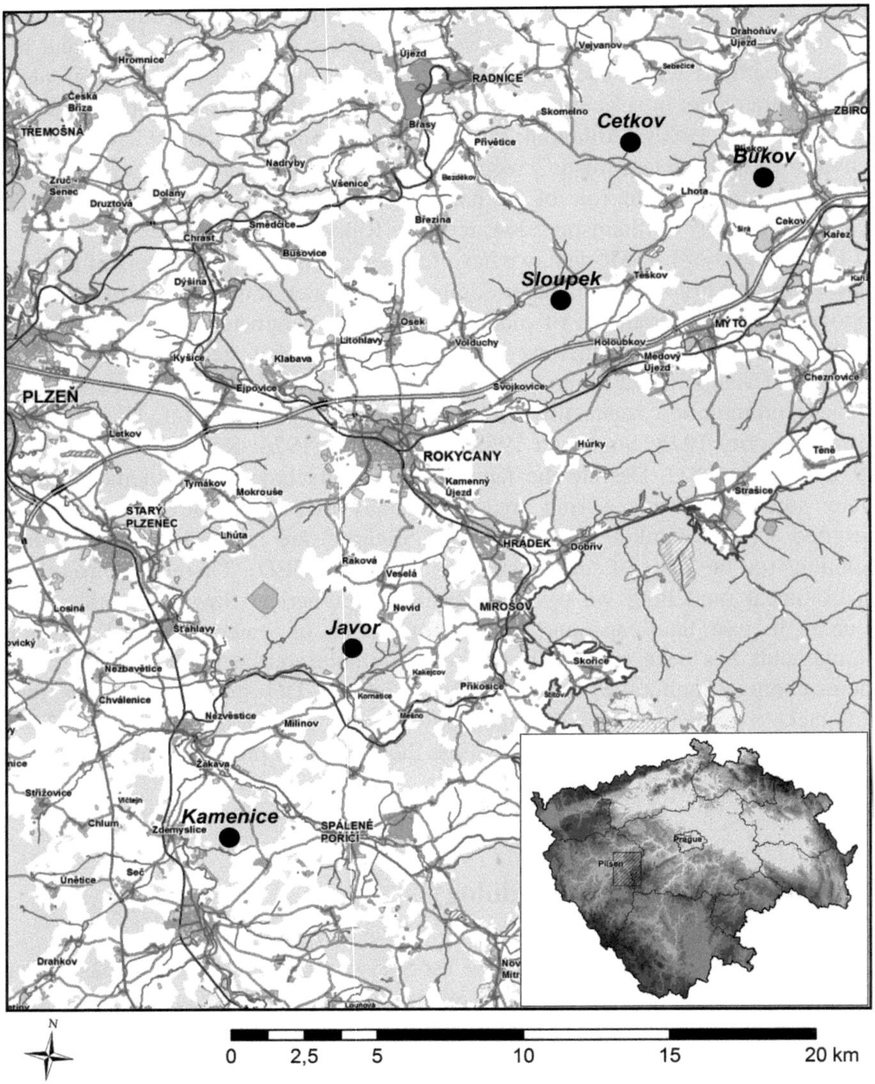

Fig. 1 Location of deserted villages mentioned in the text

Kamenice, Javor and Sloupek is dated (according to the written sources and pottery) to the first half of the 15th century; Bukov and Cetkov had more complicated settlement evolution and survived in shrunken form up to the first half of the 17th century (Rožmberský 2006, 2009). Questionable features in DTMs with more difficult interpretation have been verified by recent authors' surveys. The position and shape of the detected anthropogenic relief formations have been documented by GPS (Trimble Pathfinder ProXH in recent year) and total station (Leica TCR 407)

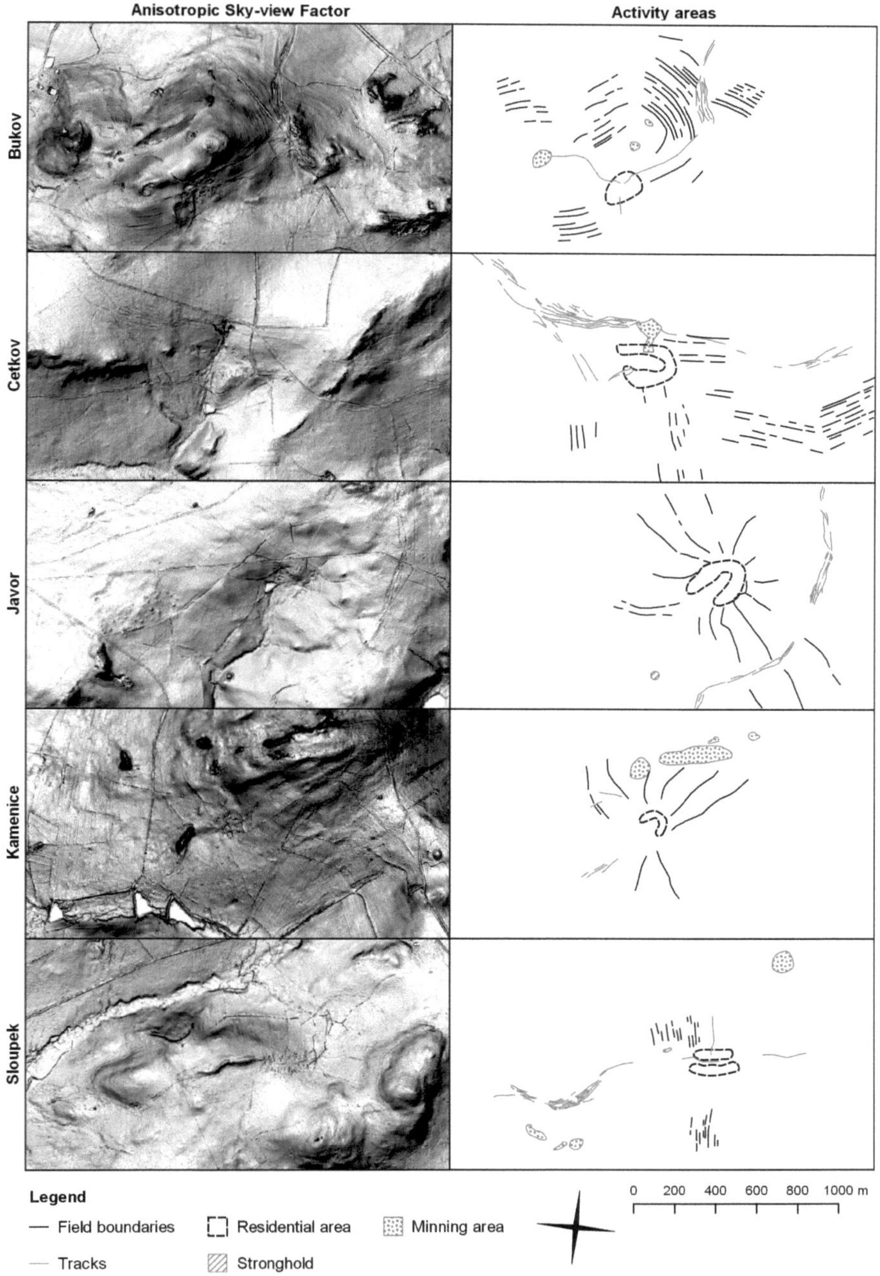

Fig. 2 DTMs of the broader areas involving deserted villages (visualised by anisotropic sky-view factor) and reconstruction of particular activity areas

and displayed on these DTMs as 'shapefile layers'. These both independent sources mutually validate each other and according to the identical morphological characteristics of other earthworks visualised by LiDAR data in the areas of deserted villages could be included to the resulting projections displaying particular activity areas (Fig. 2).

Both spatial analyses demand particular corrections of source LiDAR data. For the purpose of hydrological modelling of the past landscape, the recent features (especially roads and forest ways) had to be manually eliminated because of fundamental impact on flows and water accumulation. These "filtered" data has been applied to generation of new DEMs which constitute the bases for hydrological modelling in high resolution. It was realised in QGIS and SAGA GIS with using of SAGA Wetness Index algorithm (Böhner et al. 2002) for computing of modified Topographic Wetness Index (TWI) for three deserted villages currently without a direct attachment to the water sources. TWI is commonly used in prediction or distribution maps of archaeological sites in small scales (cf. Andresen 2008; Bevan et al. 2013). However, the calculation of TWI based on high resolution DEM can also reveals a local spatial patterns of hydrological character which can be very useful in archaeological interpretation of past land management and attachment to the water sources. Subsequently, also DEMs used for slope analysis had to be modified. The exact values of slope gradient in the areas of original fields are influenced by the later or recent features, especially by many windfalls or charcoal burning platforms (relatively dated to Early modern period) which overlay the medieval fields. Therefore, DEMs for slope analysis have been smoothed by the Low-pass filter. These corrections enable the approximation of DEMs to the original, medieval conditions without later disturbances and modifications (cultural and natural). Nevertheless, the critical evaluation of these bases is necessary. Potential transformations of the relief (e.g. extensive soil erosion) during the existence of village communities can be affected only by the systematic empirical research. This is the epiphenomenon of every research concerning the past landscape which should be overcome by testing the results in the broader geographical perspective.

3 Results

3.1 Hydrological Modelling

Hydrological modelling of the modified DTMs covers broader areas of Bukov, Kamenice and Sloupek. Due to the locations of these deserted villages on the hillsides or local saddles, the possibilities of using the surface water were limited. It is also indicated by ZABAGED (Fig. 3) where all these deserted villages are situated outside of streams. Therefore, the rainwater utilisation has been expected, as is evidenced by the presence of many small ponds or wells inside the residential areas or in their close proximity.

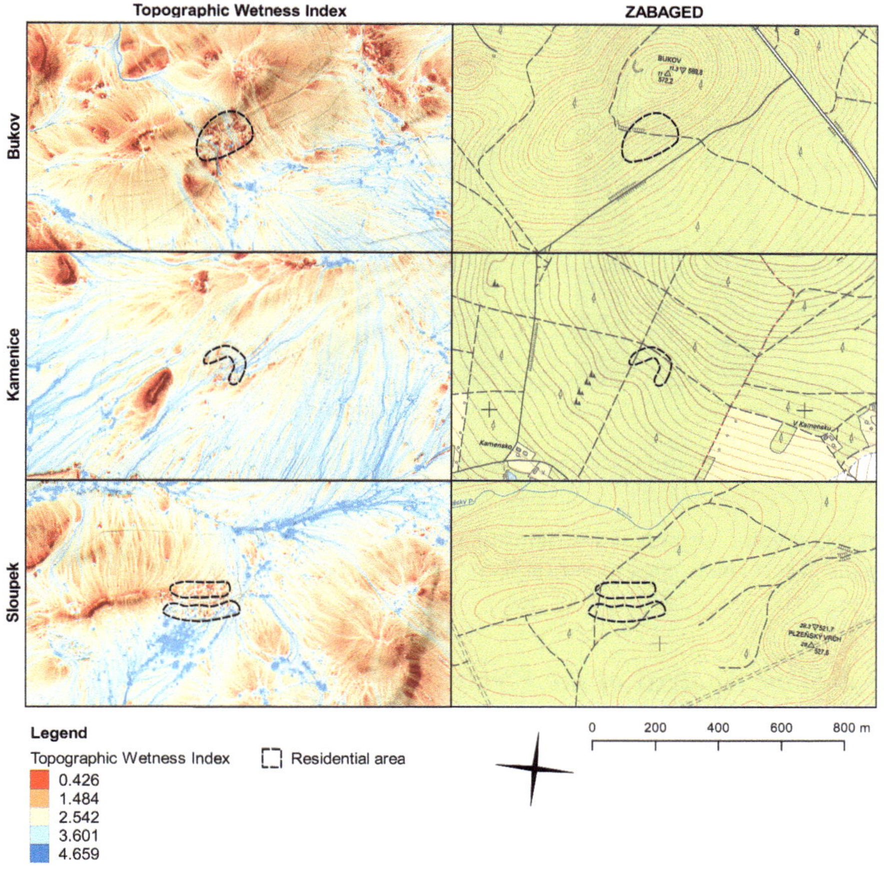

Fig. 3 Hydrological modelling on the basis of LiDAR data in the areas of three deserted villages compared with more general topography (ZABAGED)

High values of topographic wetness index indicate also the hypothetical flows or the areas with higher soil moisture just within close proximity to each of these villages (Fig. 3). In the case of Bukov and Kamenice there are apparent flows or concentrated drainage through their residential areas. Although the flow in Bukov is lesser evident, it is supplemented by the area with water accumulation in the southeast. Kamenice has several variously abundant flow accumulations in or around residential area. Deserted village Sloupek lies in the watershed with drainage to the north and south. Larger spring area is occurred in the east end of the residential area. Also the extensive area of soil moisture is situated behind the southern margin of this village. In addition, one more trend is apparent in the case of all these villages. There is certain extent of area around them with lower values

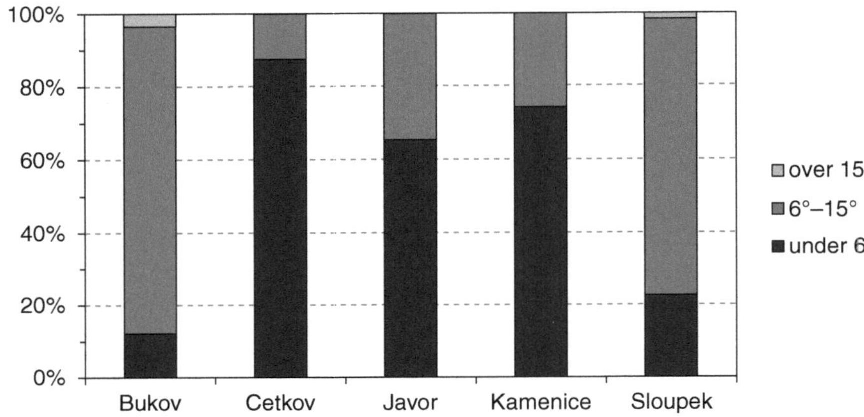

Fig. 4 Fields of deserted villages with the results of slope analysis

of topographic wetness index indicating the smaller flow accumulation or even dry areas. These areas correspond (at least partially) with the traces of ploughlands.

3.2 Slope Analysis of the Fields

Slope analysis has been done for the complete set of deserted villages. They can be divided into two groups according the slope gradient of detected segments of their fields (Fig. 4). The first group includes the villages of Cetkov, Javor and Kamenice. Their fields occupied relatively mild slopes (Mean 4.5°–5.3°). Although ploughland in Cetkov reaches the maximal value of this dataset (28°), it is represented by anomaly in DTM caused by the recent and unfiltered feature (0.09 % over 15°). The ploughlands in Javor and Kamenice also do not extend in this steep slope. The biggest part of the first group's fields is situated in areas under 6° (65–87 %). Nevertheless, particular segments fell also into steeper slopes up to 15° (12–34 %).

The second group is represented by Bukov and Sloupek. Their detected parts of fields are characterised by higher vertical articulation (Mean 8.6°–9°). Bigger portion of their ploughlands (74–87 %) was laid out in the areas which are affected by water erosion (over 6°) in greater extent (Šarapatka et al. 2002) whereas small extent is situated in mild slopes under 6° (12–22 %). The small parts of detected ploughlands in Bukov and Sloupek reached over 15° (1.6–3.4 %) which is considered as absolutely inappropriate for agriculture (Kolejka et al. 2009).

4 Conclusion

The paper presents the using of detailed terrain models generated by ALS in the archaeological research of deserted medieval or post-medieval rural settlement. These DTMs visualised by specific method (Anisotropic Sky-view factor has been used in this paper) are primarily applied to the identification and localisation of the remains of past human activities preserved as anthropogenic relief formations in woodlands. Together with the results of surface surveys they enable the reconstruction of the original settlements areas (villages with their hinterlands). However, these detailed DTMs have also broader analytical potential and it is possible to apply them in the study of deserted villages' environment.

We tested the premise involved in older literature that medieval settlement had a direct attachment to the water sources. This relationship has not been proved in the case of three surveyed deserted villages in Pilsner region by surface survey or with using of more general map basis (ZABAGED). Therefore, we applied hydrological modelling based on DTM 5G in these transects. The results demonstrate hypothetical flows or areas with higher soil moisture within residential areas or in their close proximity. All these villages were attached to water sources, their communities utilised rainfall water that was accumulated also in numerous water reservoirs or wells near the farmsteads. These outcomes extend previous statements on other suitable village locations in term of water supply. In this context, another factor can be relevant; all three pursued villages are also surrounded by the areas with smaller flow accumulation or even dry areas. Prevailing overlay with the traces of ploughlands is considered as unexpected and very interesting. The significance of this trend can be assessed only after inclusion of other deserted villages in the research. Currently we cannot exclude also the influence of post-depositional transformation in disappearance of field remains in the areas characterised by higher flow accumulation.

The lay-out of ploughlands in term of slope gradient has been only estimated to the favourable conditions in slight slopes. Specific visualisation of DTM 5G combined with surface surveys recorded field boundaries of deserted villages also in steeper slopes with extensive segments over 6°. These parts of relief can be affected by soil erosion and therefore they are considered as unsuitable for agriculture. Although we assume that only part of original ploughlands was recorded (especially in the case of Cetkov and Sloupek), gradient of slopes in fields cannot be regarded as a crucial factor in the choice of the positions for settlement areas. Using of ALS enabled us to revise the 'traditional' statements considering vertical articulation of settlement areas. We argue that the role of natural environment (concerning other parameters of physical landscape as well) in the shaping of historic landscape should be re-evaluated with the using of high-quality source data (e.g. numerous set of deserted villages, detailed map bases) and advanced spatial analyses in GIS. In this regard, their relations to cultural characteristics and human-environmental interactions in general have to be thoroughly explored.

Sources of Support This paper is the result of the project SGS-2014-057 "Airborne Laser Scanning of Landscape as Method of Archaeological Prospection".

References

Andresen J (2008) Topographic wetness index and prehistoric land use. In: Posluschny A, Lambers K, Herzog I (eds) Layers of perception. Proceedings of the 35th international conference on computer applications and quantitative methods in archaeology (CAA), Berlin, Germany, 2–6 April, Dr. Rudolf Habelt GmbH, Bonn, pp 405–410

Beresford MW (1954) The lost villages of England. Royal Geographical Society, London

Beresford MW, Hurst JG (eds) (1971) Deserted medieval village. Lutterworth Press, London

Bevan AH, Crema ER, Li X, Palmisano A (2013) Intensities, interactions and uncertainties: some new approaches to archaeological distributions. In: Bevan AH, Lake MW (eds) Computational approaches to archaeological spaces. Left Coast Press, Walnut Creek, pp 27–51

Boháč Z (1983) Vesnice v sídelní struktuře předhusitských Čech. Historická geografie 21:37–116

Böhner J, Köthe R, Conrad O, Gross J, Ringeler A, Selige T (2002) Soil regionalisation by means of terrain analysis and process parameterisation. In: Micheli E, Nachtergaele F, Montanarella L (eds) Soil classification 2001, EUR 20398 EN. The European Soil Bureau, Joint Research Centre, Ispra, pp 213–222

Čapek L (2013) Využití dat leteckého laserového skenování ke studiu středověké a novověké krajiny Velechvínského polesí, okr. České Budějovice. In: Gojda M, John J et al. Archeologie a letecké laserové skenování krajiny. Západočeská univerzita v Plzni, Plzeň

Čapek L, John J, Stolz D (2013) Příspěvek leteckého laserového skenování k poznání dvou zaniklých středověkých vesnic mezi Líšnou a Točníkem. In: Gojda M, John J et al. Archeologie a letecké laserové skenování krajiny. Západočeská univerzita v Plzni, Plzeň

Černý E (1992) Výsledky výzkumu zaniklých středověkých osad a jejich plužin: Historicko-geografická studie v regionu Drahanské vrchoviny. Muzejní a vlastivědná společnost, Brno

Chapelot J, Fossier R (1985) The village and house in the middle ages. B. T. Batsford Ltd, London

Crow P, Benham S, Devereux BJ, Amable GS (2007) Woodland vegetation and its implications for archaeological survey using LiDAR. Forestry 80:241–252

Doneus M, Neubauer W (2008) Aerial archaeology and airborne laser scanning at the iron age Hillfort Schwarzenbach-Burg. ViaVIAS 02:53–59

Gissel S, Jutikkala E, Österberg E, Sandnes J, Teitsson B (1981) Desertion and land colonization in the Nordic countries c. 1300–1600. Almqvist and Wiksell, Stockholm

Holata L (2009) Zaniklá ves Kamenice. In: Vařeka P, Rožmberský P et al. Středověká krajina na střední Úslavě I. Ing. Petr Mikota, Plzeň

Kalina T (1978) Vývoj polohy sídel v Pražské kotlině od 10. do pol. 14. stol. Historická geografie 17:311–368

Klápště J (1973) Černokostelecko jako kolonizační oblast. Historická geografie 10:123–138

Klápště J (1978) Středověké osídlení Černokostelecka. Památky archeologické 69:423–475

Kolejka J, Káňa D, Plšek V, Klimánek M, Navrátil V, Svoboda J (2009) Tématické mapy založené na digitálním modelu reliéfu. Geomorphologia Slovaca et Bohemica 2:13–27

Kuna M, Tomášek M (2004) Povrchový výzkum reliéfních tvarů. In: Kuna M et al. Nedestruktivní archeologie. Academia, Praha

Lasaponara R, Coluzzi R, Gizzi FT, Masini N (2010) On the LiDAR contribution for the archaeological and geomorphological study of a deserted medieval village in southern Italy. J Geophys Eng 7:155–163

Lennon B, Crow P (2009) LiDAR and its role in understanding the historic landscape of Savernake forest. Wiltshire Archaeol Nat Hist Mag 102:245–261

Nekuda V (1975) Pfaffenschlag. Zaniklá středověká ves u Slavonic. Blok, Brno

Pohlendt H (1950) Die Verbreitung der mittelalterlichen Wüstungen in Deutschland. Geographisches Institut der Universität, Göttingen

Pokorný O (1973) Povrchový výzkum zaniklých osad v našich zemích a některé jeho historickogeografické aspekty. Historická geografie 10:63–80

Rippon S, Wainwright A, Smart C (2014) Farming regions in medieval England: the archaeobotanical and zooarchaeological evidence. Medieval Archaeol 58:195–255

Rožmberský P (2006) Soupis zaniklých středověkých vesnic na Rokycansku. In: Vařeka P et al. Archeologie zaniklých středověkých vesnic na Rokycansku I. Ing. Petr Mikota, Plzeň

Rožmberský P (2009) Dějiny sídel na střední Úslavě. In: Vařeka P, Rožmberský P et al. Středověká krajina na střední Úslavě I. Ing. Petr Mikota, Plzeň

Šarapatka B, Dlapa P, Bedrna Z (2002) Kvalita a degradace půdy. Univerzita Palackého, Olomouc

Smetánka Z (1973) Povrchový průzkum na lokalitě Karlík u Dobřichovic. Historická Geografie 10:81–96

Smetánka Z (1978) Česká vesnice v období vzniku městských aglomerací. Archaeologia Historica 3:325–330

Smetánka Z (1988) Život středověké vesnice. Zaniklá Svídna. Academia, Praha

Smetánka Z, Klápště J (1981) Geodeticko—topografický průzkum zaniklých středověkých vsí na Černokostelecku. Památky archeologické 72:416–458

Snášil R (1976) Životní prostředí vesnických sídlišť 10.—15. století v ČSR (Nástin dosavadních výsledků). Archaeologia Historica 1:139–144

Vařeka P et al (2006) Archeologie zaniklých středověkých vesnic na Rokycansku I. Ing. Petr Mikota, Plzeň

Vařeka P, Rožmberský P et al (2009) Středověká krajina na střední Úslavě I. Ing, Petr Mikota, Plzeň

Vařeka P, Holata L, Rožmberský P, Schejbalová Z (2011) Středověké osídlení Rokycanska a problematika zaniklých vesnic. Archaeologia Historica 36:319–342

Veselá R (2006) Zaniklá vesnice Cetkov. In: Vařeka P et al., Archeologie zaniklých středověkých vesnic na Rokycansku I. Ing. Petr Mikota, Plzeň

Veselá R (2008) Zaniklá ves Javor. In: Vařeka P et al. Archeologie zaniklých středověkých vesnic na Rokycansku II. Ing. Petr Mikota, Plzeň

Williamson T (2004) Shaping medieval landscapes: settlement, society, environment. Windgather Press, Macclesfield

Zakšek K, Oštir K, Pehani P, Kokalj Ž, Polert E (2012) Hill-shading based on anisotropic diffuse illumination. In: Růžička J, Růžičková K (eds) GIS ostrava 2012—surface models for geosciences, Ostrava, 23–25 January, VŠB - Technical University of Ostrava, Ostrava, pp 283–297

Žemlička J (1978) Osídlení Litoměřicka do začátku 13. století. Historická geografie 17:65–98

DTM Impact on the Results of Dam Break Simulation in 1D Hydraulic Models

Andrea Jančíková and Jan Unucka

Abstract Dam break simulation has been quite long established module in the hydraulic modeling industrial standards, which are represented in particular by the software packages USACE HEC-RAS and DHI MIKE 11 worldwide including the Czech Republic. Coincidentally, at this level, both previously mentioned hydraulic models are using identical numerical solvers DAMBRK and WSPRO. It can be expected that the use of identical schematization of the river channels and technical objects together with the parameterization of dam body and its geometric parameters will give the comparable results of hydraulic simulations. When the same approach is applied for the reservoir and dam schematization together with its operational rules, the simulations using the identical solver DAMBRK will produce almost same results. Finally, the differences of the generated floodlakes by these particular models will be most affected by the accuracy and resolution DMT/DMR and partly by the different simulation concepts of water flow in the inundation area within each individual model. Because of the extreme situations of the "dam break" are occurring rarely (in the Czech Republic with larger water works only once), the calibration data is virtually absent. It is certainly not an argument to tell, that the simulation apparatus for HPPS and POVIS in the scope of the crisis management and planning was not prepared for these situations. A sensitivity analysis is among the crucial conditions of the successful modelling. This paper focuses particularly on the impact of DTM accuracy on dam break simulations.

A. Jančíková
Katastrální úřad, Olomouc District, Jeremenkova 110/15, 772 11 Olomouc, Czech Republic
e-mail: andrea.jancikova@cuzk.cz

J. Unucka (✉)
Czech Hydrometeorogical Institute, K Myslivně 3/2182, 708 00 Ostrava-Poruba, Czech Republic
e-mail: jan.unucka@chmi.cz; jan.unucka@vsb.cz

J. Unucka
Faculty of Mining and Geology, VSB-Technical University of Ostrava, 17. Listopadu 15/2172, 708 33 Ostrava-Poruba, Czech Republic

K. Růžičková and T. Inspektor (eds.), *Surface Models for Geosciences*, Lecture Notes in Geoinformation and Cartography,
DOI 10.1007/978-3-319-18407-4_11

Keywords Dam break analysis · Hydraulic modelling · MIKE 11 · HEC-RAS · DMT · DMR · Hloučela river · Plumlov reservoir

1 Preface

Hydraulic modelling is already among the standard tools both for the flood risk analyses and partly for the operational hydrologic forecasting in the national context together with the worldwide trends of the rainfall-runoff and hydraulic models integration. Operational Flood Forecasting Service of Czech Republic (HPPS ČR) is provided by the Czech Hydrometeorological Institute (CHMI hereinafter) in cooperation with the River boards and possibly with the other institutions of the crisis management and planning. Another impulse for the integration of hydraulic modelling methods into routine hydrological analysis is then migration of POVIS Flood Information System within the competence of the Czech Hydrometeorological Institute, that generates the obvious demand to harmonize the outcomes of the fully established rainfall modelling (hereinafter RR) with some outcomes of hydraulic modelling (hereinafter HD). It is obvious that no group of hydrological models (RR + HD) is capable of effective and accurate simulations and predictions without quality spatial/geographical data. The urgent need of accurate geographical data and consequently the quality GIS software itself from the second half of the 90 s of the last century resulted either in the integration of the GIS modules (ESRI platform and open source libraries) into the software tools for hydrological modelling or vice versa integration of selected hydrological models within the established GIS platforms. Here we can mention the integration of the EPA SWMM or SWAT models within the ESRI ArcView GIS and ArcGIS or later the integration of TOPMODEL or SIMWE rainfall-runoff models within the GRASS GIS and SAGA GIS environment. For hydraulic models the situation is not so simple due to more complicated aspects of the model schematization and simulation together with the higher amount of the manual editing, it can be concluded that the most common type of integration or rather communication with GIS software, is the existence of preprocessor and postprocessor in GIS environment. The two most widely used 1D hydraulic models and industry standards FEMA/NFIP USACE HEC-RAS and DHI MIKE 11 have the extensions HEC-GeoRAS (ESRI ArcView GIS for 3.x, ArcGIS 9.x and 10.x or ArcINFO 7.x) and MIKE 11 GIS (ESRI ArcView GIS 3.x, ArcGIS 9.x and 10.x). These two extensions are also used in this article together with the hydraulic models mentioned above. If we restrict our goals of the hydraulic modelling issues for the particular analyses of territorial impacts of special flood due to reservoir dam failure (e.g. dam break analysis), it is obvious that the digital terrain model or digital surface model (hereinafter DMT/DMR) plays the dominant role among spatial data). Parameters of their quality, aside from the obligatory requirements for DMT/DMR (see eds. Wilson and Gallant 2000), are the position and height accuracy, raster resolution and the level of singularities capture (edges, embankments, etc.). Practical experience suggests (and analyses used in this article confirm this) that the modern generation of DMT/DMR based on the laser scanning (DMR 4G/5G ČÚZK) can be considered as an important benefit for the precision of hydraulic modelling. The aim of the article was therefore to determine whether this benefit is equally strong even for the dam break analyses and whether the older generation DMT/DMR, which are still used in the Czech Republic (e.g. digital terrain models interpolated from the datasets such is DMÚ 25 or ZABAGED) are actually suitable for this type of analyses nowadays. Finally, it is necessary to emphasize that while the floods caused by natural factors (here we can classify regional floods from the stratiform precipitation, flash floods from the convective precipitation and flooding from the spring snowmelt) has a relatively solid data base for verification and calibration outputs of RR and HD modelling. On the contrary, we fortunately haven't such calibration data for dam break flood types. Such failures are not common in European countries involving Czech republic—except the accident on the water reservoir on Bila Desna river in 1916, which represent the only catastrophe of such type occurred in our country. While this complicates the verification and calibration of HD models for this type of analysis, the authors also wishes that this situation will persist also in the future. Finally, the hydraulic modelling process itself and model outputs have to be in the accordance with applicable legislation and regulations, we could refer to another works of the authors (Jančíková 2014; Unucka 2014), where the summary of this issues can be found.

2 Used Tools and Methods

As was already mentioned, this paper focuses on the analysis of the type of dam break in 1D hydraulic models and influence the accuracy of DTM/DSM on the results of these analyzes. For the actual hydraulic modeling were used industry standards USACE HEC-RAS 4.2 beta and DHI MIKE 11, 2011. Specifically, HEC-RAS 4.2 and 5.0 already integrates 1D (channel) and 2D (inundation) numerical solutions. MIKE 11 hydraulic model could be integrated with the 2D model MIKE 21c using the platform MIKE FLOOD. An important aspect for the use of 1D models (even in the context of this paper) is the fact that technical objects (e.g. dams, bridges, culverts, diverting objects, other water works) cannot be satisfactorily schematized within the 2D models and consequently there is no possibility of the proper numerical or analytical solution of their impact to the river hydraulics. Therefore, the abovementioned combination of 1D model for flow simulation in channels and the 2D model for flow simulation in inundation areas is currently (and probably in the future) the most pragmatic (Kožaná et al. 2014). Likewise, any operative communication with the RR models (e.g. HEC-HMS and MIKE SHE) is far more effective in the case of using 1D model. Both of these software products have virtually identical numeric-analytical solver for the dam body failure (NWS/FLDWAV DAMBRK) and also the solver for the influence of technical objects

(USGS WSPRO). Both models can also use a GIS preprocessor and postprocessor, specifically HEC-GeoRAS and MIKE 11 GIS ESRI ArcGIS. Part of the river Hloučela (about 12 km) and water reservoir Plumlov were schematized in these models and their preprocessors using 4th generation digital terrain model ČÚZK (hereinafter DMR 4G) and another technical objects (bridges etc.) were schematized beside the water reservoir itself. Fully automated preprocessing cannot be achieved for such situations because the schematization of the HD model including the technical objects is rather tedious task, in that respect that all required parameters cannot be obtained from the DTM. All schematization steps were therefore performed using the DMR 4G dataset and other types of DTM (interpolated by the Topo the Raster) were used only for HD model postprocessing and generation of the floodlakes. These interpolated DTMs were obtained from DMÚ 25 and ZABAGED datasets. Finally, the cell resolutions of particular digital terrain and surface models were 5 × 5 m for DMR 4G, 15 × 15 m for ZABAGED and 25 × 25 m for DMÚ 25 datasets. But it is obvious that, if there were such a variant schematization based on various abovementioned DTM datasets, the simulation of HD model and dam break analysis itself will be influenced due to the schematization errors propagation to the geometry of the longitudinal profile and cross sections. But these variations will exceed the paper extent and there is no reason using the less accurate HD model schematization nowadays. Finally, the individual DMT/DMR were compared with each other using regression analysis in ArcView GIS and IDRISI Selva (see Fig. 1). The dam break simulation results in both HD model were compared with each other, and then subjected to post-processing (analysis of the scope and depth of the flooded area) using the DMR 4G (5 × 5 m) DMT and interpolated data from the DMU 25 (25 × 25 m) and ZABAGED

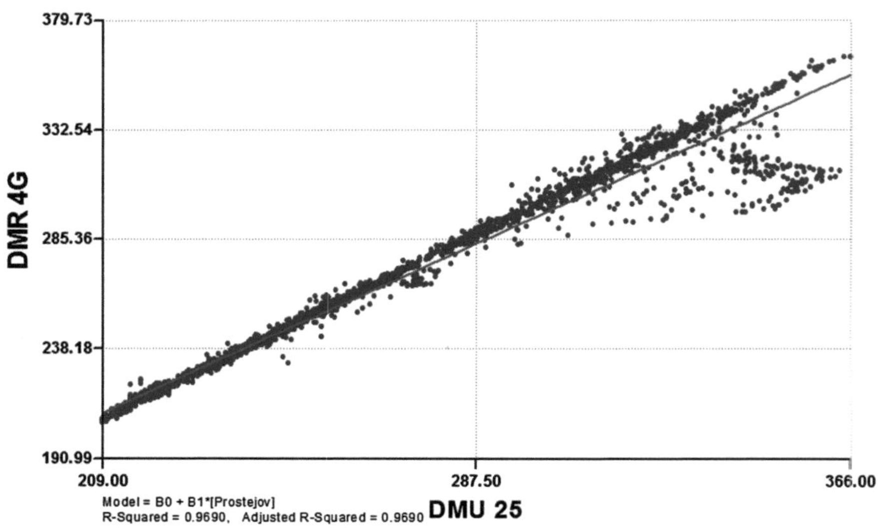

Fig. 1 Regression analysis of DTM values between DMÚ 25 a DMR 4G datasets in pilot area

(15 × 15 m). It is obvious that the grid resolution of DTM/DSM influences the extent of floodplains especially at the borders of inundation. The authors are aware of this fact and focus on the overall extent of the flooded area and floodlakes depths.

3 Brief Pilot Area Description

Hloučela River Basin is the watershed of 4th order since the number of hydrological sequence are from 4-12-01-045 to 4-12-01-057. Catchment area is 129 km^2 according to data of the Czech Hydrometeorological Institute and DIBAVOD. River Hloučela has its source in the woods of Drahanská vrchovina highlands north from the Buková village and has two sources that are spaced about 830 m as the crow flies. Hloučela is at its closing profile (total length 39 km) the right tributary of Romže (3rd order river), which then flows rightward as the Valová river onto Morava river (2nd order) near the village Uhřičice. The geological bedrock of the basin consists mainly of the Paleozoic metamorphic rocks of the Bohemian Massif, but the area of interest spans to the border of the Outer Western Carpathians, where fluvial, glacifluvial and partly eolic and lacustrine sediments of Cenozoic and Quaternary are dominant. Cambisols, gleysols and fluvisols are the most common soil types in the pilot catchment. Fluvisols are most frequented especially in the inundation areas of Hloučela and Romže rivers. Hloučela river basin belongs to moderately warm area (MT), only a small portion of the NW part of the basin lays in the cold area (CH7) and similarly the lowest parts of the catchment to the SE then belong to the warm area (T2) according to Quitt classification of climatic regions. Agricultural areas and forests are the main land use types in the catchment area (Safar 2003). Hydraulic model schematizations started above the Plumlov dam and ended on the confluence with the Romže river (0 m river stationing). Plumlov dam itself has the 9860 m stationing position in whole river schematization. Definitely, the schematization were extended above Plumlov reservoir to ensure adequate simulation of dam break due to the inflow above the Q_{100}/Q_{1000} in combination with improper manipulation on the dam outlets. Observing profile with gauge is in the category "A" of HPPS ČR with the serial number 333. Value of the average discharge Q_A is 0.580 $m^3\ s^{-1}$, Q_1 value is 7.40 $m^3\ s^{-1}$, Q_{50} value is 34.5 $m^3\ s^{-1}$ (above the Plumlov reservoir is then over 38.5 $m^3\ s^{-1}$ due to the influence of reservoir transformation effect) and Q_{100} value is 41 $m^3\ s^{-1}$ (over 47 $m^3\ s^{-1}$ above the Plumlov reservoir). For other values of N-year and m-daily Q we can refer to the pages of HPPS ČR (http://hydro.chmi.cz). Lesy ČR, s.p. is administrator of the upper part of basin and Povodí Moravy, s.p. of the lower parts of basin. Brno branch of the Czech Hydrometeorological Institute guarantees the HPPS ČR for the Hloučela river basins and its profiles, while the operational hydrological forecasting calculations are performed using the HYDROG RR model including the ensemble forecasts including ESP-ALADIN LAEF. Nowcasting forecasts are based on COTREC method in Brno and Ostrava branches of Czech Hydrometeorological Institute. Hydraulic transformation of flood waves on Morava river is provided at

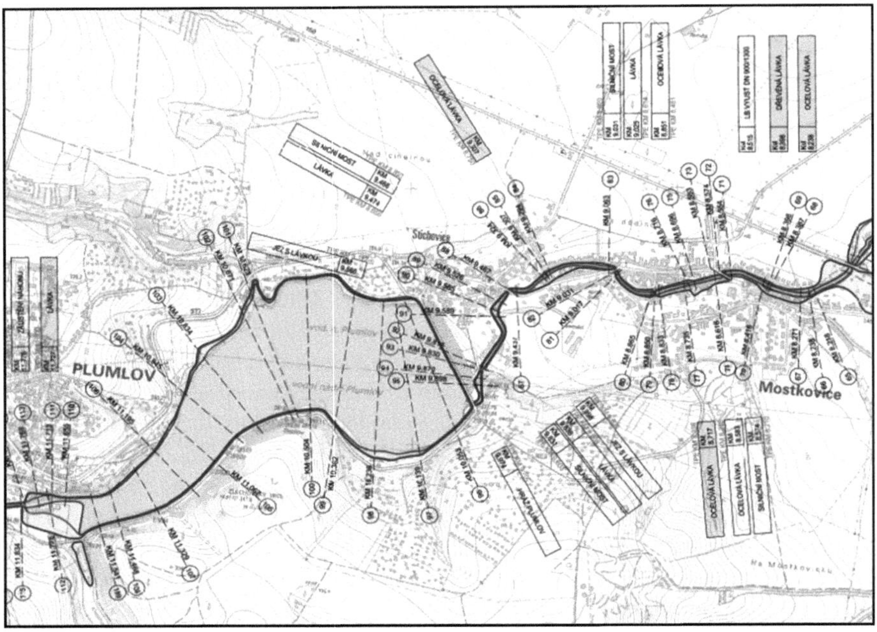

Fig. 2 Cross sections placement in HD model construction around the Plumlov reservoir

the Brno branch using MIKE 11 HD model with RR module MIKE NAM. Basic technical parameters of the Plumlov reserovoir and dam can be obtained on the website of Povodí Moravy, s.p. (http://www.pmo.cz/cz/uzitecne/vodni-dila/plumlov/) or in more detail, including the history of construction and reconstruction of the waterworks in author's work Jančíková (2014). Location of the cross sections for the schematization of HD models HEC-RAS and MIKE 11 near VD Plumlov is illustrated in Fig. 2. Calibration of models were performed for the Q_{50} and Q_{100} discharges and water levels. Fortunately, there is no observations and inundation areas evidence for the dam break situation, as was mentioned in the preface.

4 Selected Results

For detailed results of the entire cascade of the GIS preprocessing, HD modelling itself and final GIS postprocessing see the work of Jančíková (2014). For a more detailed description of the characteristics of hydraulic model HEC-RAS and MIKE 11 You can then refer to the websites of the model manufacturers, or to the work of Unucka (2014) respectively. After schematization of Hloučela river in models HEC-RAS and MIKE 11 using their particular GIS extensions the flow simulations of Q50 and Q100 were performed followed by the calibrations of hydrographs and rating curves in the channel of Hloučela river and locally the extents of floodlakes. Results correlation values between the HD models was very good (over 98.6 %), which corresponded to the entrance hypothesis of the authors that using the same input data, virtually identical schematization, simulation methods such is Bernoulli equation (HEC-RAS and MIKE 11) or the dynamic wave approximation (MIKE 11 for verification), USGS and NWS WSPRO/FLDWAV DAMBRK (HEC-RAS and MIKE 11) together with very similar mechanisms of the GIS postprocessing (for differences can again refer to the manufacturer's manuals or the SW) we can expect a high degree of results correlation of both 1D HD models. After this the dam break simulations phase were performed itself beginning in the dam crest (278.63 m above sea level) with downwards progression which stopped at the value of 271.5 m above sea level with total failure width 55 m on its upper part. The results of HD models simulations (floodlakes extents) and correlation values depths in channels and inundation are shown in Figs. 3, 4, 5 and 6.

We can characterize the actual parameters of the special flood situation due to dam break by the following parameters:

- Peak flow Qmax = 204 $m^3\ s^{-1}$,
- flood wave volume W = 4.8 million m^3,
- dam break wave celerity during the culmination raises up to 12.6 m s^{-1} just below the dam and in the first 300 m from the Plumlov water reservoir, in next kilometer downstream value drops to 1.3 m s^{-1} in the channel Hloučela (outside the channel the further decline occurs),

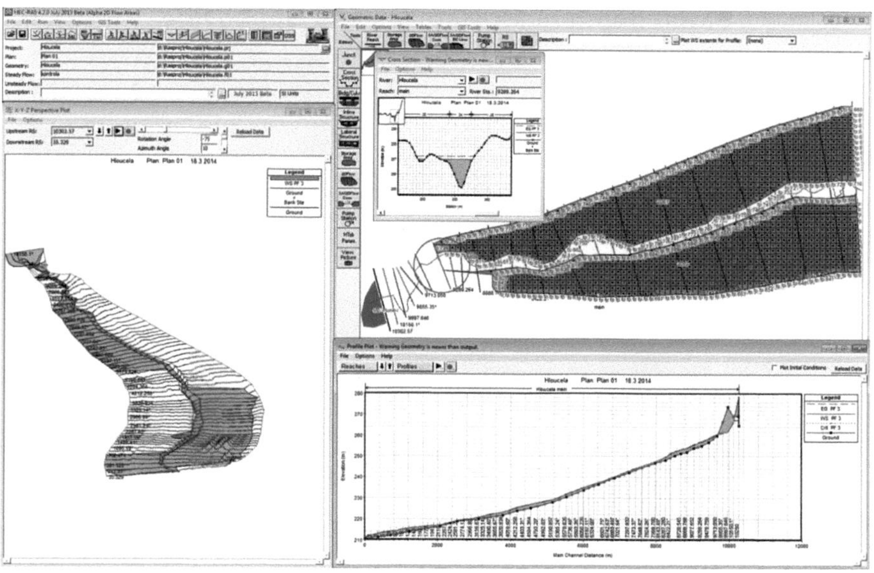

Fig. 3 Schematization of Hloučela river in HEC-RAS model for 1D and 2D flow solution

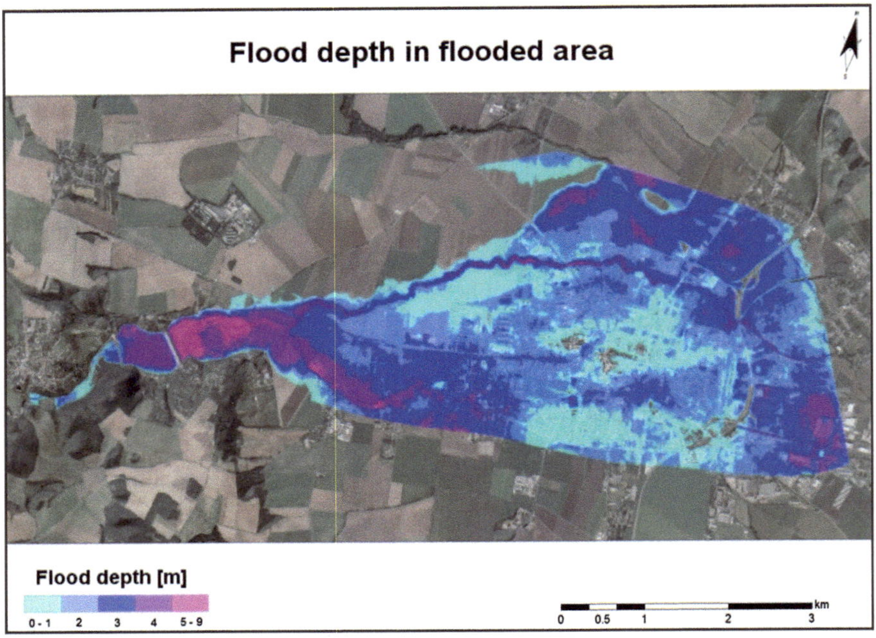

Fig. 4 Water depths and inundation are extent generated by MIKE 11/MIKE 11 GIS

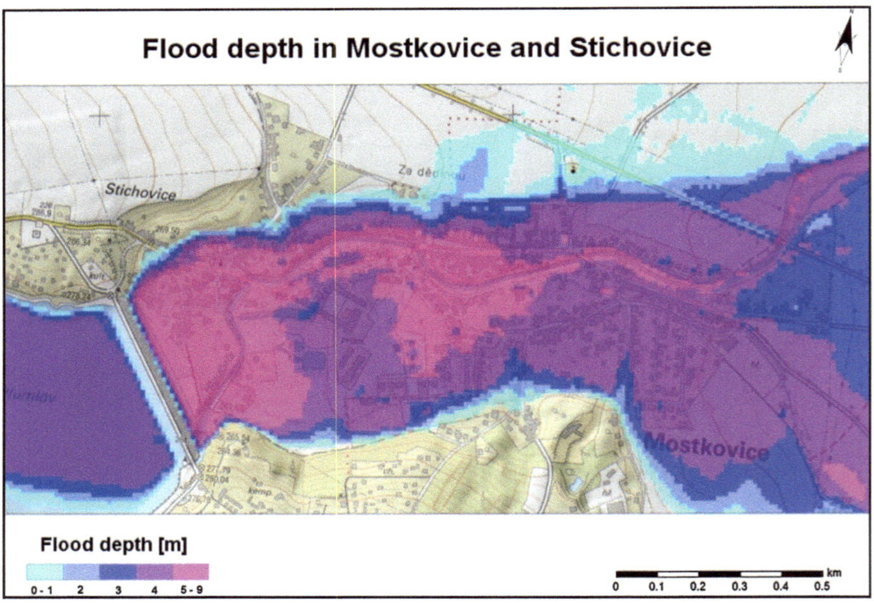

Fig. 5 Water depths and inundation are extent generated by MIKE 11/MIKE 11 GIS around the Plumlov reservoir dam

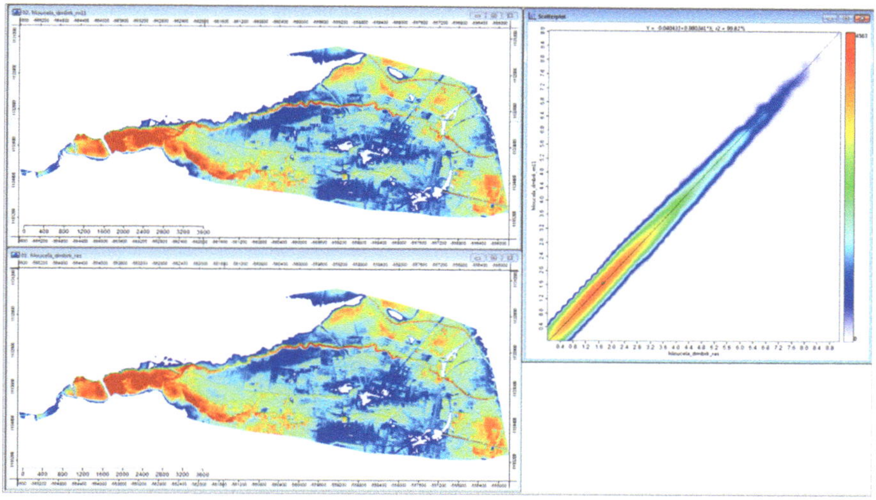

Fig. 6 Floodlakes generated by HEC-GeoRAS (*top left*), MIKE 11 GIS (*bottom left*) and regression analysis of both floodlakes in SAGA GIS (*right*)

- the speed of the break wave in the inundation area below the dam is 8 m s^{-1}, on the edges of inundation and downstream below then decreases to 0.5 m s^{-1},
- the depth below the Plumlov reservoir in the riverbed Hloučela increases up to 9 m during the culmination, in the Prostějov downtown decreases (<1 m),
- the culmination Q occurs approx. 1 h from the beginning of the dam body break,
- the maximum extent of floodplain is computed about 17.0 km^2 in HEC-GeoRAS and 17.2 km^2 in MIKE 11 GIS (these differences are due to different postprocessing when the HEC-GeoRAS are considered cross sections to the edge of considered floodplain, while such approach isn't required in the MIKE 11 GIS).

Computed floodlakes extents consequently calculated using DTMs interpolated the from DMU 25 and ZABAGED datasets indicated more significant result differences, especially in part 1.3–4.4 km from the Plumlov reservoir (cadastral evidence of Mostkovice and Prostějov) when the abovementioned parameters of the dam break wave floodplains significantly reduced and similar values depths and ranges overflowing is achieved again until cadastral area Prostějov from the station by about 6 km. According to the analysis carried out so far, there is noticeable effect complete secondary errors height grid cells DMÚ 25 (most obviously here) and ZABAGED (already small differences due to higher grid resolution and discretion terrain edges and point field dimensions during interpolation Topo to Raster). This approach has impact on the artificial "carving" of the cross sections and the height values of the banks opposite to DMR 4G dataset. The situation is summarized in the following Table 1.

Table 1 Comparison of the selected results of hydraulic modelling based on the various types of DMT for GIS postprocessing

Type of DTM	Cell size (m)	Floodlake extent (km^2)	Maximum depth (cm)
DMR 4G	5 × 5	17.2	9.2
DMÚ 25	25 × 25	14.7	10.7
ZABAGED	15 × 15	15.8	9.6

5 Discussion

It's evident from previous chapters that the HD models and the industrial standards in particular in this group of hydrologic models provide an effective analytical tool, particularly in cooperation with the GIS. Selected aspects of these mathematical models, including connectivity to GIS are well summarized in publications such are Bedient et al. (2007, 2013), Dyhouse (2007) and Di Baldassarre (2012). The data input and model structure uncertainties and consequent pitfalls of the improper use of mathematical models are discussed in Beven (2009). Technical reports and case studies are downloadable on the website USACE/HEC (including the aforementioned modules NWS/FLDWAV DAMBRK a WSPRO USGS). In this regard, it should be noted however, that in the present article presented only the simulation of scenarios during the special type of flood due the dam break event (albeit with calibrated 1D and 2D hydraulic models). As has been stated several times before, not only by the authors in previous sections of the text, hydrological and hydraulic models provide accurate results when there is an availability of the high-quality input data and vice versa, which Starý called the "data crisis" (in Jandora et al. 2002). A virtually identical conclusions in the next case study is also presented by the team of authors Kourgialas and Karatzas (2013) in the catchment areas of the island of Crete or in the Elbe basin case studies (Chaterjee et al. 2008). In both cases, the platform DHI MIKE was obviously proven in different physical-geographical and hydrological conditions. Similarly, the potential of HD model HEC-RAS tested for similar analyses in the work of Johnson et al. (1999). For detailed verification and validation of models it would be necessary to perform hydrometric measurements during the special flood event occurrence. However measurement during extreme discharge conditions brings significant error and data inaccuracy using both classical methods (hydrometric propeller) or new technologies based on ADCP. Finally, it is in fact hardly technically feasible not only on the level of measurement itself, but also on the level of rating curve construction, which represents further complication of the HD model calibration. Generally, the situation can be summarized by saying that, if the hydraulic models give satisfactory results during the floods caused by natural factors, we can expect that equal simulation apparatus can generate similarly good results for the special flood events simulations assuming the use of high-quality input data and verified tools (industrial standards). An interesting contribution to the debate is the article of USACE/HEC developers Ackermann and Brunner (2005) or Xiong (2011).

6 Conclusion

We can conclude that the phenomenon of the dam break events and simulation of this type of special flood represents a challenge both for geoinformatics and for hydroinformatics and hydrological modeling. Input hypothesis of the direct impact of DTM quality and resolution was confirmed with the utilization of GIS analyses and mathematical modelling. The greatest spatial accuracy of hydraulic model schematization can be achieved by using the LIDAR based DTM/DSM (DMR 4G ČÚZK) in combination with geodetic measurements of thalwegs and cross-sections. An interesting aspect will be the comparison with the next generation of digital terrain/surface model DMR 5G. It is interesting not just in the case of special flood events simulation, but also for the spatial and temporal reconstruction already occurred flood events caused by the natural factors. Finally, it is clear that the implementation of the so-called Flood Directive (2007/60/EC) was prepared responsibly and passed without significant problems in the Czech Republic at the level of methods, tools and input data. So we can expect that there will be further development of methods HPPS CR for which Czech Hydrometeorological Institute is responsible in cooperation with the River Boards and other institutions involved in crisis management and planning. Data of State Administration of Land Surveying and Cadastre (ČÚZK) then form an integral part of the critical data base. At the level of mathematical models we can already hardly expect fundamental changes, industrial standards such are USACE HEC-RAS and DHI MIKE 11 show their quality at the level of a separate HD modeling and even by the integration possibilities with other tools for RR modeling (HEC-HMS, MIKE SHE etc.), urban hydrology simulations (MIKE URBAN, EPA SWMM) or hydrogeological modeling packages (MODFLOW, FEFLOW).

Acknowledgments The authors would like to thank to the project TAČR č.TA04021123 "New methods of the surface water bodies morphology measurements and their utilization in landscape and emergency planning" without its support realization of this article would not have been possible. We thanks also to Czech Hydrometeorological Institute, Morava River Basin Board, and State Administration of Land Surveying and Cadastre for the support not only at the level of the data base for the model building and simulations.

References

Ackermann CT, Brunner GW (2005) Dam failure analysis using HEC-RAS and HEC-GeoRAS. 8 p. USACE/HEC

Bedient PB, Huber WC, Vieux BC (2007) Hydrology and floodplain analysis, 4th edn. Prentice Hall, London, 795 p. ISBN: 978-0131745896

Bedient PB, Huber WC, Vieux BC (2013) Hydrology and floodplain analysis, 5th edn. Prentice Hall, London, 815 p. ISBN: 978-0-273-77427-3

Beven KJ (2009) Environmental modelling: an uncertain future? Routledge, London, 310 p. ISBN: 978-0-415-46302-7

Chaterjee C, Förster S, Bronstert A (2008) Comparison of hydrodynamic models of different complexities to model floods with emergency storage areas. Hydrol Process 22(24):4695–4709

Di Baldassarre G (2012) Floods in a changing climate. xiv, 105 p. ISBN 11-070-1875-7

Dyhouse GR et al. (2007) Floodplain modelling using HEC-RAS. Bentley Institute Press, Watertown, 696 p. ISBN 978-1-934493-02-1

Jančíková A (2014) Simulace dopadu poruchy hrázního tělesa VD Plumlov. Diplomová práce. 92 p. VŠB-TU Ostrava

Jandora J, Stara V, Starý M (2002) Hydraulika a hydrologie. n. Brno University of Technology, Faculty of Civil Engineering, Institut of Water Structures: Fakulta stavební, VUT v Brně, Akademické nakladatelství CERM, s.r.o. Brno, pp. 1–186. ISBN: 80-214-2204-1

Johnson GD, Strickland MD et al. (1999) Quantifying impacts to riparian wetlands associated with reduced flows along the Greybull River, Wyoming. In Wetlands, Vol 19, Issue 1 pp:71–77

Kourgialas NN, Karatzas GP (2013) A hydro-economic modelling framework for flood damage estimation and the role of riparian vegetation. In: Hydrological Processes, vol 27, Issue 4, pp:515–531. doi:10.1002/hyp.9256

Kožaná B, Štěrba O, Unucka J et al. (2014) Příspěvek k možnostem stanovení vlivu lužního lesa na tlumení povodňových vln s využitím 1D a 2D hydraulických modelů a GIS. 7 p. In: Zprávy lesnického výzkumu, vol. 59, 2/2014. ISSN: 0322-9688

Šafář J (ed) (2003) Olomoucko. Vyd. 1. Praha: Agentura ochrany přírody a krajiny ČR, 454 p. ISBN 80-860-6446-8

Unucka J (2014) Environmentální modelování 1. Skriptum PřF OU. 209 p. SVZZ CZ.1.07/2.3.00/35.0053 & PřF OU

Xiong YF (2011) A dam break analysis using HEC-RAS. 9 p. J Water Resour Prot 3:370–379

Wilson JP, Gallant JC (eds) (2000) Terrain analysis. In: Principles and Applications. Wiley, London, 479 p. ISBN: 978-0471321880

Dasymetric Mapping of Population Distribution in Serbia Based on Soil Sealing Degrees Layer

Nikola Krunić, Branislav Bajat and Milan Kilibarda

Abstract This paper outlines a methodology used to disaggregate a census population in order to more accurately determine the population distribution over a regional area or a state scale. Data regarding population distributions are usually accessible at the level of individual census designation places and are usually mapped as aggregated polygons by the choropleth method with the assumption of a homogeneous distribution of population within a cartographic unit. In contrast, dasymetric mapping provides a more reliable view into the allocation of inhabitants, which can be of significant importance when estimating population distributions. Coupling this methodology with the GIS environment and a free open access database of soil sealing facilitates the acquisition of population surface models for human and urban geography applications.

Keywords Dasymetric mapping · Soil sealing · Statistical layer

1 Introduction

In the Republic of Serbia, data are usually presented on the level of census designation places by the aggregation of data obtained by censuses performed by individual districts. Demographic census data are mapped as statistical surfaces

N. Krunić (✉)
Institute of Architecture and Urban and Spatial Planning of Serbia, Bul. Kralja Aleksandra 73, 11000 Belgrade, Serbia
e-mail: nikola@iaus.ac.rs

B. Bajat · M. Kilibarda
Faculty of Civil Engineering, Institute for Geodesy and Geoinformatics, University of Belgrade, Bul. Kralja Aleksandra 73, 11000 Belgrade, Serbia
e-mail: bajat@grf.bg.ac.rs

M. Kilibarda
e-mail: kili@grf.bg.ac.rs

K. Růžičková and T. Inspektor (eds.), *Surface Models for Geosciences*, Lecture Notes in Geoinformation and Cartography,
DOI 10.1007/978-3-319-18407-4_12

(DeMers 1999) and most often presented on choropleth maps (Harvey 2008), i.e. maps, which consist of a number of individual estimated uniform surfaces that are separated by clear and crisp boundaries. The major drawback of the choropleth method is that it requires an often invalid assumption of a homogeneous distribution of population within a spatial unit that is used as cartographic unit (Tobler 1979). The techniques we use in order to create a more realistic depiction of population density is called "dasymetric mapping", which in fact corresponds to the term "downscaling" that is more commonly used term in environmental studies.

Mennis (2003) defines dasymetric mapping as an areal interpolation technique that uses additional/ancillary data to assist in the areal interpolation process in order to produce a finer distribution of population. A dasymetric map depicts a statistical surface, most commonly population density, as a set of simply connected smaller spatial units such that variation within each unit is minimized and the unit boundaries approximate the steepest escarpments of the surface (Mennis 2009). A comprehensive review and the explanation of different dasymetric techniques is given in Maantay et al. (2007). Most of these techniques are based on combining different spatial databases and layers with official aggregated polygons of population. Land cover represents the most widely used spatial database of ancillary data in dasymetric mapping. Recent research in the area has examined the performance of alternative sources of ancillary data including: imperviousness, road networks, and nighttime lights (Zandbergen and Ignizio 2010).

A variety of dasymetric databases with coarse resolution are already available on Internet. Recent efforts are focused on producing refinement population surfaces suitable for small areas analysis (Maantay et al. 2007; Zandbergen 2011; Leyk et al. 2014).

A current example of areal interpolation analysis at the city level uses a soil sealing (imperviousness) database with adjusted weights as an ancillary predictor associated with Urban Morphological Zones built on Corine Land Cover classes.[1] Our motivation for the research presented here is to explore the possibility of accommodating this method for settlements with small or middle spatial cover. In particular, our research focuses on finding weighted coefficients representing the ratio among terrain elevation and soil sealing values and, on the other hand, population density.

1.1 World Gridded Demographic Data Sets

One of the examples for mapping the spatial population distribution at a global level is a the LandScan Global Population Database (LSGPD), which has been developed by the Oak Ridge National Laboratory (ORNL 2003) within the Global Population

[1]http://www.eea.europa.eu/data-and-maps/figures/data/urban-morphological-zones-2000-umz2000-f1v0.

Project for the needs of assessing the population vulnerability after disastrous events (Dobson et al. 2000). The LandScan database is a dasymetric model of population distribution in a grid format and is based on accumulated census data obtained from countries that provide a 24-h ambient population estimate. The resolution of the LandScan database is of 30 arc seconds by 30 arc seconds (grid size 1 × 1 km).

The Special Report on Emission Scenarios (SRES) is a similar project showing the assessment and projection of spatial distribution as a long-term projection of world population for the period 1990–2100 (Bengtsson et al. 2006). A database with a similar purpose is the "The Gridded Population of the World: Future Estimates, 2015" (GPW2015) database, which has a resolution of 2.5 min by geographic coordinates and is based on a projection of population for 2015 (Balk and Yetman 2004). Population estimates were projected to the year 2015 using the same simple extrapolation methods as the Gridded Population of the World version 3 (GPWv3) and prior GPW databases (Deichmann et al. 2001) GPWv3 (CIESIN and CIAT 2005) was produced under Global Rural-Urban Mapping Project (GRUMP) and it represents a 30 arc-second grid depicting population density over the entire world. However, all of these publicly available population densities are inappropriate for smaller scale demographic studies, analyses for microlocations and for large scale social, economic and environmental applications.

1.2 Experiences in Dasymetric Mapping in Serbia

The very first experience in dasymetric mapping in Serbia dealt with mapping the population change index (PCI), which represents the ratio of change in the number of inhabitants at certain location for an observed period of time between two censuses (Bajat et al. 2011a). The area of interest in this study was the Region of Južno Pomoravlje and the obtained dasymetric models were based on spatial regression (multiple regression, regression kriging and geographical weighted regression).

The eastern part of the Republic of Serbia, which is actually known as the Timok Region, was picked for more complex research concerning the mapping of PCI (Bajat et al. 2011b) and population density (Bajat et al. 2011c) as well. Population change indexes for the period of 1961–2002 have been mapped based on population census data, while data on the planned population forecast have been used for the period of 2002–2027. Dasymetric mapping of PCI was performed using geographically weighted regression along with roads network, digital terrain model, and soil sealing database. Census data of the year 2002 were disaggregated by CORINE 2000 land cover data (CLC) (Nestorov and Protić 2009) and were used as ancillary predictors needed for the Three-Class dasymetric Method (Eicher and Brewer 2001).

The possibility of using a dasymetric method for the purpose of assessing "daily"or "ambient population"and "night" population in the territory of Vojvodina has been presented in the paper written by a group of authors (Krunić et al. 2011). Their study demonstrated how three land use classes from the CORINE 2006 land cover spatial database for Serbia can be used to generate a basic layer for disaggregation.

Bajat et al. (2013) recently presented a methodology for rescaling soil imperviousness data by using a spatial database of building height typology in order to develop a dasymetric model that might be used for estimating a population distribution in urban areas.

2 Data and Methodology

2.1 Population Data

The Republic of Serbia covers an area of 88,361 km^2 and in accordance to the 2011 census it has a population of 7,120,666 (no data for the Autonomous Province of Kosovo and Metohija) and a population density of about 92 inhabitants per km^2. Due to methodological harmonization, the estimated population of Serbia in 2006 (SORS 2012) was used in this study. A total of 160 municipalities (comparable to the lower Local Administrative Units—LAU level 1, of the Nomenclature of Units for Territorial Statistics-NUTS), of which 16 belonged to the City of Belgrade, were encompassed within this study. Census data was not officially available for the 29 municipalities in Kosovo and Metohija. The total estimated population in the year 2006 was 7,411,982 inhabitants. The distribution of population density visualised as a choropleth map is given in Fig. 1.

The distinct spatial heterogeneity of population distribution highlighted in Fig. 1 is partially the consequence of terrain relief (Fig. 2), but can also be attributed to recent demographic trends and urbanization processes. The population distribution is generalized concentrated around the Novi Sad-Belgrade-Niš axis, in surroundings of major cities (Subotica, Kragujevac, Leskovac) and along the major river valleys (in Zapadna Morava axis Užice-Čačak-Kraljevo-Kruševac). Depopulation with the accompanying negative consequences is represented in most municipalities, especially in peripheral and mountainous areas. The extreme values of population density are expressed by the ratio of the total population and the total territory of the municipality and are given in Table 1.

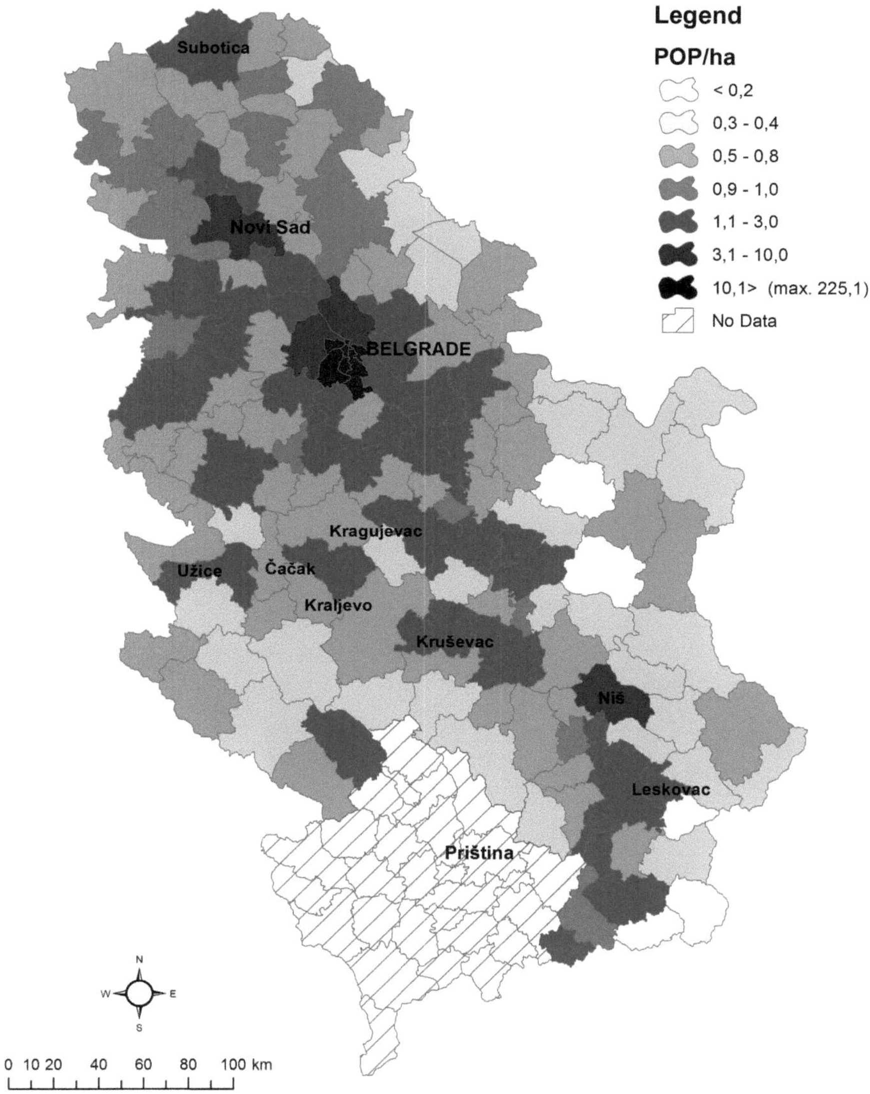

Fig. 1 Choropleth map of population density in Serbia for the year 2006

2.2 Digital Elevation Model (DEM)

The ASTER GDEM global model of Earth's surface, which was published in the year 2009, was used to generate a digital terrain model. The ASTER GDEM data are distributed by NASA in a GeoTIFF format with a spatial resolution of 30 m and estimated height accuracy of 20 m.

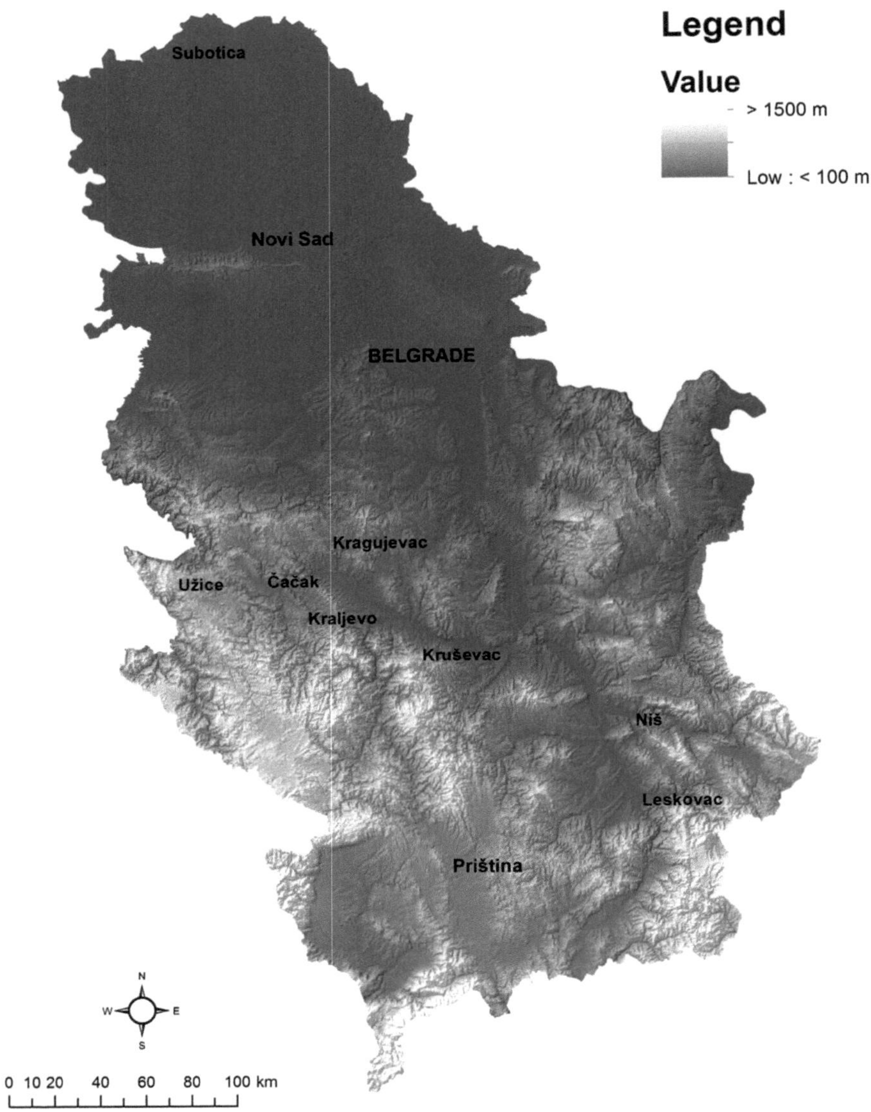

Fig. 2 DEM map of Serbia based on ASTER GDEM

Table 1 Population density statistics

Analysed municipalities	Min density (pop/km^2)	Max density (pop/km^2)	Mean density (pop/km^2)	Standard deviation
160	6.5 (Crna Trava, Leskovac, Southern Serbia)	23097.5 (Vračar, City of Belgrade)	397.3	2050.16

2.3 Soil Sealing Database (SSD)

This database represents yet another free open access database available via Internet.[2] It indicates the land surfaces which have, due to anthropogenic impact, become sealed areas (Burghardt 2006). As such, they directly reflect the percentage of built-up land given in the scale from 0 to 100. The database is developed by the European Environment Agency (EEA) and is available in two spatial resolutions of 20 and 100 m, respectively. The database with resolution of 100 m has been selected for the purpose of this research.

2.4 Methodology

A dasymetric population density map for municipalities in Serbia in 2006 was produced and was based on an adapted degree of soil sealing raster layer and terrain altitude:

$$\mathrm{G_D} \int SS, A \tag{1}$$

where GD represents Dasymetric density, SS represents soil sealing and A represents altitude of terrain.

The proposed methodology for dasymetric mapping is a modified formula for obtaining the estimate of population in municipalities as defined in the technical report.[3]

In the first step of the procedure, the total population of the municipality was allocated by the sum of the values of soil sealing scaled with $\mathrm{K_A}$ and $\mathrm{K_{SS}}$ coefficients. The "dasymetric coefficient" D was calculated as follows:

$$D_m = \frac{\mathrm{POP_m 2006}}{\sum \mathrm{K_{Am}} + \sum \mathrm{K_{SSm}} + \sum \mathrm{SS_m}} \tag{2}$$

where: $\mathrm{POP_m 2006}$ represents the total population of the municipality in 2006, $\mathrm{K_{Am}}$ and $\mathrm{K_{SSm}}$ represent total values of weight coefficients within the municipality and $\sum \mathrm{SS_m}$ represents the total value of soil sealing for the municipality m.

The resulting coefficient D has a unique value for each municipality. Afterwards we calculated the number of people P_S in each cell (pixel) of the statistical layer taking into account the unique value of $\mathrm{D_m}$ for each municipality.

[2] http://www.eea.europa.eu/data-and-maps/data/eea-fast-track-service-precursor-on-land-monitoring-degree-of-soil-sealing.

[3] http://www.eea.europa.eu/data-and-maps/figures/soil-sealing-in-the-capitals/explanation-of-the-methodology/download.

$$P_S = D_m \times (\mathrm{K_A} + \mathrm{K_{SS}} + \mathrm{SS}), \text{given that} \sum\nolimits_m P_S = \mathrm{POP_m}2006 \quad (3)$$

The obtained statistical layer was converted in raster with resolution of 100 m. Finally, the population density was calculated by the number of inhabitants for each raster cell divided by its area (density per 1 ha).

3 Initial Results

In order to simplify the preparation of a "statistical layer", the soil sealing raster was converted into a vector layer such that each pixel of the soil sealing grid was attached to the corresponding centroid vector layer (point). In further steps the vector layers were filtered so that the values of "0" and values greater than "100" (errors that arise due to the sensitivity of the sensor recording) were removed. All of the points not belonging to the settlements area were then removed using additional vector layers including agriculture, infrastructure, natural areas and water surfaces (i.e. all CORINE CLC classes excluding values: 111, 112 and 121). The total number of soil sealing centroids was reduced from about 8.8 million to approximately 410,000 after filtering.

Thereafter, data referring to terrain altitude was associated to the statistical layer such that the altitude value of the GDEM grid cell was associated to the centroid. Thus, the mean GDEM value for municipalities given in Table 3 represents the average height of all the settlements in each municipality rather than the average altitude of the total territory of each municipality.

The statistical layer was then overlaid on the vector layer of municipal boundaries, which contains information about the number of inhabitants in 2006 (SORS 2012). Weight coefficients K_A and K_{SS} (Eq. 2) were estimated taking into account total population, statistical parameters for GDEM and Soil sealing given in Tables 2 and 3.

The final result of dasymetric mapping is presented in Fig. 3. The obtained dasymetric map was mapped by the *plotGoogleMaps* tool developed and designed in the open source R software language, which is utilized for the automatic creation

Table 2 Related municipalities statistics and estimated K_A weights

Altitude (m)	Number of municipalities	Total population (year 2006)	Total count of soil sealing points	Mean soil sealing values	Weight (K_A)
≤200	89	4,668,622	263,040	33.8	3
201–600	62	2,505,888	135,147	21	2
≥601	9	237,472	11,239	14	1
∑/Mean	160	7,411,982	409,426	28.4	

Table 3 Related terrain altitude statistics and estimated K_{ss} weights

Soil sealing degree	Mean GDEM values	Sum of soil sealing values	Mean soil sealing values	Weight (K_{SS})
≤10	251	663,122	5	1
11–20	214	1,091,685	15	2
21–50	167	4,070,699	33.9	3
51–100	145	5,812,949	68.9	4
∑/Mean	198	11,638,455	28.4	

of web maps by combining the users' data and Google Maps layers (Kilibarda and Bajat 2012). The map of population density in Serbia is available in HTML format at the web page URL: http://osgl.grf.bg.ac.rs/PopDensSerbia2006.html.

Better insight into the results can be achieved by overlaying the obtained population grid with existing orthophoto and topographic maps (Fig. 4).

The mean average percentage error (MAPE) was used to appraise the performance of the proposed model:

$$MAPE = \frac{100\,\%}{n}\sum_{t=1}^{n}\left|\frac{A_t - F_t}{A_t}\right| \tag{4}$$

where A_t is the actual (Pop2006) and F_t is the forecast value (dasymetric) of population within a settlement.

Percentage errors have the advantage of being scale-independent and are frequently used to compare forecast performance in cases where data are considerably different in size, or across series, if they are scale-dependent. For entire validation set consisting on 16 settlements with different number of inhabitants (min. 3100; max. 150,000) and spatial extension, the obtained MAPE value is 14.8 %, which indicates that proposed model is appropriate for resolving spatial heterogeneity of population density on the municipality scale.

4 Final Considerations

Without a doubt, increasingly accessible GIS packages that are intended for modelling of spatial data open up new possibilities in processing and visualization of demographic data. The option of combining demographic and other spatial databases, such as soil sealing database, is particularly important in expanding the versatility of such packages. Our research furthers the possibility of applying dasymetrically modelled spatial distributions of population for vastly important tasks such as in spatial and urban planning, accident risk assessment, hazard control, environmental protection, and socio-economic disciplines, etc. It is reasonable to expect an increased interest in both academic and professional application of this method, thereby considerably advancing methods and models that can ultimately

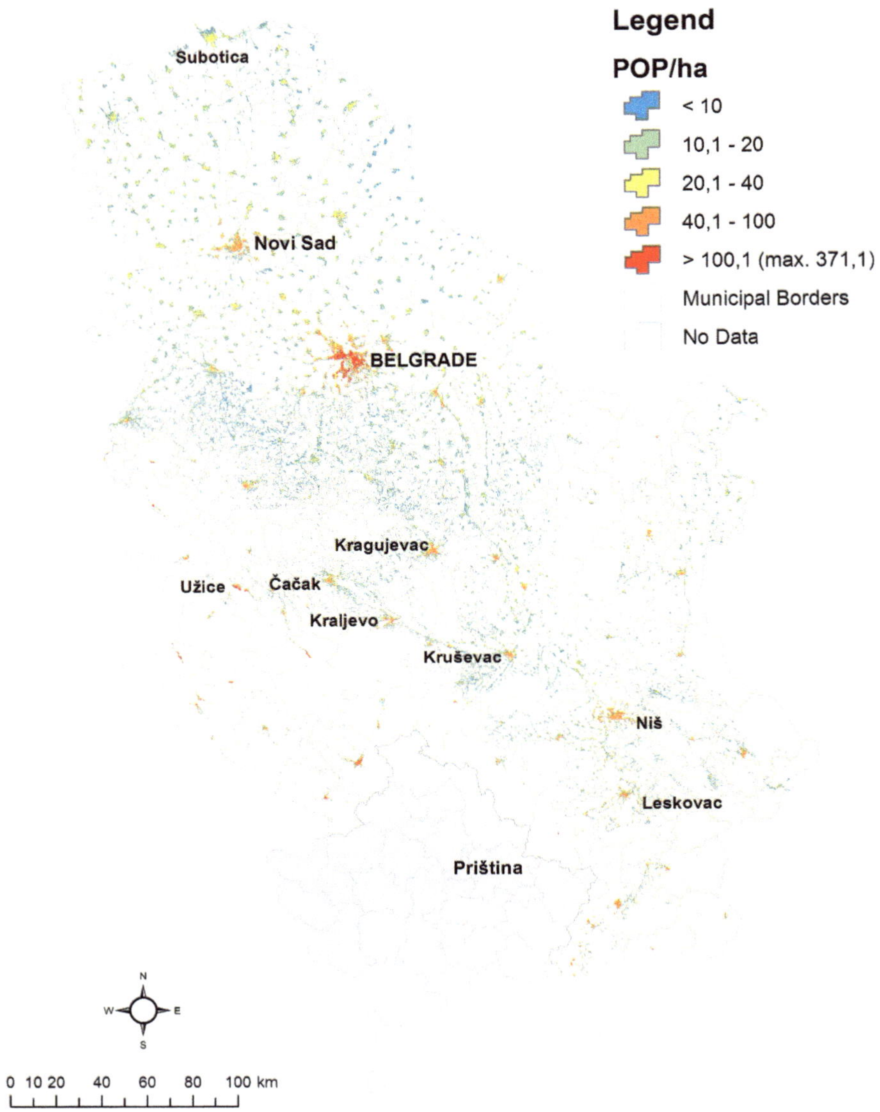

Fig. 3 Dasymetric map of population density in Serbia

reach higher levels of research that investigate coverage of the space, phenomena and processes in the Republic of Serbia.

The main drawback of CLC data, which is often used in similar dasymetric models, is that there is a lack of recognition of small anthropogenic areas in natural environments (Gallego 2010). Such areas in Serbia include dispersed settlements

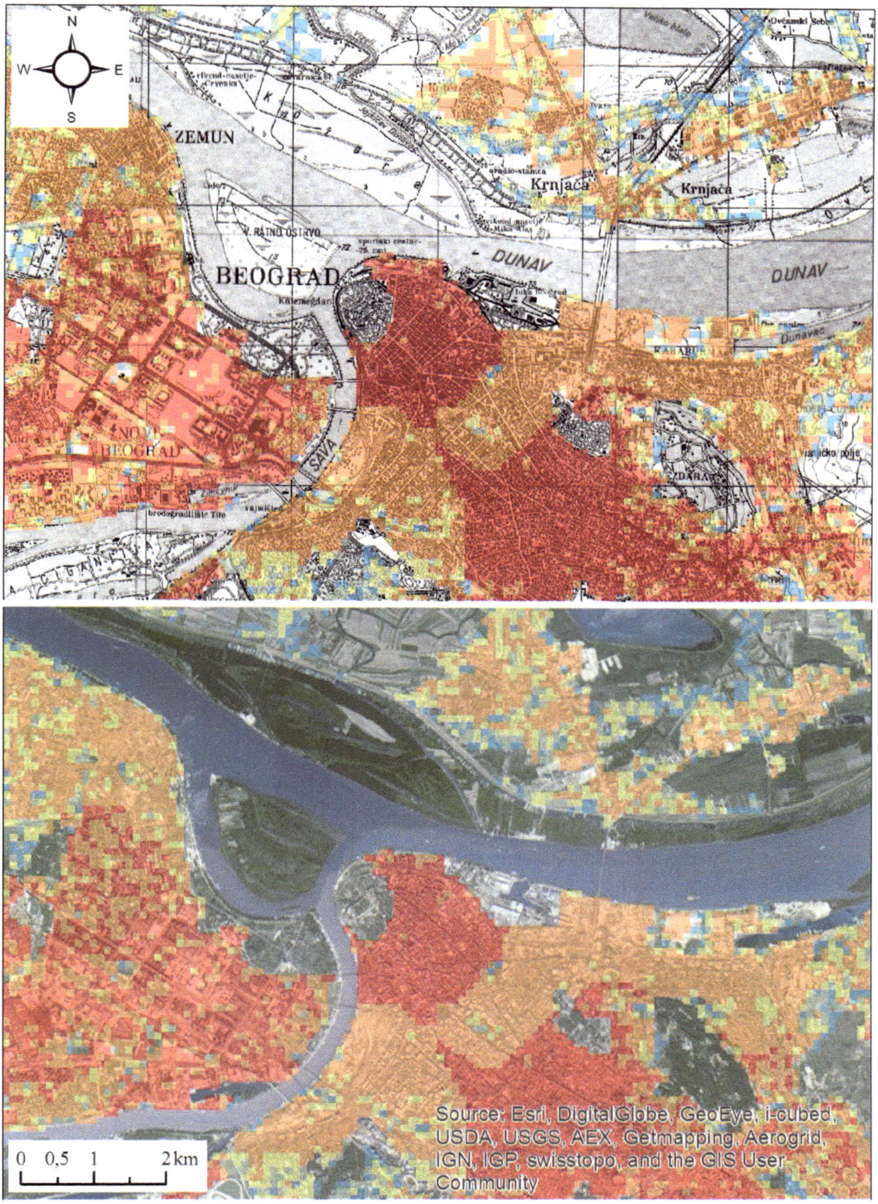

Fig. 4 Dasymetry of Belgrade City centre, compared with topographic map scale 1:100000 (*above*) and Satellite image (*beneath*)

and high-altitude (600 m above sea level) settlements with small area coverage. This problem can be resolved using the soil sealing database.

On the other hand, an arising problem of this treatment is that the soil sealing detects anthropogenic surface regardless of whether they are residential areas, infrastructure or industrial areas. In addition, it is not possible to determine whether high SSD values represent parking plot or the block of high-storey buildings with flat roofs. At the end, sensors used for soil sealing acquisition are sensitive to the reflection of water surfaces and snow. For that reason it is necessary to filter the original grid, which might be accomplished automatically.

Brief validation of the initial model was conducted and it gives satisfying mean average percentage error (MAPE) of 14.8 %. The next step should be additional validation of the obtained results, further calibration of K_A and K_{SS} coefficients and model dissemination through the web service.

Acknowledgments The paper represents the result of research carried out on projects No. III 47014 "The role and implementation of the National spatial plan and regional development in renewal of strategic research, thinking and governance in Serbia", TR 36036 "Sustainable spatial development of Danube area in Serbia" and TR 36035 "Spatial, ecological, energy and social aspects of settlements' development and climate changes—interrelationships" financed by the Ministry of Education and Science of the Republic of Serbia.

References

Bajat B, Krunić N, Kilibarda M, Samardžić-Petrović M (2011a) Spatial prediction of human population change. In: CD proceedings of 14th AGILE international conference on geographic information science, Utrecht, Netherlands, 18–21 April 2011

Bajat B, Krunić N, Kilibarda M, Samardžić-Petrović M (2011b) Spatial modelling of population concentration using geographically weighted regression method. Zbornik radova Geografskog Instituta "Jovan Cvijić", 61:151–167

Bajat B, Krunić N, Kilibarda M (2011c) Dasymetric mapping of spatial distribution of population in Timok region. In: Proceedings of international scientific conference professional practice and education in geodesy and related fields, Kladovo, Serbia, pp 30–34, 24–26 Jun 2011

Bajat B, Krunić N, Samardžić Petrović M, Kilibarda M (2013) Dasymetric modelling of population dynamics in urban areas. Geodetski Vestnik 57(4):777–792

Balk D, Yetman G (2004) The global distribution of population: evaluating the gains in resolution refinement, documentation for GPWv3. CIESIN, Columbia, Palisades, NY

Bengtsson M, Shen Y, Oki T (2006) A SRES-based gridded global population dataset for 1990–2100. Popul Environ 28(2):113–131

Burghardt W (2006) Soil sealing and soil properties related to sealing. Geol Soc Lond Spec Publ 266:117–124

CIESIN, CIAT (2005). Gridded population of the world, VERSION 3 (GPWv3). In: Socioeconomic data and applications center (SEDAC), Columbia University, Palisades, NY. Available at http://sedac.ciesin.columbia.edu/gpw

Deichmann U, Balk D, Yetman G (2001) Transforming population data for interdisciplinary usages: from census to grid. Available at: http://sedac.ciesin.org/gpw-v2/GPWdocumentation.pdf

DeMers MN (1999) Fundamentals of geographic information systems, 2nd edn. John Wiley, New York

Dobson JE, Bright EA, Coleman PR, Durfee R, Worley BA (2000) Landscan: a global population database for estimating populations at risk. Photogram Eng Rem S 66:849–857

Eicher CL, Brewer CA (2001) Dasymetric mapping and areal interpolation: implementation and evaluation. Cartography Geogr Inf Sci 28:125–138

Gallego FJ (2010) A population density grid of the European Union. Popul Environ 31:460–473

Harvey F (2008) A primer of GIS: fundamental geographic and cartographic concepts. The Guilford Press, New York

Kilibarda M, Bajat B (2012) PlotGoogleMaps: the R-based web-mapping tool for thematic spatial data. Geomatica 66:37–49

Krunić N, Bajat B, Kilibarda M, Tošić D (2011) Modelling the spatial distribution of vojvodina's population by using dasymetric method. Spatium 24:45–50

Leyk S, Buttenfield BP, Ruther M (2014) Dasymetric refinement for improved temporal small area analysis. In: Extended abstract proceedings of the GIScience 2014, Vienna, Austria, pp 42–46, 23–26 Sept

Maantay JA, Maroko AR, Herrmann C (2007) Mapping population distribution in the urban environment: the cadastral-based expert dasymetric system (CEDS). Cartogr Geogr Inf Sci 34(2):77–102

Mennis J (2003) Generating surface models of population using dasymetric mapping. Prof Geogr 55(1):31–42

Mennis J (2009) Dasymetric mapping for estimating population in small areas. Geogr Compass 3(2):727–745

Nestorov I, Protić D (2009) CORINE land cover mapping in Serbia. Građevinska knjiga, Beograd

ORNL (2003) Integrated analysis of mitigation and adaptation as responses to concerns about impacts of global climate change, ORNL Working paper, June 2003

Statistical Office of the Republic of Serbia (SORS) (2012) Natural changes of population in the Republic of Serbia. Statistical Office of the Republic of Serbia, Belgrade

Tobler WR (1979) Smooth pycnophylactic interpolation for geographical regions. J Am Stat Assoc 74:519–530

Zandbergen PA, Ignizio DA (2010) Comparison of dasymetric mapping techniques for small-area population estimates. Cartogr Geogr Inf Sci 37(3):199–214

Zandbergen PA (2011) Dasymetric mapping using high resolution address point datasets. Trans GIS 15(s1):5–27

Geographic Information System for Flood Hazard Analysis and Early Warning Using Numerical Weather Predictions

Marcin Kulawiak, Andrzej Chybicki, Lukasz Markiewicz and Marlena Jankowska

Abstract Floods are a phenomenon which causes significant losses of lives as well as property damage, which in turn severely impact the local economy. The nature of a flooding has been well researched, and several methods of flood hazard simulation have been established over the last decades. The current legislation in the European Union requires the Member States to create, maintain and update flood risk and hazard maps, as well as identify possible sources of flooding. In the Tricity area in Northern Poland, the historical records indicate that one of the most common causes of flooding are high amounts of rainfall which cause a sudden surge of the water levels in the Vistula River estuary. This was the reason for the tragic flood of 2001, which caused damages worth millions and affected thousands of people. Over the last decade, the City invested millions in rebuilding the infrastructure and creating flood prevention and early warning measures. The paper presents a Geographic Information System which aims to provide early warning as well as flood risk assessment in semi-real time. For this purpose the system is integrated with a weather forecast model and a flood simulation model. The simulation results are automatically made available to interested parties via a user-friendly Web-based client.

Keywords GIS · Flood · Modeling · Responsibility · Analysis · Integration · Liability

M. Kulawiak (✉) · A. Chybicki · L. Markiewicz
Department of Geoinformatics, Faculty of Electronics, Telecommunication and Informatics, Gdansk University of Technology, Narutowicza Street 11/12, 80-233 Gdansk, Poland
e-mail: Marcin.Kulawiak@eti.pg.gda.pl

A. Chybicki
e-mail: Andrzej.Chybicki@eti.pg.gda.pl

L. Markiewicz
e-mail: Lukasz.Markiewicz@eti.pg.gda.pl

M. Jankowska
Department of Civil Law and Private International Law, Faculty of Law and Administration, University of Silesia, Bankowa Street 11b, 40-007 Katowice, Poland
e-mail: Marlena.Jankowska@interia.pl

K. Růžičková and T. Inspektor (eds.), *Surface Models for Geosciences*,
Lecture Notes in Geoinformation and Cartography,
DOI 10.1007/978-3-319-18407-4_13

1 Introduction

Application of Geographic Information Systems (GIS) to mapping and analysis of flood zones has a long history. In 1996, Muzik (1996) applied a GIS to flood modelling for the purpose of describing the connectivity of links in a watershed flow network and to calculate distances and travel times to the watershed outlet for various intermediate points within the watershed. Two years later, GIS was used by Townsend and Walsh (1998) who have integrated a Digital Elevation Model (DEM) with photographs obtained via remote sensing in order to identify potential flood areas. In the next decade, Islam and Sado (2000) have integrated land cover category and elevation height data within a GIS in order to rank them by flood affected frequency, creating a new type of a flood hazard map. Later Werner (2001) used GIS to show that in certain conditions a two-dimensional flood model may be successfully replaced by a simplified single-dimensional model. Four years later Knebl et al. (2005) proposed a framework for integrating GIS and hydrological modelling for the purpose of researching flash floods in Texas. More recently, Tran et al. (2009) demonstrated the significance of GIS and flood mapping in reducing the risk and losses caused by floods, while Dabrowski et al. (2009) applied GIS to real-time integration and analysis of data from several aquatic sensors. In 2011 Qi and Altinkar have presented a GIS-based decision support system for two-dimensional simulation of possible outcomes of a flood hazard under defined uncertainty conditions (Qi and Altinkar 2011). Two years later Chau et al. (2013) have applied DEM analysis via GIS to estimate the potential impact of floods on agricultural land in Vietnam. In the same year Tehrany et al. (2013) used GIS to compare several different methods of predicting areas which are susceptible to flooding based on a test-case in Malaysia, while Kulawiak and Lubniewski (2014) presented a method of integrating a flood simulation module within a GIS dedicated to protecting city residents and infrastructure. Recently Kulkami et al. (2014) have shown a web based GIS framework for organizing spatial datasets and running a flood simulation model within this framework. A common characteristic of the aforementioned solutions is that all of them form dedicated solutions to static problems, however none of them constitutes an operational system.

The threat caused by flooding has been long recognized internationally. On 17 March 1992 in Helsinki the United Nations adopted the Convention on the Protection and Use of Transboundary Watercourses and International Lakes, which in Article 7 obliges the Parties to support appropriate international efforts to establish rules, criteria and procedures in the field of responsibility and liability (United Nations Economic Commission for Europe 1992). In Europe, reacting to the increasing threat of flooding [213 major damaging floods, causing damages worth approximately EUR 52 billion (Ludlow 2011)], on 26 November 2007 the European Commission has issued Directive 2007/60/EC on the assessment and management of flood risks. The Directive aims to establish a framework for the assessment and management of flood risks for the purpose of reducing their consequences. It requires all Member States to identify possible sources of floods on their territories and perform their flood risk assessments, taking into account factors such as potential climate changes. After performing the risk assessments, each Member State has been made responsible for creation, updating and publishing the following documents (European Commission 2007):

- preliminary flood risk assessment results—indicating areas where there is a significant risk of flooding, or a future occurrence of a significant risk of flooding is likely (Article 4 and 5 of the Directive),
- flood hazard maps—identifying the areas that may be affected by flooding with low, medium or high probability (Article 6 of the Directive).
- flood risk maps—showing the potential adverse consequences of flooding, including the number of affected inhabitants, impacts on local economy, and vulnerable installations which may release environmentally hazardous substances (such as sewage treatment plants) (Article 6 of the Directive),
- flood risk management plans covering all aspects of flood risk management, including prevention, protection and preparedness. These may include flood forecasts and early warning systems, promotion of sustainable land use practices, improvement of water retention as well as controlled flooding of certain areas (Article 7 and 8 of the Directive).

The Directive has been transposed into Polish law in Polish Water Law, the legal act as of 5 January 2011, that entered into force on 18 March 2011 (Water Law 2011). The law introduced the concept of flood risk management including prevention, protection and readiness. The creation, management and update of flood hazard and risk maps has been made, pursuant to Article 88a (1), 88c and 88f of the aforementioned act, the general responsibility of the Minister of the Environment and the President of National Water Management Authority as well the local administration (for most flood hazard sources) and the directors of maritime offices (for flood hazards originating from the sea). These maps are meant to be used e.g. for the development of urban plans, making decisions on the location of public purpose investments, zoning permits etc. (Verdict 2011). The protection of the citizens, including creating operational plans of flood prevention as well as calling and withdrawing flood alerts, relies mainly on local authorities (voyt, mayor, president of the city) (cf. Article 31a of the Polish legal act as of 08.03.1990 on commune self-government). In the City of Gdansk, the Gdansk University of Technology Department of Geoinformatics has been working on a Geographic Information System for the assessment of risk to the City's Critical Infrastructure. In addition to its functionalities in the field of maintaining an inventory of natural resources as well as spatial planning, the GIS has been integrated with a module dedicated to flood risk assessment.

2 Processing Digital Elevation Model Data in Gis

One of the most important types of data used for determining flood zones is a Digital Elevation Model, which represents the surface of the analysed area in a three-dimensional space. In general, a DEM is described by a set of points defined by coordinates x, y and z, and the actual three-dimensional surface is obtained by interpolation of these coordinates. The physical distance between these points (measured horizontally) represents the horizontal resolution of the model (Zhang and Montgomery 1994). Nowadays the most popular methods of obtaining a DEM include LIDAR scanning as well as stereographic satellite and aerial photography. These methods produce a dense mesh of three-dimensional points which describe the shape of the investigated terrain (Hodgson et al. 2003). After the data has been collected, it is converted into one of the most popular formats of DEM information storage. These include (Kreveld 1997):

- GRID—in this format, the points of a DEM are stored in the form of a georeferenced matrix. Every cell of the matrix is given geographical coordinates, and the value attribute of the cell represents the height of the given geographical point (usually measured above water level). The main disadvantage of this format is its fixed resolution, which makes it unfit for accurate representation of intricate details such as terrain discontinuities.
- Triangulated Irregular Network (TIN)—in this format, the three-dimensional surface is represented by an irregular mesh of triangles of different sizes. Unlike GRID, TIN preserves a variable density of measurement points in order to better represent areas of high vertical variance. This allows to optimize the size of the files containing the DEM by removing unnecessary vertices, while preserving a desirable resolution of the model. As a disadvantage, the format is much more complex to process, which limits its practical applications.

A DEM describes the three-dimensional shape of the analysed terrain, without which it would not be possible to determine the potential areas vulnerable to flooding in the event of rising water levels. The analysis of the interpolated surface in three dimensions allows to determine its impact on certain hydrological processes involved in flooding. In particular, the application of a DEM in flood modelling requires the calculation of the following topographical characteristics (Li et al. 2005):

- Slope—denotes the change in height per unit of length in the direction of steepest descent. It may be expressed as a percentage or an angle at which a surface is inclined to the horizontal.
- Aspect—determines the horizontal direction which the slope faces. It is defined by its azimuth, or angle measured clockwise between the direction of the north and the direction of maximum gradient.
- Curvature—describes the shape of the slope. Most commonly it is measured independently in the surface parallel to the slope direction (Profile curvature) and perpendicular to the slope direction (Planar curvature).

However, a successful application of a DEM for the purpose of flood modelling requires the definition of additional parameters, such as (Zekster and Everett 2000; Robock 2014), (Bonia et al. 2007; Silvestro and Rebora 2014):

- Soil moisture—a quantitative measure describing the effect of terrain on hydrological processes, commonly used along with the hydrological model describing water flow in an area. The value of soil moisture for a given cell is often defined as the ratio between its actual and maximum water storage capacity.
- Watershed—defines an area which is characterized by the fact that all of the water which passes through it is drained through a single outlet. In case of a rainfall, the amount of water which passes through this outlet is then proportional to the amount of rain fallen on the entire watershed area.
- Hydrogeological conditions—these pose an important element of precise flood modelling in areas which exhibit large groundwater reservoirs, and thus are susceptible to flooding caused by the rise in groundwater level.
- Land use/land cover—producing flood risk maps requires knowledge regarding the location of potentially affected people and installations. The ongoing changes in land use are one of the most important factors increasing flood risk in Central Europe (De Roo et al. 2003).

The presented study applies a Digital Elevation Model to the analysis of flood risks in the Vistula Marshlands, located in Northern Poland. The region, formed around the Vistula River estuary, is particularly at risk from flooding by its positioning in the depressed areas close to the shores of the Baltic sea. The Marshlands are constantly threatened by river floods caused by water masses from the South and Centre of the country, Coastal flooding caused by storm surges, as well as drain and sewer flooding caused by rising sea water levels.

At the same time, the area's close proximity to both the Baltic sea as well as Poland's greatest river has caused some of the country's vital companies to be located there. In particular, the city of Gdansk houses both the Lotos refinery and a Phosphate Fertilizer Plant. Other important objects in the area include the "Lipce" water source and the sewage treatment plant "East", as well as national road No. 7 and the Tricity—Warsaw railway line. Figure 1 shows a DEM of Northern Poland where areas located below sea level are marked in dark green. The Vistula Marshes are clearly visible in the centre of the image, directly below the Gulf of Gdansk.

3 Numerical Weather Modeling

A numerical weather prediction (NWP) model is a software package which main function is to predict the weather using numerical methods basing on initial and boundary conditions retrieved from global atmospheric observations. The software utilises mathematical models of the atmosphere and the hydrosphere in order to generate weather predictions based on current weather data. The NWP results can

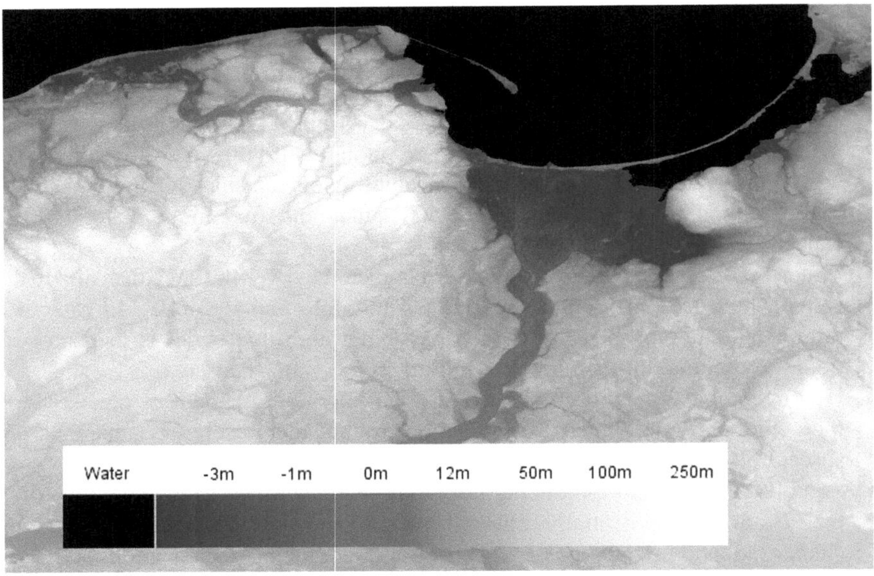

Fig. 1 Digital elevation model of Northern Poland. Areas below sea level are marked in *dark green*

be applied in various fields, including energy potential prediction for wind farms, agriculture, environment protection and for operational forecasting. Weather Research and Forecast (WRF) model utilized in the presented research uses Advanced Research WRF (ARW) physical core developed by the National Center for Atmospheric Research (NCAR) (Klemp et al. 2007) which incorporates a non-hydrostatic model. It supports acquisition and processing of initialization data, model execution, output data processing and result data visualization.

The applied WRF numerical weather prediction is embedded in virtual processing environment managed by Department of Geoinformatics in Faculty of Electronics, Telecommunication and Informatics of Gdansk University of Technology. It generates 48-h forecasts for the area of Poland with 4 km × 4 km spatial resolution (Chybicki et al. 2011). The system delivers a number of parameters describing the state of Earth atmosphere and processes like precipitation, wind, pressure and others. For historical datasets, developed architecture also allows for high-resolution 1 km × 1 km reanalysis that can be performed on demand. Sample results of precipitation generated by NWP are presented by Fig. 2.

Another valuable source of meteorological observational data are weather radars (also called Doppler radars) which are commonly used in meteorology and real-time weather forecasting to detect and locate precipitation and its intensity. Such data can be additionally analysed in order to estimate future motion of weather fronts, which is precipitation saturation along with temperature contrast, and determine its type (rain, snow, sleet, hail etc.). Weather products generated by radars provide full 2D coverage

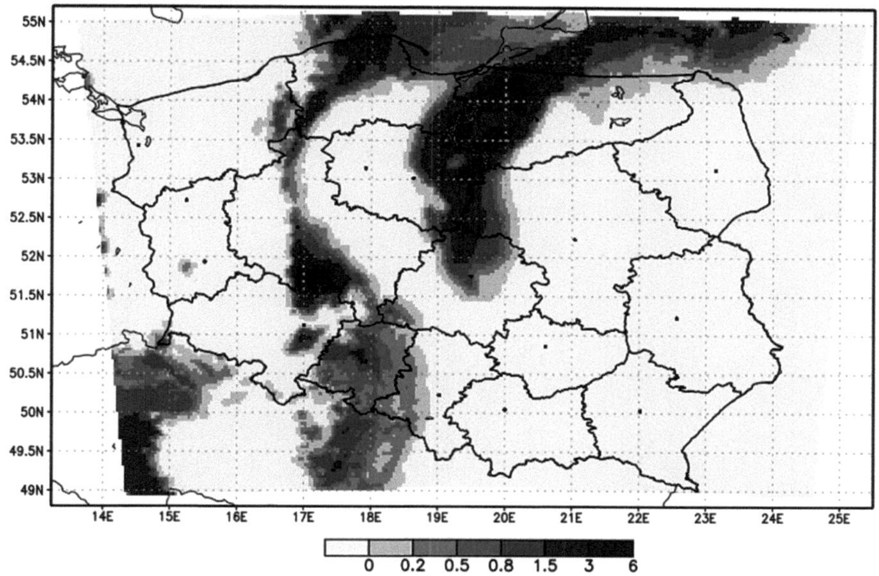

Fig. 2 Spatial distribution of precipitation (mm) field generated by WRF model of KSG-Met for 10 UTC on 3rd of September 2010 (historical real analysis data sample)

of an investigated area, including real-time information on weather conditions. The properties of precipitation (intensity, movement direction, precipitation type) can be determined by analysis of the shape of the echo reflected towards the receiver antenna. Sample results of spatial distribution of rainfall volume generated by Doppler radars for the area of Poland are presented in Fig. 3.

4 The System

The presented Web-based GIS for prediction and modelling of flood hazards integrates data from an operational weather forecast model and a flood simulation system in order to dynamically produce flood hazard maps for potentially affected areas. The system uses a modular structure, where every type of functionality has been enclosed in a dedicated module. The architecture of the system is presented in Fig. 4.

The Radar Data Transformation module analyses the data from Meteorological Radars and uses it to discover potentially hazardous weather conditions. The actual rainfall estimates are performed by the dedicated Numerical Weather Prediction module. When the estimated level of rainfall for the City of Gdansk crosses a certain threshold, the information is passed to the Rainfall Data Adaptation module, which starts the procedure of flood hazard mapping. The Flood Modelling module uses a Digital Elevation Model (produced by the provincial geodetic unit with a

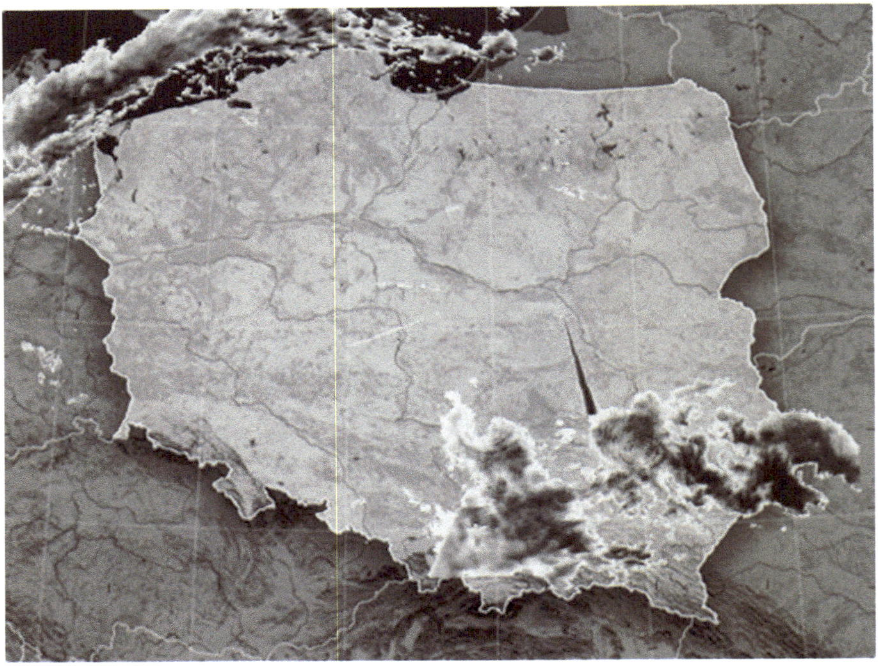

Fig. 3 Sample weather radar image showing heavy precipitation in the north and south Poland, 6 June 2010 00:00 UTC+2

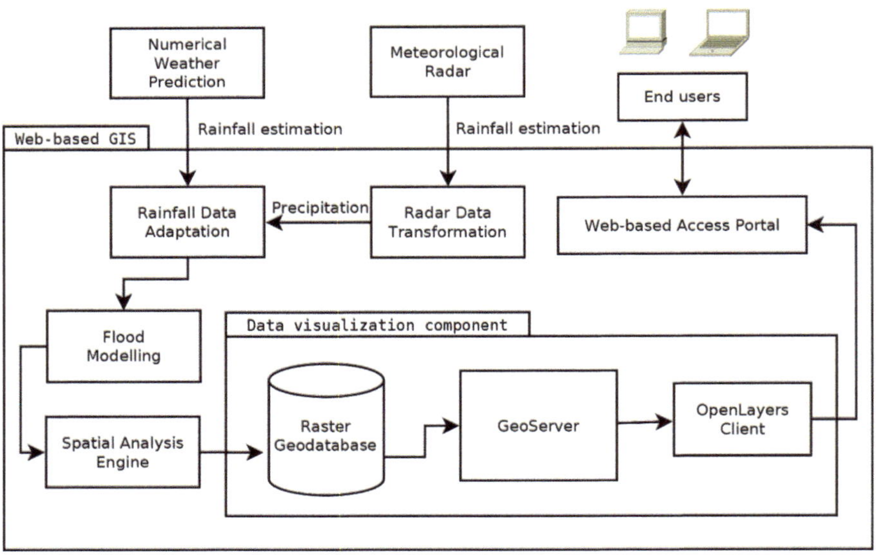

Fig. 4 Architecture of the GIS for prediction and modelling of flood hazards

resolution of 0.5 m) for the Tricity area to perform an analysis of the Slope and Aspect of the DEM in the rainfall area in order to estimate the amount of water which will pass through the Vistula River Watershed. The applied simulation model assumes a worst-case scenario in which the rainfall precipitation rate is greater than soil infiltration capacity. Under such conditions all rainfall water over the analysed area directly contributes to surface runoff, which allows the model to estimate the maximum value of water level in the River estuary. This information is passed to the Spatial Analysis Engine which produces thematic layers showing the extent of the estimated water overflow. The generated flood zones are registered in the Raster Geodatabase, which is used by the Map Server component. For serving thematic layers the system uses GeoServer, which is an Open Source map server with excellent support for open standards of spatial data exchange such as Open Geospatial Consortium's Web Feature Service and Web Map Service. The latter is used to serve thematic layers to the Web GIS client, which uses the Open Source OpenLayers JavaScript library to provide the end users with rich GIS functionality from any modern Web browser, without the need for additional plugins. The client enables end users to view the simulation results in a geographical context, overlaid on background data showing terrain and city maps.

5 Results

The presented system works in semi-real time, analysing weather forecasts and simulating the effects of increased rainfall over an area. With a high enough level of rainfall, the water level in the riverbed could rise significantly. This has been the case in 2001, when a rainfall of 90 mm per square meter caused the water levels in the entire Vistula estuary to rise by several meters. In consequence several river banks have been broken, causing a devastating flooding of the entire City of Gdansk. Over the last decade, the Province administration has made great efforts to rebuild and improve the river banks and associated infrastructure. Figure 5 presents the result of a simulation produced by the system with regards to the 2001 rainfall values, but using the recently produced DEM.

The results of the simulation indicate that the rebuilt infrastructure should be capable of receiving the increased amount of water, at least initially (the model does not take into account the effects of increased water pressure on the banks).

The month of August 2010 has shown some moderate precipitation for the City of Gdansk area. The recorded values of rainfall in millimetres per hour were not particularly high, but the long duration of rain caused considerable increases in recorded water levels. Figure 6 shows the spatial distribution of precipitation over Poland for 3rd of August 2010, as retrieved from the WRF model.

After several hours of rainfall, the water level in the Vistula estuary rose by more than 100 cm, however at the end of the day the recorded water levels were well below warning threshold. Figure 7 presents an image generated in response to the rainfall estimations of 03.08.2010 gathered between 9:00 and 10:00 UTC.

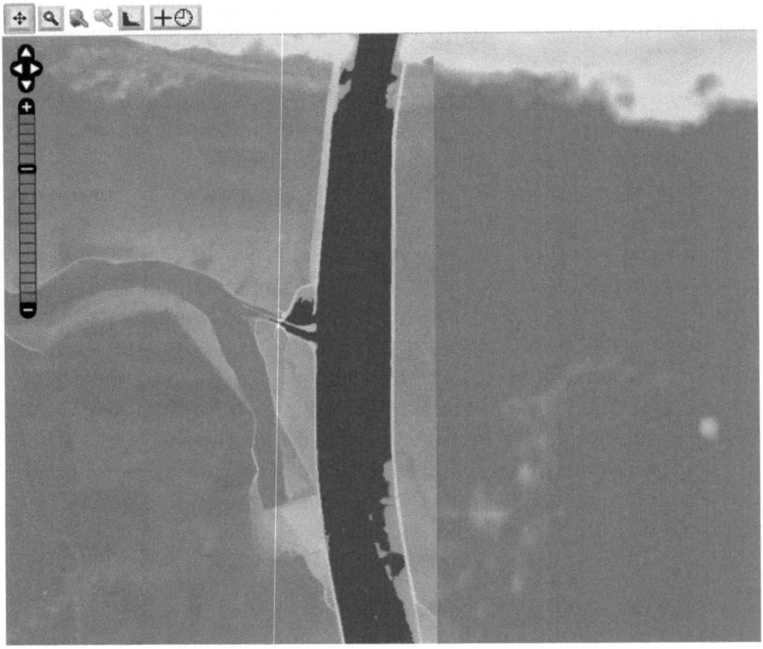

Fig. 5 Simulated results of rising the Vistula river water level by 3 m

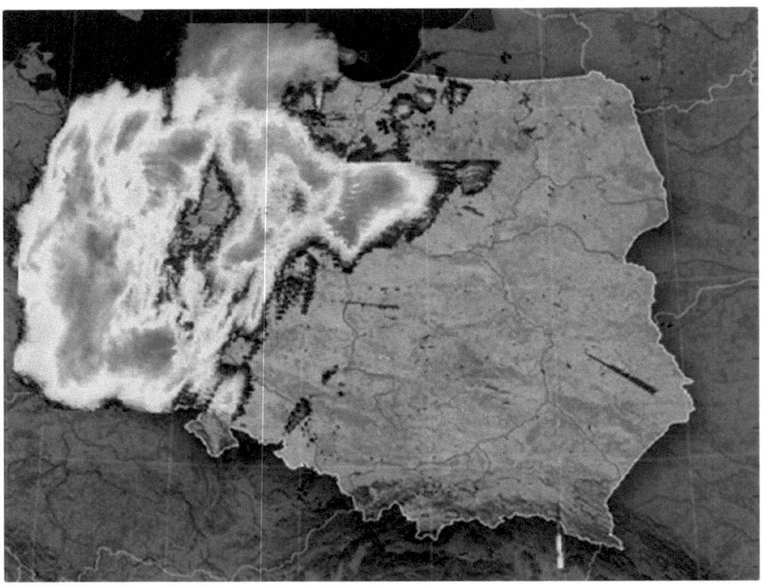

Fig. 6 Spatial distribution of precipitation predicted by WRF model on the 3rd of August 2010. Accumulation period 9-10 UTC

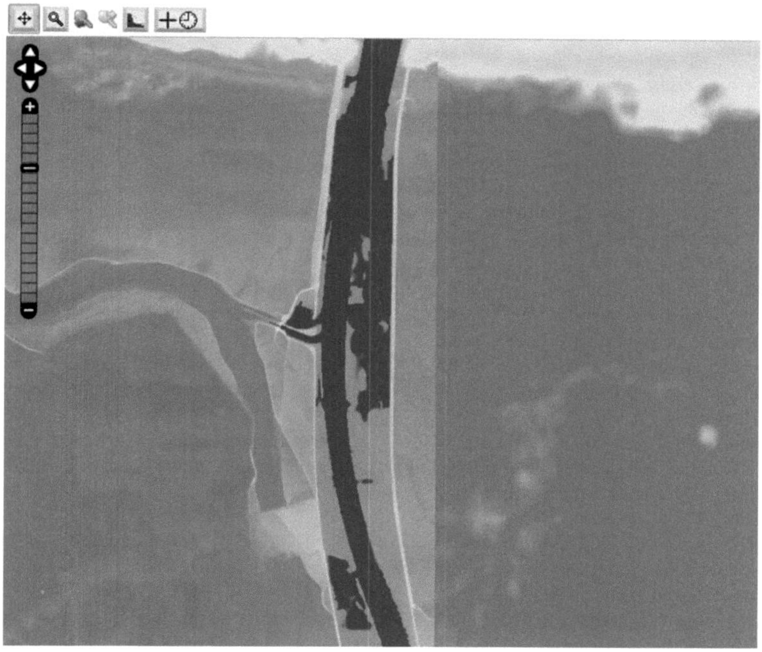

Fig. 7 Simulated results of rising the Vistula river water level by 0.8 m

As it can be observed, the rainfall forecast did not cause any danger of flooding. This was also the case in reality, as the increased level of water was still contained within the river banks.

As concerns more recent events, fortunately the analysed weather forecasts have not shown any considerable danger of flooding the city.

6 Conclusions

The paper presents a Geographic Information System dedicated to flood area modelling in the Vistula Marshes region of Northern Poland. The system is integrated with the WRF weather model for the purpose of automatic prediction of water level changes. This combination enables the system to dynamically estimate possible flooding regions and produce informative maps in semi-real time.

Acknowledgments The Digital Elevation Model used for the purposes of visualizing elevation of Northern Poland was produced using Copernicus data and information funded by the European Union—EU-DEM layers.

The weather radar presented in the research data was provided by Institute of Meteorology and Water Management National Research Institute, Poland.

References

Bonia G, Ferrarisa L, Giannonic F, Rotha G, Rudaria R (2007) Flood probability analysis for ungauged watersheds by means of a simple distributed hydrologic model. Adv Water Resour 30 (10):2135–2144

Chau VN, Holland J, Cassels S, Tuohy M (2013) Using GIS to map impacts upon agriculture from extreme floods in Vietnam. Appl Geogr, 41:65–74

Chybicki A, Lubniewski Z, Kaminski L, Bruniecki K, Markiewicz L (2011) Numerical weather prediction—data fusion to GIS systems and potential applications. GIS Odyssey ISBN HIZ 953-6129-32-9

Dabrowski J, Kulawiak M, Moszynski M, Bruniecki K, Kaminski L, Chybicki A, Stepnowski A (2009) Real-time web-based GIS for analysis, visualization and Integration of marine environment data. Lecture notes in geoinformation and cartography, Springer, pp 278–288

De Roo A, Schmuck G, Perdigao V, Thielen J (2003) The influence of historic land use changes and future planned land use scenarios on floods in the Oder catchment. Phys Chem Earth, Parts A/B/C 28(33):1291–1300

European Commission (2007) Directive 2007/60/EC of the european parliament and of the council of 23 October 2007 on the assessment and management of flood risks. Available at: http://eur-lex.europa.eu/LexUriServ/LexUriServ.do?uri=OJ:L:2007:288:0027:0034:EN:PDF. Accessed on 10.10.2014

Hodgson ME, Jensen JR, Schmidt L, Schill S, Davis B (2003) An evaluation of LIDAR-and IFSAR-derived digital elevation models in leaf-on conditions with USGS Level 1 and Level 2 DEMs. Remote Sens Environ 84(2):295–308

Islam MM, Sado K (2000) Flood hazard assessment in Bangladesh using NOAA AVHRR data with geographical information system. Hydrol Process 14(3):605–620

Klemp JB, Skamarock WC, Dudhia J (2007) Conservative split-explicit time integration methods for the compressible nonhydrostatic equations. Mon Weather Rev 135(8):2897–2913

Knebl MR, Yanga Z-L, Hutchisonb K, Maidmentc DR (2005) Regional scale flood modeling using NEXRAD rainfall, GIS, and HEC-HMS/RAS: a case study for the San Antonio River Basin Summer 2002 storm event. J Environ Manage 75(4):325–336

Kreveld M (1997) Digital elevation models and TIN algorithms. Algorithmic foundations of geographic information systems. Springer, Berlin Heidelberg, pp 37–78

Kulawiak M, Lubniewski Z (2014) SafeCity—A GIS-based tool profiled for supporting decision making in urban development and infrastructure protection. Technol Forecast Soc Chang 89:174–187

Kulkarni AT, Mohanty J, Eldho TI, Rao EP, Mohan BK (2014) A web GIS based integrated flood assessment modeling tool for coastal urban watersheds. Comput. Geosci Volume 64:7–14

Legal act as of 08.03.1990 on commune self-government. Journal of Laws of 2013, Item 594 with further changes. Current revision as of 30.11.2014.

Li Z, Zhu Q, Gold C (2005) Digital terrain modeling: principles and methodology. CRC Press, Boca Raton

Ludlow D (ed) (2011) Mapping the impacts of natural hazards and technological accidents in Europe. European environment agency technical report No 13/2010. Available at: http://www.eea.europa.eu/publications/mapping-the-impacts-of-natural Accessed on 10.10.2014

Muzik I (1996) Flood modelling with GIS-derived distributed unit hydrographs. Hydrol Process 10(10):1401–1409

Qi H, Altinakar MS (2011) A GIS-based decision support system for integrated flood management under uncertainty with two dimensional numerical simulations. Environ Model Softw 26 (6):817

Robock A (2014) Hydrology, floods and droughts|soil moisture. encyclopedia of atmospheric sciences, 2nd edn, pp 232–239

Silvestro F, Rebora N (2014) Impact of precipitation forecast uncertainties and initial soil moisture conditions on a probabilistic flood forecasting chain. J Hydrol, Part A 519:1052–1067

Tehrany MS, Pradhan B, Jebur MN (2013) Spatial prediction of flood susceptible areas using rule based decision tree (DT) and a novel ensemble bivariate and multivariate statistical models in GIS. J Hydrol 504:69–79

Townsend PA, Walsh SJ (1998) Application of remote sensing and GIS in geomorphology. Geomorphology 21(3–4):295–312

Tran P, Shaw R, Chantry G, Norton J (2009) GIS and local knowledge in disaster management: a case study of flood risk mapping in Viet Nam. Disasters 33(1):152–169

United Nations Economic Commission for Europe (1992) Convention on the protection and use of Transboundary watercourses and International lakes done at Helsinki on 17 March 1992. Available at: http://www.unece.org/fileadmin/DAM/env/water/pdf/watercon.pdf. Accessed on 19.10.2014

Verdict (2011) Verdict of Voivodship administrative court in Kraków as of 18 May 2011, ref. no II SA/Kr 1205/2010, lexPolonica no 2788444

Werner MGF (2001) Impact of grid size in GIS based flood extent mapping using a 1D flow model. Phys Chem Earth, Part B: Hydrol, Oceans Atmos, 26(7–8):517–522

Water Law (2011) Legal act as of 05.01.2011. Journal of Laws of the Republic of Poland, Item 145 with further changes

Zektser IS, Everett LG (2000) Groundwater and the environment: applications for the global community. CRC Press, ISBN: 978-1566703833

Zhang W, Montgomery DR (1994) Digital elevation model grid size, landscape representation, and hydrologic simulations. Water Resour Res 30(4):1019–1028

Effect of DEM Inaccuracy on Precision of Satellite InSAR Results

Milan Lazecký, Matúš Bakoň and Ivana Hlaváčová

Abstract This short paper summarizes the effects of DEM errors influencing precision of deformation monitoring results using satellite SAR interferometry and presents basic mathematical approaches to quantify DEM accuracy needed for proper estimation of SAR interferometry parameters. While in case of appropriately large SAR dataset for Multi-Temporal Interferometry (MTI) processing the height difference is usually well estimated, still DEM accuracy should be carefully considered in case of using low number of images—and especially in case of processing differential InSAR with only one image pair. Depending on the interferogram pair combination characteristics, especially parameter known as perpendicular baseline (B_{perp}) describing the distance between satellite positions during their image acquisitions, the interferogram is more or less prone to DEM accuracy errors. In case of higher B_{perp}, a relatively high quality DEM is necessary to remove stereoscopic effect apparent in interferogram in order to distinguish interferogram phase element caused by temporal terrain deformation.

Keywords InSAR · DEM · Accuracy · Deformation monitoring

M. Lazecký (✉)
IT4Innovations, VSB-TUO, 17. listopadu 15, 708 33 Ostrava-Poruba, Czech Republic
e-mail: milan.lazecky@vsb.cz

M. Bakoň
STU Bratislava, Radlinského 11, 813 68 Bratislava, Slovakia
e-mail: matus.bakon@gmail.com

I. Hlaváčová
CVUT Prague, Thákurova 7, 166 29 Praha 6, Czech Republic
e-mail: hlavacova@insar.cz

M. Lazecký · M. Bakoň · I. Hlaváčová
UTAD Vila Real, Quinta de Prados, 5000-081 Vila Real, Portugal

K. Růžičková and T. Inspektor (eds.), *Surface Models for Geosciences*,
Lecture Notes in Geoinformation and Cartography,
DOI 10.1007/978-3-319-18407-4_14

1 Introduction

Satellite Synthetic Aperture Radar Interferometry (InSAR) is a valuable technology currently being used mainly for generation of high quality DEM models—projects including global models as SRTM (2000) or WorldDEM (2014, continuously updated), and for monitoring of terrain deformations such as subsidence, landslides, structure deformations, effects of earthquakes or volcano activity. Satellite missions, as TanDEM-X, can deliver data allowing extracting DEM models with resolution of few meters and height estimation accuracy generally in the order of few meters. In terms of deformation monitoring, satellite InSAR is able to deliver information of near-vertical movements of observed objects in millimetric or even better precision (Ferretti et al. 2007). To achieve such a precision, however, the majority of non-deformation effects influencing phase of the radar wave must to be filtered out—including the phase induced by topography changes. Generally, the phase change ($\Delta\varphi$) between two SAR images can be ascribed as due to displacement in time, height difference, atmosphere-caused delay and various noise—as described by Eq. 1.

$$\Delta\varphi = \Delta\varphi_{displacement} + \Delta\varphi_{height} + \Delta\varphi_{atmosphere} + \Delta\varphi_{noise} \tag{1}$$

2 Height Effects in Differential SAR Interferometry

A rate of influence of topography in satellite SAR interferogram can be described using parameter called Height ambiguity (H_{amb}) that is a height difference corresponding to a 2π phase difference (a fringe). This parameter is directly corresponding to image pair parameters, like perpendicular baseline (B_{perp}) between satellite positions at the time of image acquisition—a larger B_{perp} yields more sensitivity of interferogram to height—see Eq. 2.

$$H_{amb} = \frac{\lambda}{2} \cdot \frac{R \sin\theta_{inc}}{B_{perp}} \tag{2}$$

where λ… satellite SAR carrier wavelength, θ_{inc}… incidence angle of SAR line of sight, R… distance of satellite from observed terrain.

So, the effect of topography depends on the B_{perp} of computed interferogram. With a very small perpendicular baseline (depending to applied SAR system, for example less than ~50 m in case of Envisat), meaning the satellite is scanning both images from almost the same spot in some time delay, there will be a very small or negligible stereoscopic effect. This is an optimal configuration for deformation monitoring. Higher effect of topography is visible in higher rates of B_{perp}—Fig. 1 shows this effect. To remove the topographic phase in interferograms, an appropriated DEM,

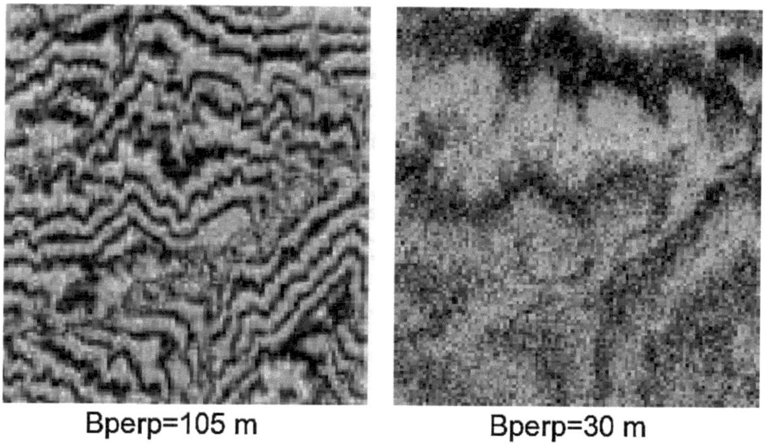

Fig. 1 Topography effects in Envisat interferograms with B_{perp} = 105 m (*left*) and B_{perp} = 30 m (*right*). Images contain same hilly area (Kampes et al. 2008)

converted to SAR coordinates as a simulated phase with the same H_{amb} as in the interferogram, has to be used.

Any DEM inaccuracy will cause a residual topographic signal ($\Delta\varphi_{res,topo}$) that can be described as modified from (Hlaváčová 2008) at Eq. 3:

$$\Delta\varphi_{topo}^{res} \simeq \frac{4\pi}{2}\left(\frac{B_{perp} + \Delta B_{perp}}{R\sin\theta_{inc}}\Delta H_p\right) \tag{3}$$

where ΔB_{perp}… error in local B_{perp} estimation due to height error ΔH_P. Value of ΔB_{perp} is for practical purposes often neglectable, however it can yield significant impact in case of detecting displacements at steep slopes.

Such value of $\Delta\varphi_{res,topo}$ can be misinterpreted as signal caused by displacement $\Delta\varphi_{displacement}$ [rad]. This equals a value of detected deformation in SAR line of sight (D_{LOS} [m]) given by the relation in Eq. 4 (Hanssen 2001).

$$D_{LOS} = \lambda \cdot \Delta\varphi_{displacement}/4\pi \tag{4}$$

In case of Envisat at B_{perp} = 100 m, the error of H_P = 5 m will yield $\Delta\varphi_{res,topo}$ = 0.3 rad (i.e. error in deformation estimation is neglectable: D_{LOS} = 1.5 mm) but in case of higher B_{perp} = 600 m, the same DEM error H_P would yield $\Delta\varphi_{res,topo}$ = 2 rad (i.e. D_{LOS} = 9 mm). This indicates that it is not necessary to use a very precise DEM for topography removal in shorter B_{perp} configurations, especially in areas with not very high terrain gradients. Currently active SAR missions tend to minimize B_{perp} by establishing precise satellite orbital tracks. Different situation can occur in case of huge displacements where millimetric-decimetric sensitivity of InSAR phase can be strongly affected by outdated DEM. Different satellite SAR systems have different

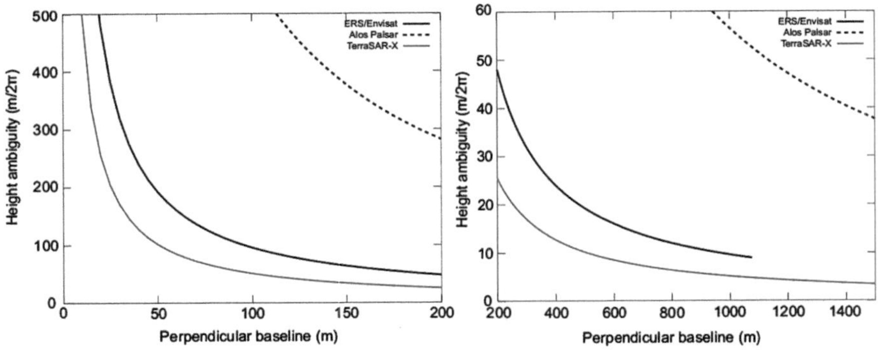

Fig. 2 Relation between H_{amb} and B_{perp} of selected popular SAR missions, adapted from (Lazecký 2011)

sensitivity to height errors—relation between B_{perp} and H_{amb} is demonstrated for three most popular SAR systems in Fig. 2: Envisat (C-band SAR), Alos (L-band SAR) and TerraSAR-X (X-band SAR).

3 Issues Within Application of Popular DEMs in InSAR

SRTM DEM model (from year 2000) still being the most used for topography removal in InSAR, with a resolution of ~90 m (in Europe) and a vertical accuracy that should be better than 15 m (with some 90 % coverage within a tolerance of 5 m). Such a model was considered sufficient for most of the InSAR uses and more reliable than other global projects as Aster GDEM (with significantly lower vertical accuracy). Since September 2014, the SRTM (2000) of 1 arcsecond resolution (~30 m in Europe) is available (Simon 2015). Obviously, two properties should be considered in DEM selection for InSAR topography phase removal: DEM precision in terms of accuracy and resolution (see Fig. 3) and whether DEM is up-to-date with regards to InSAR images timeline (see Fig. 4).

4 DEM Inaccuracies in Multitemporal InSAR

A continuous processing of InSAR of various sites within last two decades has shown high potential of InSAR data for geological, geodetic or urban needs. With the new generation of high spatial resolution radar satellites, as TerraSAR-X or COSMO Sky-Med, the InSAR focus started to be directed for monitoring bridges, dams and other urban structures and, therefore, complement conventional structural monitoring techniques due to, among others, its capability to measure small and slow deformation phenomena and the high density of stable points that can be

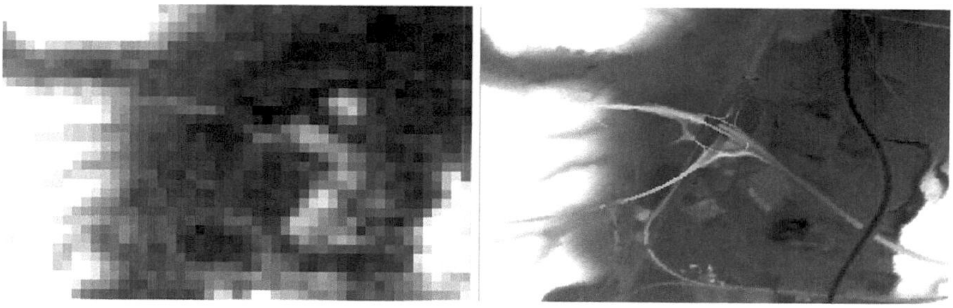

Fig. 3 Resolution difference between SRTM (*left*) and custom DEM based on Czech national geodetic base ZABAGED (*right*), used for InSAR monitoring of highway bridge deformations (Lazecký et al. 2014)

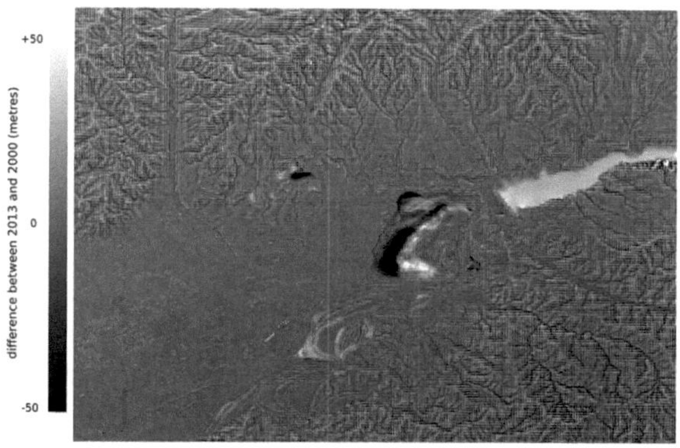

Fig. 4 Difference between DEM from Pleiades (from year 2013) and SRTM (from year 2000) over Angren, Uzbekistan—difference in landslide areas goes up to 50 m affecting InSAR detection ability (Lazecký et al. 2013)

detected. Thanks to the usage of multiple images, it is possible to estimate both deformation trends in time and height, relatively precisely, using information about correlation of SAR phase changes and B_{perp}. Depending on quality of dataset (especially in terms of number of input images and their B_{perp} configuration), as well as on quantitative values of major phase elements $\Delta\varphi$, it is possible to accurately estimate height from MTI, supposing the height difference of processed points is in reasonable range (a coarse DEM is helpful to allow such proper range).

Let's consider that the most reliable estimation of any major $\Delta\varphi$ element is in case where this element doesn't exceed critical value corresponding to $\lambda/4$ (i.e. $\Delta\varphi = \pi$) between neighboring measurements in the MTI dataset, based on (Hanssen 2001). If we consider two major elements: due to linear deformation in time $\Delta\varphi_{defo}$ and due to height $\Delta\varphi_{height}$, we can coarsely suppose that maximally reliable estimation will happen if both of them don't exceed $\pi/2$ in most of interferograms. A maximal $H_{amb,max}$ is at the same time representing maximal DEM error possible to correct in such dataset—this $H_{amb,max}$ then is simply computable using Eq. 5 (where $B_{perp,max}$ can be selected as the highest B_{perp} from the majority of image pairs, e.g. from 25–75 % quantiles of the dataset).

$$H_{amb,max} = H_{amb}\left(B_{perp,max}\right)/4 \tag{5}$$

For example using TerraSAR-X dataset of 11 images with majority of B_{perp} spreading between quantiles −164 and 107 m, the $H_{amb,max}$ = 6 m means that DEM errors of up to 6 m will be reliably corrected. Note that this estimation is simplified and condition of critical value of $\Delta\varphi = \pi$ was chosen as rather strict, however can help in analysis of basic quality of SAR dataset and need of input DEM quality for MTI processing.

Once a DEM error is already estimated in MTI, it is possible to evaluate its accuracy using standard deviation of phase residuals $\delta\Delta\varphi$. One should take into account the number of input images (*M*) and compute standard deviation error due to the estimated height ($\delta\Delta h$) using Eq. 6, based on (Perissin 2006). Such reliability analysis describes goodness of fitting of height estimation model for the input data.

$$\delta_{\Delta h} = \sqrt{\left(\frac{\lambda R \sin\theta}{4\pi}\right)^2 \frac{\delta^2_{\Delta\varphi}}{M\delta^2_{B_{perp}}}} \tag{6}$$

In the same example of TerraSAR-X dataset, described further in (Lazecký et al. 2015), a DEM error was estimated for selected point ID 2274 to be Δh = 7.5 m. Phase residuals after removal of modelled parameters varied with a standard deviation of $\delta_{\Delta\varphi}$ = 0.4 rad. Therefore standard deviation of estimated height was calculated to be $\delta_{\Delta h}$ = 0.58 m -> the height was estimated as $\Delta h = 7.5 \pm 0.58$ m.

5 Conclusions

This paper summarizes issues and gives directions in analysis of height accuracy needed for proper InSAR processing for deformation monitoring. The own idea of estimation of needed DEM accuracy is basically delivered as term $H_{amb,max}$ in Eq. 5. It can be said from the authors experience that in case of having dataset spreading 20 or more images, the theoretical limit of $\lambda/4$ for reliable estimation is neglected—depending on site-to-site experience, it can be summarized that dataset with higher number of images (especially with more spread perpendicular baselines) is usually not prone to most of DEM errors. The results from Hong-Kong dataset processing using 73 TerraSAR-X images (Perissin 2013) reveals the potential to perform heights estimation even of very high skyscraper buildings using MTI processing. Still, DEM accuracy should be carefully considered in case of using low number of images—and especially in case of processing differential InSAR with only one pair of images—the sensitivity for DEM errors is estimable using Eq. 3.

Acknowledgments This work was supported by the European Regional Development Fund in the IT4Innovations Centre of Excellence project (CZ.1.05/1.1.00/02.0070). This paper has been elaborated in the framework of the project New creative teams in priorities of scientific research, Reg. No. CZ.1.07/2.3.00/30.0055, supported by Operational Programme Education for Competitiveness and co-financed by the European Social Fund and the state budget of the Czech Republic and is partially supported by the European Fund for Regional Progress—FEDER (Fundo Europeu de Desenvolvimento Regional) through the project BI/COMPETE/38732/UTAD/2014 entitled "RemotWatch—Alert and Monitoring System for Physical Structures". This work is also the result of implementation of the project: the National Centre of Earth's Surface Deformation Diagnostic in the area of Slovakia, ITMS 26220220108 supported by the Research and Development Operational Programme funded by the ERDF and the grant No. 1/0642/13 of Slovak Grant Agency VEGA. This work was also supported by the program Safety and Security Research of Ministry of Interior, Czech Republic in the Disaster management support scenarios using geoinformation technologies project (VG20132015106).

References

Ferretti A, Savio G, Barzaghi R, Borghi A, Musazzi S, Novali F, Prati C, Rocca F (2007) Submillimeter accuracy of InSAR time series: experimental validation. IEEE Trans Geosci Remote Sensing 45(5):1142–1153. doi:10.1109/TGRS.2007.894440

Hanssen RF (2001) Radar interferometry: data interpretation and error analysis. Dordrecht, Kluwer Academic Publishers, 328 p. ISBN 0-7923-6945-9

Hlaváčová I (2008) Interferometric stacks in partially coherent areas. Ph.D. thesis. ČVUT Praha, Prague 170 p

Kampes BM, Hanssen RF, Perski Z (2008) Doris user's manual and technical documentation. TU Delft, 166 p

Lazecký M (2011) Monitoring of terrain relief changes using synthetic aperture radar interferometry: application of sar interferometry techniques in a specific undermined Ostrava-Karviná district. Ph. D thesis, VŠB-TU Ostrava, Ostrava

Lazecký M, Bláha P, Khasankhanova G, Minchenko V (2013) Monitoring of landslide hazard in selected areas of Uzbekistan. ESA living planet 2013 symposium, 9–13 Sep 2013, Edinburgh

Lazecký M, Perissin D, Bakoň M, Sousa JJM, Hlaváčová I, Real N (2015) Potential of satellite InSAR techniques for monitoring of bridge deformations. JURSE Lausanne, 30.3.-1.4.2015, 4 p. (in print)

Lazecký M, Rapant P, Perissin D, Bakoň M (2014) Deformations of highway over undermined Ostrava-Svinov area monitored by InSAR using limited set of SAR images. CENTERIS 2014, 15–17 Oct 2014, Troia, Portugal. Procedia Technology. doi: 10.1016/j.protcy.2014.10.107

Perissin D (2006) SAR super-resolution and characterization of urban targets: an advanced analysis on the physical nature of synthetic aperture radar permanent scatterers through ERS and envisat interferometry. VDM Verlag Dr. Müller. 144 p. ISBN 978-3639294170

Perissin D (2013) Permanent scatterers in the hong kong harbor. Available at: https://engineering.purdue.edu/~perissin. Accessed on 24 Feb 2015

Simon (2015) SRTM-1 (1 Arc second) now available with large global coverage. Digital Geography. Available at: http://www.digital-geography.com/srtm-1-1-arc-second-now-available-large-global-coverage/. Accessed on 23 Feb 2015

Creating of DSM Based on RPAS Measurement and Accuracy Testing

Karel Pavelka and Jaroslav Šedina

Abstract At the Czech Technical University in Prague, a project for using RPAS in photogrammetry, archaeology and the monitoring of the biosphere was launched in 2011. The project's aim was to develop, test and verify simple and low-cost technology for the monitoring and modelling of small areas such as archaeological digs, historical objects, dumps, mines or and small orthophoto projects. The creation of a digital surface model from photos and its accuracy is one of main tasks in all projects. The idea of this contribution is not only to introduce today's popular unmanned aerial systems (UAS, UAV or RPAS), but to provide a critical analysis of the newly-created high resolution digital surface model (DSM). The introduction is focused on general information and legislation of RPAS in the Czech Republic and other countries. The quality of DSM's is discussed based on the data processing of three case projects in Agisoft Photoscan. The use of control points in the bundle adjustment seems to be a problem in this software. In many cases, the DSM is distorted and we tried to find a better method.

Keywords DSM · RPAS · Accuracy · UAV · Quality control · Agisoft photoscan

1 Introduction

UAV (unmanned aircraft vehicle), nowadays called UAS (unmanned aircraft systems) in the U.S. or RPAS (remote piloted aircraft systems) are gaining importance in the mapping and monitoring tasks of our environment. Nowadays, for control and piloting, RPAS are equipped with sophisticated micro-instruments such as

K. Pavelka · J. Šedina (✉)
Faculty of Civil Engineering, Department of Geomatics,
Czech Technical University in Prague, Thakurova 7, 166 29 Prague 6, Czech Republic
e-mail: jaroslav.sedina@fsv.cvut.cz

K. Pavelka
e-mail: pavelka@fsv.cvut.cz

K. Růžičková and T. Inspektor (eds.), *Surface Models for Geosciences*,
Lecture Notes in Geoinformation and Cartography,
DOI 10.1007/978-3-319-18407-4_15

IMU, gyroscopes, GNSS receivers, wireless image insights, wireless controls, automatic stabilization, flight planners, etc. Benefits can be found by using RPAS, but there are a lot of misconceptions about RPAS regulations (McGovern 2014). In the European Union, there is still a lack of legislation—as each member tries to regulate separately, which increases bureaucracy, leading to problems with the use of RPAS. However, there are issues as well as benefits with RPAS (EU, Enterprise and industry 2015). Safety seems to be a big problem. Examples include RPAS control measures due to accidents, threats to air traffic, privacy protection, spying and terrorism. In April 2014 The European Commission adopted the slogan "A new era for aviation—Opening the aviation market to the civil use of RPAS in a safe and sustainable manner". This motto expresses the Commission's views on how to address civil drone (also known as remotely piloted aircraft systems—RPAS) operations in a European level policy framework which will enable the progressive development of the commercial drones market while safeguarding the public interest.

RPAS can provide not only photographic data, but also other data types like multispectral (with NDVI capability) or thermal data too (depending on sensors and type). (Pavelka and Řehák 2012; Jon et al. 2013; Řezníček and Straková 2013). Bigger RPAS can be equipped with more complex and expensive instruments like laser scanners or hyperspectral scanners. The RPAS method of acquisition combines the benefits of close range and aerial photogrammetry (Eisenbeiss 2009). As a result, a higher resolution and mapping precision can be obtained over compact and possibly less accessible areas (e.g. mountains, moors, swamps, dumps, small natural reserves, dangerous areas). RPAS has the capability to create quality outputs, but before we consider RPAS as a fully functional and accurate technology in mapping, it is necessary to test the quality and accuracy of the outputs—for example, orthophoto and DSM (Pavelka et al. 2014). This can be done using comparative projects, based on the analysis of measured elements. There are a lot of RPAS types with different hardware quality, and each one's processing software and quality is often unique. For this reason, it is necessary to standardize the terminology and define parameters for each category, or to issue certificates of quality (e.g. by certified research institutes). The legislation and the defining of the quality of devices, software and outputs is the task of nearest future.

2 Used Drone

Our research is, in this case, focused on DSM accuracy that was created using low cost RPAS from images taken using an ordinary camera or a camera modified for near infrared range. We have at our disposal the eBee drone. The eBee (a good price-performance ratio for basic university research) drone consists of a styropor airframe, with an electrically powered engine with a pusher propeller, battery (4200 mA), GNSS and IMU unit, digital camera, radio-modem and control system. For flight control, an ordinary notebook computer with special software is used.

With this software it is possible to easily manage all the necessary project information like area (based on free Google maps and other maps or orthophotos), pixel resolution, image overlapping, flight time, distances etc. The typical flight altitude is 100–200 m, with a flight time of up to 40 min and typical documented area of approximately 1 km^2; pixel size can be set and reaches normally 3–6 cm (with small compact Canon camera 14MPix; there are two possibilities—one classic camera for visible spectral range and one modified camera (for near infrared range). Primary outputs from a set of taken photos (typically hundreds of photos) are digital surface models (DSM) and orthophotos (Marčiš 2013; Pavelka and Řezníček 2011). Via two flights over the same area and the changing of cameras, it is possible to obtain vegetation indexes (NDVI). Data processing to an ortophoto and DSM can be done using additional software; we use Agisoft Photoscan software (Fig. 1).

Fig. 1 EBee drone (*left*) and view on central part of NNR (Božídarské rašeliniště) with a control point (*right*)

Fig. 2 Relative deformation of DSM caused by adjustment (optimize) using control points in comparison with DSM created using adjustment (optimize) without control points, NNR project (VIS camera)

3 Case Projects

Three different types of terrain were used for DSM accuracy testing.

3.1 National Natural Reserve (NNR)

Small NNR ("Božídarské rašeliniště") lies in the Karlovy Vary region (Czech Republic) and covers an extensive complex between the municipalities of Boží Dar and Hřebečná along the Krušné hory (Ore mountains) mountain ridge. It has an area of about 1000 ha at an elevation of approximately 1000 m above sea level (Božídarské rašeliniště 2014). The subject of the protection is peat bog as well as mountain peaty meadows, which merge into heathlands. RPAS was used here for biodiversity and vegetation growth monitoring in three periods (May, June and August). These give us the possibility to analyse vegetation growth influence. Signalized control points (white painted wood crosses and plastic boards measuring 60–80 cm) and random object points, measured with the precise GNSS Leica Viva (precision in position cca 1 cm with using online correction of CZECHPOS net), were used for the testing of DSM accuracy (control points are used for transformation—it is photogrammetric technical term; primarily these points are geodetically measured in terrain and then secondary on images). This area is non-homogenous, consisting of uneven terrain, swampy areas, and small water bodies with discontinuous vegetation. It is problem to define solid terrain, because it is partly swamp and the compact vegetation cover has been changed during time (three flights in different season over the mapped area with the same parameters). The photographed area has about 1.2 km^2 with a square shape; the flight height was 165 m, pixel size 5.5 cm. A steady wind was blowing (a problem with flight stability for a small and light drone). Both cameras (one for visible and one for near infrared range) were used separately (Pavelka et al. 2014).

3.2 Historical Field Fortification

In the neighbourhood of nice historical city Litoměřice, there are still visible ramparts from the Prussian—Austrian war in the 19th Century. This was a field forward fortification, but has never been used in battle and later disappeared because of agricultural activities. Some parts are detectable by their terrain signatures, visible on shaded DSMs. By the documentation and research of these relics, we tested the quality of the DSM using directly measured profiles and points using the GNSS Leica Viva with cm accuracy. 180 object points and 5 signalized control points (concentric circular signals with a diameter 15 cm) were measured on a relative flat agriculture area after the harvest by vegetative rest in August (terrain was dry with very small and sparse vegetation). The photographed area has about 50 ha with a rectangular shape; the flight height was 130 m, with a pixel size of 4 cm. Both cameras were used. For fortification monitoring, only a 150 × 150 m^2 area was analyzed (within control points).

3.3 Unknown Historical Construction in Landscape

A small, well known village Lipany in Czech history lies 40 km east of Prague. Near the village is a probably medieval linear formation (mound), now hardly recognizable after several cultivations of the local fields. This mound was located using RPAS and verified by ground penetrating radar. The measuring of an excessive amount of control points give us a good possibility to test the quality of DSMs based on the taken photos. The photographed area has about 50 ha with a very elongated rectangular shape 550 × 100 m). The flight altitude was 130 m, with a pixel size of 4 cm. Only one camera for NIR range was used. The terrain was photographed after a harvest (stubble field, 15 cm and partly with subsequent vegetation), and was relatively flat and dry. 6 signalized control points (square black and white signals on printing paper 20 × 30 cm) and 15 object points were measured with the GNSS Leica Viva (the same instrument as in above mentioned two case projects).

4 Accuracy Testing

In an ordinary case, the quality of outputs can be indicated by deviations on control points, which were not included in the general adjustment. The accuracy of automatically created orthophotos and DSM from RPAS depends on many parameters. First the accuracy is understandably the quality of hardware (RPAS stability, vibrations, camera type and calibration, lens, sensor and consequential image quality etc.) Secondary it is the quality, position and number of control points, image overlapping and flight height, next the result depends strongly on used software for data processing.

We use Agisoft Photoscan software which operates in steps. It is necessary to explain, that Agisoft itself is not pure photogrammetrical software, but it uses image correlation principles (image based modelling). Firstly, it computes a sparse point cloud (used for camera orientations), for DSM secondary, a dense point cloud, and finally creates the orthophoto or textured 3D model). By this process, the solution is not computed as a photogrammetrical model, but as a used image based modelling technology (computer vision). From a mathematical point of view, the process and adjustment of all photos with control points are fine (Marčiš 2013; Remondino 2011). With low cost RPAS, there is a large amount of approximately perpendicular (right to oblique) images (which images a small part of object only), taken by common non photogrammetric cameras. The internal orientation is unstable during process; lens distortion is usually very high. It reasons, that the system deals with lots of unknowns and many degrees of freedom

1. It was discovered that Agisoft PhotoScan creates a distorted DSM by a fully automated process of adjustment with ground control points. This distortion of the DSM occurs in the phase of creating sparse clouds, when the bundle adjustment is done (with the function optimize). Distortion is caused by the control points; the software minimizes deviations on ground control points. This hard criterion causes changes in the external orientation elements of the camera and the DSM is deformed as shown in Fig. 3. The deformation of the model is not easily detectable and deviations on ground control points are minimal. For example, received maximum deviation of the ground control points was 0.007 m by Lipany project only (Table 1 and Fig. 2).

Fig. 3 Point clouds comparing for NIR and VIS range by high vegetation season (NNR 3)

Table 1 Control point coordinates deviations after adjustment in Agisoft PhotoScan software, Lipany project; very good result

Project lipany NIR				
Point	dX (m)	dY (m)	dZ (m)	δ (m)
1	−0.005	0.001	−0.002	0.005
2	0.000	−0.004	0.001	0.004
3	0.005	0.007	0.000	0.009
4	0.001	−0.004	0.000	0.004
5	−0.001	−0.001	0.000	0.001
6	0.000	0.000	0.000	0.001

δ—positional deviations of ground control point
dX, dY, dZ—coordinate difference calculated ground control point with GNSS measurements

Comparing of two models with the software CloudCompare the mean remoteness of 0.309 m with a standard deviation of 0.158 m has been found. Table 2 shows the variations of the ground control points that were not included in the adjustment (*optimize*) and were left as quality control object points. Deviations are largest in the Z coordinate and take values up to one meter. Results for images in the visible spectral range show greater variation than in the images in the near-infrared range.

Table 2 Control points coordinate deviations omitted stepwise from the adjustment (using function *optimize*)

	NIR				VIS			
Point	dX (m)	dY (m)	dZ (m)	δ (m)	dX (m)	dY (m)	dZ (m)	δ (m)
	NNR flight 1							
1	0.058	0.013	−2.026	2.027	0.243	−0.574	−16.216	16.228
7	−0.001	0.073	0.573	0.577	0.000	0.120	−3.284	3.286
8	−0.017	−0.053	−0.013	0.057	−0.069	−0.093	1.546	1.551
9	0.048	0.114	−0.108	0.164	0.185	0.185	−2.770	2.782
	NNR flight 2							
1	−0.026	0.004	0.126	0.129	−0.076	0.019	−0.042	0.089
2	0.085	0.060	−0.196	0.222	0.126	0.038	0.010	0.132
7	−0.032	0.067	−0.152	0.169	0.065	−0.007	−1.119	1.121
8	−0.061	−0.068	−0.201	0.221	−0.069	−0.053	0.879	0.883
9	0.210	0.065	0.393	0.450	0.194	0.099	−1.248	1.267
	NNR flight 3							
1	0.028	−0.009	−0.343	0.344	−0.020	−0.020	−0.278	0.279
2	0.086	−0.043	0.478	0.488	0.066	0.115	0.316	0.342
7	0.079	0.088	0.487	0.501	−0.020	0.139	−0.523	0.542
8	−0.084	−0.042	0.095	0.134	−0.046	−0.087	0.933	0.938
9	0.190	0.074	−0.220	0.300	0.125	0.162	−1.363	1.378
	Litoměřice							
1	−0.014	−0.008	0.097	0.098	0.014	0.001	0.055	0.057
2	−0.021	0.011	−0.126	0.128	−0.054	−0.037	−0.112	0.130
3	−0.008	−0.001	0.100	0.101	0.053	−0.032	0.084	0.104
4	−0.033	0.007	−0.125	0.129	−0.010	−0.005	−0.124	0.125
5	0.010	−0.001	0.028	0.030	0.002	0.013	0.081	0.082
	Lipany							
1	−0.037	0.001	−0.054	0.066				
2	0.000	−0.012	0.013	0.018				
3	0.011	0.019	0.004	0.022				
4	0.007	−0.028	0.013	0.032				
5	−0.009	−0.002	−0.043	0.044				
6	−0.005	0.001	0.033	0.033				

δ—positional deviations of ground control point
dX, dY, dZ—coordinate difference calculated ground control point with GNSS measurements

2. The second variant of the DSM in the required coordinate system is the adjustment of the sparse cloud (optimize) without ground control points. Thus, a model is created that optimizes the elements of interior and exterior orientation and the DSM retains its internal accuracy. The usage of ground control points can be model placed into the DSM coordinate system (using function update). Deviations of control points are shown in Table 3. In the NNR project VIS point

Table 3 Deviations of control points after transformation (function *update*)

	NIR				VIS			
Point	dX (m)	dY (m)	dZ (m)	δ (m)	dX (m)	dY (m)	dZ (m)	δ (m)
	NNR flight 1							
1	−0.025	0.026	0.001	0.036	−0.053	0.022	−0.005	0.058
7	0.017	−0.018	0.015	0.029	0.038	−0.054	0.003	0.066
8[a]	−0.012	−0.059	−0.042	0.073	−0.040	−0.121	1.687	1.692
9	0.020	0.051	0.023	0.059	0.014	0.033	−0.001	0.036
	NNR flight 2							
1	0.001	−0.030	0.060	0.067	−0.037	0.006	0.236	0.239
2	0.070	0.122	−0.042	0.147	−0.006	0.030	−0.260	0.262
7	−0.162	0.010	0.058	0.173	−0.002	−0.098	−0.105	0.144
8[a]	−0.054	−0.043	−0.075	0.102	−0.071	−0.050	1.032	1.036
9	0.123	−0.051	0.061	0.146	0.063	0.044	0.026	0.081
	NNR flight 3							
1	−0.012	0.023	−0.112	0.115	−0.023	−0.021	0.211	0.213
2	0.030	0.042	0.117	0.128	0.010	0.022	−0.179	0.181
7	−0.023	−0.012	0.125	0.128	0.032	−0.014	−0.101	0.107
8[a]	−0.091	−0.061	−0.160	0.194	−0.058	−0.104	1.097	1.103
9	0.104	0.025	0.124	0.164	0.000	0.016	0.020	0.025
	Litoměřice							
1	0.003	0.005	0.004	0.007	−0.011	0.009	−0.017	0.022
2	−0.011	0.003	0.005	0.012	−0.028	−0.015	−0.052	0.061
3	−0.012	−0.022	0.004	0.025	0.026	−0.022	0.000	0.034
4	0.018	−0.013	0.002	0.022	0.008	0.020	−0.055	0.059
5	0.000	0.027	−0.011	0.029	0.009	0.012	0.112	0.113
	Lipany							
1	−0.040	0.013	−0.105	0.113				
2	−0.002	0.012	0.034	0.036				
3	0.017	0.001	0.034	0.038				
4	0.027	−0.015	0.018	0.035				
5	−0.004	−0.002	−0.063	0.063				
6	−0.010	−0.023	0.080	0.084				

δ—positional deviations of ground control point
dX, dY, dZ—coordinate difference calculated ground control point with GNSS measurements
[a]Point 8 was not used as a control point by transformation of DSM model (based on images in visible range)

8 was omitted from the adjustment; it has a systematic error in elevation. Point 8 was placed on isle lying in the small water body (peat lake).
The absolute orientation of the model in the S-JTSK coordinate system is performed using control points, which are calculated from oriented images by spatial intersection. These points need not lie in the plane of DSM. Due to errors, points can be calculated out of the plane of DSM. However these points only serve for the transformation of DSM into the coordinate system. The elevation deviation of these points is then a systematic error of DSM in neighborhood of these points. Comparing transformations (*update*) in Agisoft PhotoScan, it was found that the results of this transformation are worse than results of seven-parameters Helmert transformation, according to the criteria $\Sigma vv \rightarrow \min$. Algorithm and the conditions of transformations in Agisoft Photoscan are unclear (Table 4).

3. To create a DSM without distortion based on optional transformation into the coordinate system (JTSK), the following approach was chosen. The sparse point cloud was adjusted (optimize) without ground control points. The dense point cloud was created; after this a polygonal mesh was generated. Raster DSM and orthophoto were exported. From these products model X, Y, Z coordinates of control points were subtracted. These have been transformed into S-JTSK by seven-parameters Helmert transformation. Obtained coordinate deviations of control points are presented in Table 5.

Table 4 Differences of control points coordinates from adjustments in Agisoft PhotoScan and adjustments using Helmert seven-parameters transformation compared with control points coordinates acquired by GNSS (Leica Viva)

	Agisoft PhotoScan				7-parameter Helmert transformation			
Point	dX (m)	dY (m)	dZ (m)	δ (m)	dX (m)	dY (m)	dZ (m)	δ (m)
1	0.001	−0.030	0.060	0.067	−0.007	0.028	−0.049	0.057
2	0.070	0.122	−0.042	0.147	−0.073	−0.127	0.028	0.149
7	−0.162	0.010	0.058	0.173	0.153	−0.007	−0.010	0.154
8	−0.054	−0.043	−0.075	0.102	0.050	0.048	0.092	0.115
9	0.123	−0.051	0.061	0.146	−0.124	0.058	−0.060	0.149
Σ (m)	−0.023	0.009	0.062	0.635	0.000	0.000	0.000	0.624
Σvv (m^2)	0.088				0.085			

δ—positional deviations of ground control point
dX, dY, dZ—coordinate difference calculated ground control point with GNSS measurements
Σ—the sum of coordinates corrections
Σvv—the sum of the squares corrections in the project

Table 5 Deviations of model control points coordinates after Helmert transformation

	NIR				VIS			
Point	dX (m)	dY (m)	dZ (m)	δ (m)	dX (m)	dY (m)	dZ (m)	δ (m)
	NNR flight 1							
1	0.040	−0.013	−0.001	0.042	0.049	0.004	0.001	0.049
7	−0.034	0.033	−0.008	0.048	−0.052	0.027	−0.001	0.058
8[a]	0.002	0.016	0.022	0.027	0.033	0.148	−1.807	1.813
9	−0.009	−0.035	−0.013	0.038	0.003	−0.031	0.001	0.031
	NNR flight 2							
1	−0.050	0.038	0.066	0.091	0.020	0.012	−0.229	0.230
2	−0.077	−0.144	−0.047	0.170	0.022	−0.048	0.152	0.161
7	0.204	−0.030	−0.056	0.214	0.000	0.065	0.094	0.115
8[a]	0.066	0.043	0.087	0.117	0.073	0.056	−1.047	1.051
9	−0.144	0.093	−0.049	0.179	−0.042	−0.029	−0.017	0.054
	NNR flight 3							
1	0.049	0.027	0.021	0.060	0.135	0.208	−0.197	0.316
2	−0.060	−0.071	−0.018	0.095	−0.041	−0.252	0.133	0.288
7	0.010	0.004	−0.041	0.042	0.015	0.063	0.080	0.102
8[a]	0.106	0.027	0.092	0.143	0.017	0.136	−1.000	1.010
9	−0.104	0.013	−0.054	0.118	−0.109	−0.018	−0.016	0.111
	Litoměřice							
1	0.002	−0.013	0.001	0.013	−0.013	−0.017	0.020	0.029
2	0.023	−0.015	−0.014	0.031	0.055	0.001	0.037	0.066
3	0.011	0.030	0.004	0.032	−0.022	0.038	0.010	0.045
4	−0.019	0.000	−0.012	0.023	−0.003	−0.002	0.034	0.034
5	−0.017	−0.001	0.021	0.027	−0.017	−0.020	−0.101	0.105
	Lipany							
1	0.022	0.029	0.090	0.097				
2	0.013	−0.034	−0.003	0.037				
3	−0.015	−0.029	−0.100	0.106				
4	−0.032	0.014	−0.004	0.035				
5	0.003	−0.012	0.100	0.100				
6	0.009	0.032	−0.082	0.089				

δ—positional deviations of ground control point
dX, dY, dZ—coordinate difference calculated ground control point with GNSS measurements
[a]Point 8 was not used as a control point by transformation of DSM model (based on images in visible range)

5 Accuracy Testing

For outputs accuracy testing, the results of Helmert seven-parameters transformation were used. During testing it was found that the positioning accuracy is affected by vegetation growth. Projects with low vegetation growth are summarized in

Table 6 Calculation of standard deviation X, Y and standard position errors in projects with minimal vegetation cover

	Projects with minimum of vegetation											
	NNR flight 1				Litoměřice				Lipany			
	NIR		VIS		NIR		VIS		NIR			
Point	dX	dY	dX	dY	dX	dY	dX	dY	dX	dY		
1	0.040	−0.013	0.049	0.004	0.002	−0.013	−0.013	−0.017	0.022	0.029		
2					0.023	−0.015	0.055	0.001	0.013	−0.034		
3/7	−0.034	0.033	−0.052	0.027	0.011	0.030	−0.022	0.038	−0.015	−0.029		
4/8	0.002	0.016			−0.019	0.000	−0.003	−0.002	−0.032	0.014		
5/9	−0.009	−0.035	0.003	−0.031	−0.017	−0.001	−0.017	−0.020	0.003	−0.012		
6									0.009	0.032	σ_X (m)	σ_Y (m)
σ_X/σ_Y (m)	0.031	0.030	0.050	0.029	0.018	0.018	0.031	0.023	0.020	0.029	0.032	0.026
σ_{XY} (m)	0.031		0.041		0.018		0.028		0.025		0.028	

dX, dY—coordinate difference calculated ground control point with GNSS

σ_X, σ_Y—standard deviation in X and Y of control points coordinates

σ_{XY}—standard deviation

Table 7 Calculation of standard deviation X, Y and standard position errors in projects with dense vegetation cover

	Projects with full-grown vegetation									
	NNR flight 2				NNR flight 3					
	NIR		VIS		NIR		VIS			
Point	dX	dY	dX	dY	dX	dY	dX	dY		
1	−0.050	0.038	0.020	0.012	0.049	0.027	0.135	0.208		
2	−0.077	−0.144	0.022	−0.048	−0.060	−0.071	−0.041	−0.252		
7	0.204	−0.030	0.000	0.065	0.010	0.004	0.015	0.063		
8	0.066	0.043			0.106	0.027				
9	−0.144	0.093	−0.042	−0.029	−0.104	0.013	−0.109	−0.018	σ_X (m)	σ_Y (m)
σ_X/σ_Y (m)	0.137	0.092	0.030	0.050	0.084	0.041	0.103	0.193	0.097	0.111
σ_{XY} (m)	0.117		0.041		0.066		0.154		0.103	

dX, dY—coordinate difference calculated ground control point with GNSS
σ_X, σ_Y—standard deviation in X and Y of control points coordinates
σ_{XY}—standard deviation

Table 6. Table 7 analyzes projects with grown vegetation. For projects without or with low vegetation the standard position error was determined as 0.028 m on ground control points. Projects with grown vegetation have significantly worse results—standard position error reaches 0.103 m. The resulting standard position errors in the coordinates (Tables 7 and 8) are squared weighted average of the

Table 8 Calculation of standard height deviation in projects with dense vegetation and low vegetation cover

	Projects with minimum of vegetation					Projects with full-grown vegetation			
	NNR flight 1		Litoměřice		Lipany	NNR flight 2		NNR flight 3	
	NIR	VIS	NIR	VIS	NIR	NIR	VIS	NIR	VIS
Point	dZ	dZ	dZ	dZ	dZ	dZ	dZ	dZ	dZ
1	−0.001	0.001	0.001	0.020	0.090	0.066	−0.229	0.021	−0.197
2			−0.014	0.037	−0.003	−0.047	0.152	−0.018	0.133
3/7	−0.008	−0.001	0.004	0.010	−0.100	−0.056	0.094	−0.041	0.080
4/8	0.022		−0.012	0.034	−0.004	0.087		0.092	
5/9	−0.013	0.001	0.021	−0.101	0.100	−0.049	−0.017	−0.054	−0.016
6					−0.082				
σ_Z (m)	0.016	0.001	0.014	0.058	0.084	0.070	0.168	0.059	0.145
	σ_Z without vegetation (m)				0.051	σ_Z with vegetation (m)			0.115

dX, dY—coordinate difference calculated ground control point with GNSS
σ_Z—standard height deviation

standard position errors of individual projects, where the weight is the number of control points in the project. Another finding was that the positional accuracy of control points does not depend on the type of image (visible range and the near infrared range).

The same applies to height standard deviations; for projects with low vegetation the height standard deviations was determined as 0.051 m and by projects with grown vegetation the height standard deviations was determined as 0.115 m. Standard height deviation is computed by a weighted squared mean where the weight is the number of points in the project.

6 Quality Control

In the Litoměřice project 180 points in four profiles were measured using precise GNSS (Leica Viva); in the Lipany project only 8 points in a single profile were measured. Height differences between GNSS measurements and point height determined from DSM in the Z coordinate are shown in Table 9. By comparing the results in the near infrared and visible range in the Litoměřice project, it is evident that the model created from images taken by a camera for near-infrared range is better.

Onwards the standard deviation of the model with a mean equal to 0 is computed. For the near infrared range the standard deviation increases from 0.028 to 0.073 m, and for the visible range the standard deviation increases from 0.083 to 0.186 m. This higher standard deviation would better reflect the determination of the height of the DSM—primarily it is unknown a constants by which the DSM is shifted in the Z coordinate. Thus obtained data should have bias error equal to the average height error Z (ϕ_Z) and a random component equal σ_Z (Table 10).

Table 9 Height differences between GNSS measurements and point height determined from DSM

	Litoměřice		Lipany
	NIR	VIS	NIR
ϕ_Z (m)	−0.067	−0.165	0.000
σ_Z (m)	0.028	0.083	0.095
σ_Z ($\phi_Z = 0$) (m)	0.073	0.186	0.095
Min (m)	−0.142	−0.308	−0.173
Max (m)	−0.003	0.100	0.120
Number of points	180	180	8

ϕ_Z—mean deviation in Z coordinate
σ_Z—standard deviation in Z coordinate

Table 10 Height differences between GNSS measurements and point height determined from DSM (NNR case project, three flights in different season)

Cover	NIR			VIS			
	Peat	Grassland	Solid obj.	Peat	Peat*	Grassland	Solid obj.
	NNR 1						
ϕ_Z (m)	−0.169	−0.242	0.143	−0.308	0.161	0.818	0.331
σ_Z (m)	0.076	0.058	0.048	0.919	0.387	0.311	0.198
σ_Z ($\phi_Z = 0$) (m)	0.189	0.256	0.167	0.972	0.422	0.899	0.420
Min (m)	−0.279	−0.375	0.059	−2.063	−0.613	0.372	0.111
Max (m)	−0.044	−0.115	0.180	0.762	0.762	1.477	0.513
	NNR 2						
ϕ_Z (m)	−0.262	−0.312	0.014	−0.088	0.238	0.460	0.055
σ_Z (m)	0.147	0.074	0.056	0.643	0.297	0.296	0.091
σ_Z ($\phi_Z = 0$) (m)	0.306	0.330	0.058	0.650	0.385	0.559	0.110
Min (m)	−0.611	−0.424	−0.063	−1.372	−0.353	0.034	−0.053
Max (m)	0.007	−0.166	0.079	0.603	0.603	1.061	0.130
	NNR 3						
ϕ_Z (m)	−0.060	−0.592	−0.144	−0.042	0.318	0.381	0.283
σ_Z (m)	0.136	0.116	0.085	0.691	0.257	0.249	0.172
σ_Z ($\phi_Z = 0$) (m)	0.149	0.621	0.182	0.692	0.418	0.465	0.360
Min (m)	−0.389	−0.736	−0.273	−1.414	−0.280	−0.013	0.063
Max (m)	0.147	−0.372	−0.039	0.623	0.623	0.817	0.443
Number of points	20	17	5	20	15	17	5

ϕ_Z—mean deviation in Z coordinate
σ_Z—standard deviation in Z coordinate

7 Conclusion

The RPAS technology is very popular nowadays. It will be interesting for many customers by providing new and efficient ways to manage both large projects and common 2D, 3D or special measurements using thermal, hyperspectral or other instruments. Manufacturers of hardware and software speak in superlatives, and RPAS becomes a fashion trend (just for the event Intergeo 2014 in Berlin there were 70 examples of RPAS). The rapid, effective, ecological and, in many cases, low-cost data acquisition capability is indisputable. However, quality, precision and comparability of results are not quite the best. The main result of our projects is the precision research of DSMs based on RPAS photographic data joined with geodetically (GNSS) measured control and object points. Adjustments (optimize), together with control points, distort the DSM, because while it is minimizing deviations on ground control points, elements of exterior orientation of images become deformed. Using transformation (update) with control points the DSM can be transformed into the frame coordinate system. These control points can be burdened with measuring errors (positioning in images) and their height can thus be determined incorrectly. Height that is incorrectly measured or computed ground control points causes a systematic error of the DSM in the close area of such a point. Transformation *update* is not described (black box), Helmert seven-parameters transformation gives similar results and it is the least squares method. A dependent positional and height accuracy to grown vegetation was detected. When testing the accuracy of the DSM, it was found that the deviation between GNSS measurement and the point height from the DSM consists of a systematic component and a random component. The size of both components was significantly smaller in the near infrared model. The optimal data processing procedure seems to be the adjustment (*optimize*) of the sparse cloud without control points and a gradual filtration of the sparse cloud points to form a rigid model followed by the creation of the dense cloud (and from this the generation of a polygon model). When generating the polygon surface model (*Surface type*) the *Height field* setting was chosen. This method allows one to generate meshes with some simplification, and the process takes considerably less time than with the setting *Arbitrary*. Consequently, it is possible to generate an orthophoto or raster DSM. From orthophotos, it can be deducted that model X, Y (with an approximate transformation to S-JTSK which is given by the transformation (*update*) in Agisoft PhotoScan) and Z coordinate can be obtain from the DSM. Control points can be transformed into S-JTSK by Helmert seven-parameters transformation. The transformation key can further serve to transform the dense cloud and the polygon DSM in the S-JTSK.

It is necessary to remark, that used software is not exactly photogrammetric. It is based on image correlation (*image based modelling*) and all parameters (lens distortion, camera calibration and image exterior orientation) are computed in one step using tie points searched by image correlation method and control points. That is why the final model created from many images is distorted—mathematically it is correct, but from photogrammetric point of view it is questionable. Used software minimizes deviations on control points and changes exterior and interior orientations. It is necessary to introduce limits for photogrammetric variables before adjustment, otherwise global optimization procedure changes according to its algorithm parameters to suit task notwithstanding the reality. In NNR case project a systematic error in visible range was found (extreme variables are located in the middle of an area near control point Nr.8). Strong dependence in positional and height accuracy was also discovered with vegetation growth. This was confirmed by results of Litoměřice and Lipany projects, where minimum vegetation was present. Standard deviations were 28 mm (x, y) and 51 mm (z) with minimum vegetation, 103 and 115 mm with high vegetation season (NNR project). Differences on object points between GNSS measurement and measurement on DSM have a systematic error from vegetation and random error. Better results give near infrared images.

Acknowledgments This paper project was supported by a grant of the Ministry of Culture of the Czech Republic NAKI DF13P01OVV002 (New modern non-invasive methods of cultural heritage objects exploration).

References

Božídarské rašeliniště National Nature Reserve (2014) http://www.cittadella.cz/europarc/index.php?p=index&site=NPR_bozidarske_raseliniste_en. Accessed 1 Feb 2015

Dan McGovern (2014) Busting UAS myths. http://www.sparpointgroup.com/Blogs/Disruptive-Perspective/Busting-UAS-myths/. Accessed 1 Feb 2015

Eisenbeiss H (2009) UAV photogrammetry. Zurich, DISS. ETH No. 18515. ETH Zurich. http://www.igp-data.ethz.ch/berichte/Blaue_Berichte_PDF/105.pdf. Accessed 1 Feb 2015

EU, Enterprise and industry, Remotely Piloted Aircraft Systems (RPAS) http://ec.europa.eu/enterprise/sectors/aerospace/uas/index_en.htm. Accessed 1 Feb 2015

Jon J, Koska B, Pospíšil J (2013) Autonomous airship equipped with multi-sensor mapping platform. ISPRS Int Arch Photogrammetry Remote Sens Spat Inf Sci XL-5/W1(40):119–124

Marčiš M (2013) Quality of 3D models generated by SFM technology. Slovak J Civil Eng 11 (4):13–24

Pavelka K, Řehák M (2012) Using of UAV for photogrammetry and thermal imaging. In: Proceedings of ACRS 2012 Asian Association of Remote Sensing (AARS), Chang Watana Road, Bangkok, vol 1, pp 589–601

Pavelka K, Řezníček J (2011) New low-cost automated processing of digital photos for documentation and visualization of the cultural heritage. Geoinformatics CTU FCE 6(6):245–258. ISSN 1802-2669. http://www.geoinformatics.fsv.cvut.cz

Pavelka K, Řezníček J, Faltýnová M, Matoušková E (2014) RPAS as a tool for the monitoring of a natural reserve. In: 14th international multidisciplinary scientific geoconference SGEM 2014, informatics, geoinformatics and remote sensing, proceedings, vol III. Sofia STEF92 Technology Ltd., pp 291–298

Remondino F (2011) UAV photogrammetry for mapping and 3D modelling: current status and future perspectives. http://www.isprs.org/proceedings/XXXVIII/1-C22/papers/remondino.pdf. Accessed 1 Feb 2015

Řezníček J, Straková H (2013) Documentation of dumps and heaps by use of UAV. In: 13th international multidisciplinary scientific geoconference SGEM 20134, informatics, geoinformatics and remote sensing, proceedings vol I. Sofia STEF92 Technology Ltd., pp 151–158

The Use of Real and Fictitious Surfaces for Territorial Distribution Assessment of Given Geographic Phenomenon

Aleš Ruda and Jaromír Kolejka

Abstract Considering the usability of real or fictitious surfaces, we analyze their formation both in terms of purpose and input data. Relief is understood as a major distribution factor of processes in the landscape. This concept is related to both continuous (e.g. air temperature, rainfall) and a discrete (various forms of land use) phenomena. While at continuous phenomena simulated surfaces of terrain distributed phenomena respect the configuration of the relief directly, simulated surfaces reflect rather discrete phenomena. Although largely shaped by terrain they differ from the relief itself, not regarding their different nature. A paper discusses within physical and human geography the issue of raster surfaces making. In case of real surfaces, authors use for calculation known regression dependences of air temperature and precipitation on various relief parameters regarding deeper intention of drought risk assessment. This serves as an evidence of dependences of given parameters on digital elevation model which results in higher accuracy opposite to fictitious surfaces generated using spatial analyst or statistic. On the example of Focal Statistic tool (Neighborhood Statistic) and IDW interpolation method differences towards the real state (e.g. concentric shapes in IDW generated surfaces) and too smoothed and generalized areas (using different size of moving window) are described and discussed. When defining the areas of post-industrial landscape authors came to the conclusion that not very often recommended IDW interpolation does not make so big inaccuracies as compared with Focal Statistic tool.

Keywords Raster surfaces · DEM · Real surfaces · Fictitious surfaces · Neighborhood statistics

---

A. Ruda (✉)
Faculty of Regional Development and International Studies, Department of Regional Development and Public Administration, Mendel University in Brno, Zemědělská 1, 61300 Brno, Czech Republic
e-mail: ruda@mendelu.cz

J. Kolejka
Faculty of Education, Department of Geography, Masaryk University, Poříčí 7, 60300 Brno, Czech Republic
e-mail: jkolejka@centrum.cz

K. Růžičková and T. Inspektor (eds.), *Surface Models for Geosciences*, Lecture Notes in Geoinformation and Cartography,
DOI 10.1007/978-3-319-18407-4_16

1 Introduction

All human activities are spatially located. When needed to express the level of dependence or to derive further information surface models are being used. The easily understood is digital elevation model (DEM) concerning elevation data. Although surfaces can be represented by raster and vector data, raster representation consisting of pixels (elementary cells) containing real-world information provides a suitable representation for further processing. Basically, within raster data we may work with continuous, thematic (discrete) or picture data. When working with surfaces, we use continuous data which have been measured as a specific value in a fixed point in space. Many other raster surfaces are derived from DEM and then may be depicted in a two-dimensional (planar view) or can be generated and modeled with other image sources for 3D perspective views. The question arises, why is DEM so important for surface model creation? In fact, the terrain plays an important role as a distributional factor of many terrain related issues when studying the landscape on choric level. Compare the importance of the terrain as a key factor for precipitation and temperature distribution illustrating the water—energy balance, forming landslides, modeling water runoff from watershed, etc. Spatial analyses often works with raster surfaces indirectly related to the terrain. It means that although terrain basically influences the distribution of human activities (considering the least consumption of energy) following analyses uses geostatistically based tools to derive the prevailing distribution of given phenomena in wider space (for instance Neighborhood statistics). These outputs may be generalized and smoothed and do not have to fit the real distribution as it is influenced by the terrain. Considering the aim of geography studying both natural and human activities generated raster surfaces are based on given relations. When deriving natural characteristics dependent on the terrain accurate DEM is necessary. Following surfaces derived from DEM (directly slope or aspects characteristics or indirectly solar insolation) represent a work with continuous data and provide almost a real state close to the landscape. Working with human activities or distribution of objects created by men pushes us in case of further analysis to create fictitious surfaces best fitting studying phenomena and thus representing almost discrete data. Using several tools for data processing differs in physical and human geography although we can find certain convergence. Because of two faces of geography (physical and human) two different surfaces creations and following derivations are discussed.

2 Importance of Surface Modeling

The terrain as a firm natural landscape surface plays the dominating role in the distribution of many various natural phenomena, as well as human products. This role is based on its critical on the water, energy and material distribution. The exposition of slopes to solar radiation, wind, within a combination of the Earth's gravitation and sea elevation form local climate defined by behavior of air phenomena in the course of a standard time unit. Each time unit depends on the primary distribution of energy as the principal source of movement of matter. The available energy, its amount and space/time distribution lays in the background of regularly repeated processes that cause the origin and following consequent existence of terrain, soil and vegetation forming procedures. Geomorphological processes form the terrain features, kind of rock weathering. The final product is the typical and logical combination of terrain, rock, soil, water, warm and vegetation parameters of each site. Such combination of parameters represents a subject of human interests—in the background of human interest is the maximum profit and minimum expenditure of time, energy and matter. This is why the terrain plays the dominating role in the spatial distribution of natural and human phenomena.

In terms of research applications, it is necessary to underline the role of GIS. A number of available tools can be used for surface map creation. Surface maps are widely used in many branches because of their capability to express continuous values. The issue of surface modeling was discussed many times. Smith et al. (2006) solved the importance of the implication of DEM resolution in soil mapping. Arrouays et al. (2014) and McKenzie et al. (2000) confirmed that terrain attributes derived from DEM are related to many landscape properties. On the other hand Openshaw (1983) worked with fictitious surface and found out that the property of aggregated spatial data generating population surface can result in very different local area distributions.

3 Raster Surfaces Modeling Directly Dependent on DEM

Authors used DEM to derive directly dependent surfaces (air temperature and precipitation surface) for geographical drought modeling. Air temperature model considered maximum daily air temperature and precipitation model is based on identification water conditions in the Hustopeče study area (Fig. 1).

3.1 Air Temperature Surface in Hottest Period

Climatologists and other scientist building air temperature model come out from existing time series data and consider various factors and parameters expressed in equation. Accuracy of the model depends on the scale and also data availability. In many cases interpolation method is used (DeGaetano and Belcher 2007). The spatial interpolation of temperature is influenced by elevation therefore accurate DEM is needed. Jarvis and Stuart (2001) consider elevation as the strongest covariate for estimating temperature which was confirmed in numerous studies (e.g. Johnson et al. 2000; Kurtzman and Kadmon 1999). Willmott and Matsuura (1995)

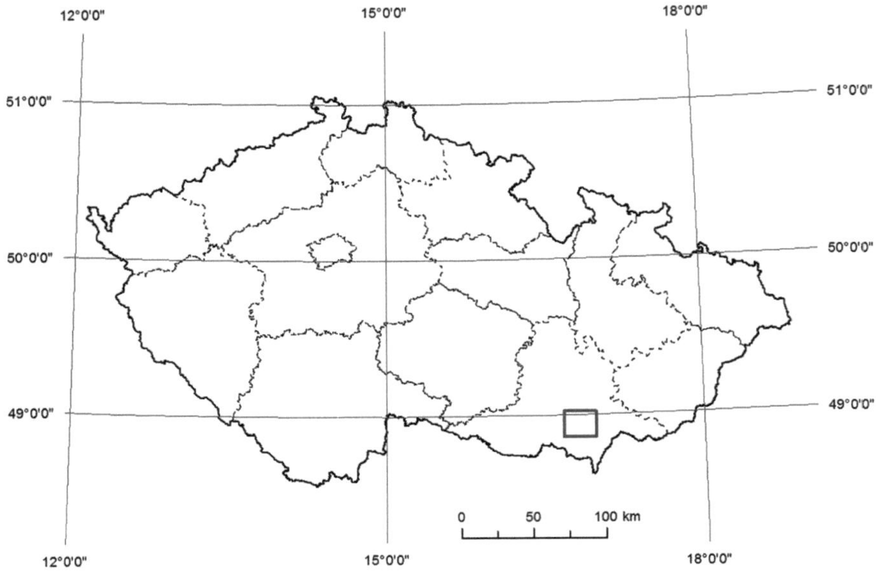

Fig. 1 Hustopeče study area

reduced each temperature to the sea level using average environmental lapse rate and then interpolated these values to DEM which is very similar to our approach.

Further discussed air temperature model considering maximum daily temperature in study area was proposed for geographical drought risk assessment. With regard to the accuracy of the model of temperature change with altitude in the area of interest or in its vicinity meteorological stations with permanent data collection were identified. Four key stations (Kobylí, Nemochovice, Staré Město and Luhačovice) were used to collect temperature data. Their maximum daily air temperatures were analysed and from the time series the period (drought period) from 18.7 to 9.8 with prevailing tropical daytime temperatures (above 30 °C) and minimal rainfall was identified. This period served as a basis from which an arithmetic mean was calculated for each station and subsequently used for the calculation of temperature changes related to elevation.

For this purpose a cartographic model (Fig. 2) based on the work of Vahalík and Klimánek (2013) and reflecting the change in air temperature with altitude respecting slope and aspects was created. DEM was created using Topo to Raster tool (ArcGIS for Desktop 10.1) working with hydrologically corrected algorithm. As an input data we used contours (5 m interval) derived from a topographic map in scale of 1:25,000, important 158 elevation points taken from topographic map in scale of 1:10,000, water stream, water bodies and known relief barriers. As the model shows, the arithmetic mean of the maximum daily temperatures for mentioned stations were with regard to their altitude subjected to regression analysis. From the linear regression model a regression equation of dependence

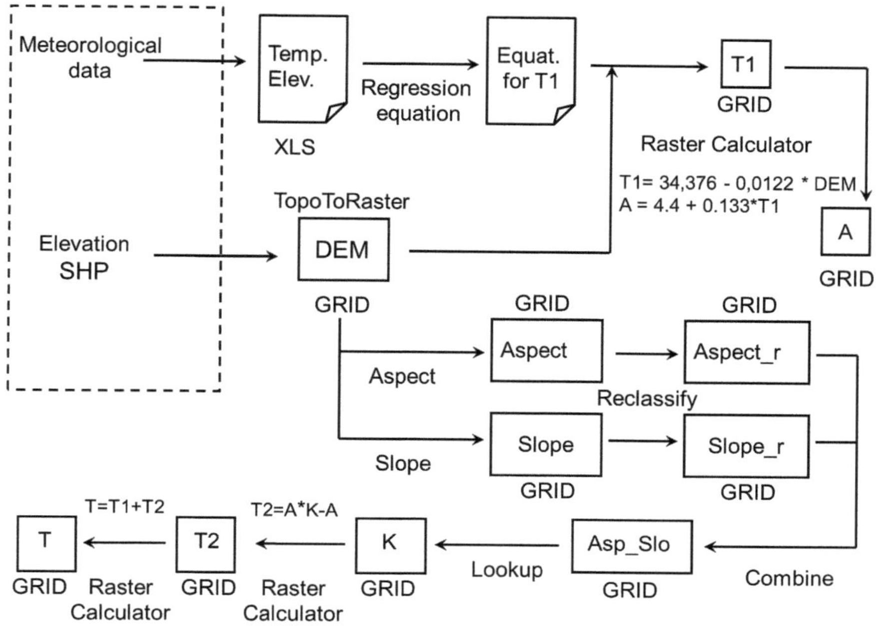

Fig. 2 Flow chart of air temperature model

−0.0122x + y = 34,376 was used for further processing. Using map algebra a raster image bearing vertical temperature gradient in relation to digital model elevation (layer T1) was created. The equation T1 enables regression estimation of temperatures related to an elevation, but do not reflecting local (zonal) temperature trend. In this case, temperature normality (a proportion between the actually measured temperature and the temperature detected by calculating a regression equation) can be used as input data. Interpolating the values of individual temperature normality and following multiplying with T1 raster results in zonal correction for applied regression relationship. Using experimental equation T2 (Vašků 1971) to refine calculated temperature (T1), the resulting potential temperature is given by the sum of the calculated values for the equation of T1 and T2 (Fig. 3). For this purpose raster layer combining reclassified values of mentioned elevation characteristics was created and further served to derive raster layer bearing information of the coefficient of relative insolation (Table 1) calculated by Vašků (1971). The coefficient reflects the influence of slope and aspect of the terrain on the final temperature value. For mid-attitudes it confirms that areas on southern slopes with slope between 40 and 50° are most insolated and opposite on northern slopes of the same slope interval are least insolated. The results of air temperature model were verified by comparing temperature averages in given stations during chosen period with temperature values in raster surface and we found out 95 % accordance respecting the first decimal place.

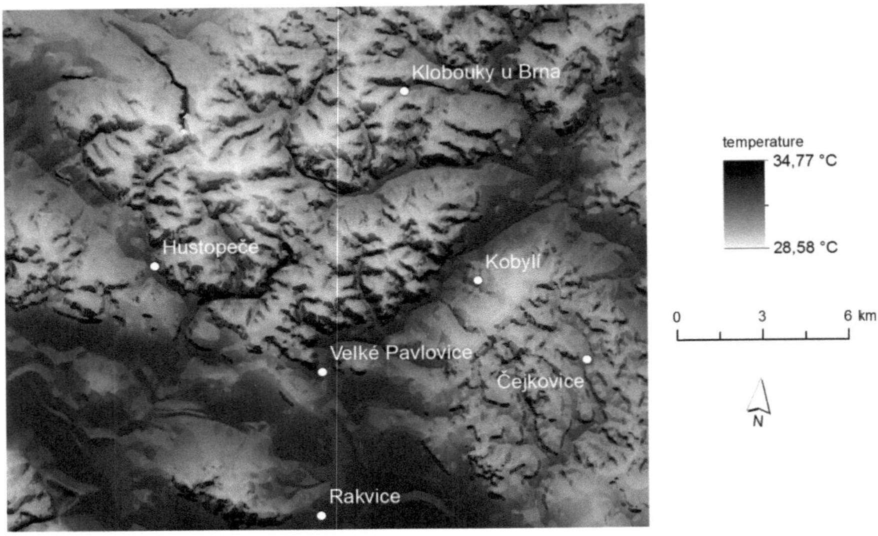

Fig. 3 Visualization of air temperature surface results of Hustopeče area

Table 1 Derivation of coefficient of relative insolation reflecting aspect and slope (Vašků 1971)

Slope								
K	(1) 0°–5°	(2) 5°–10°	(3) 10°–15°	(4) 15°–20°	(5) 20°–25°	(6) 25°–30°	(7) 30°–40°	(8) 40°–50°
(1)S	1.05 (04)	1.38 (03)	1.17(06)	1.22 (08)	1.26 (31)	1.31 (34)	1.34 (35)	1.37 (40)
(2) SE, SW	1.04 (05)	1.10 (02)	1.16 (01)	1.20 (07)	1.24 (30)	1.26 (33)	1.28 (32)	1.30 (39)
(3) E, W	1.02 (09)	1.06 (20)	1.09 (21)	1.11 (14)	1.12(27)	1.12 (26)	1.10 (29)	1.07 (38)
(4) SW, NW	1.00 (10)	1.02 (11)	1.01 (15)	1.00 (17)	0.99 (22)	0.97 (24)	0.92 (28)	0.84 (36)
(5) N	0.99(13)	1.00 (12)	0.98 (16)	0.96 (18)	0.93(19)	0.87 (23)	0.81 (25)	0.75 (37)
Aspect								

3.2 Precipitation Surface

Spatial distribution of precipitation using interpolation method can be found in many studies (e.g. Kurtzman and Random 1999; Mitas and Mitasova 1988). In case we have a lot of meteorological stations and not very variable relief we can use this approach. In much difficult cases results are not very accurate because interpolation methods do not consider other properties of the landscape (Marquínez et al. 2003). Spreen (1947) and Smith (1979) informed that precipitation generally increases with elevation and in workes of Goodale (1998), Wotling (2000) and Weisse and Bois (2001) and others are highlighted relationships between precipitation and various topographic variables such as altitude, latitude, continentality, slope, orientation or exposure Although other authors used many stations a different time series (e.g. Marquínez et al. 2003) in our case without many data collections and in relatively small area we used one station inside and other stations near the study area.

In our study in contex of geographical drought assessment we developed a model considering relationships between precipitation and a range of topographic variables. Similarly, the distribution of precipitation is also influenced by the characteristics of the relief. In addition to the current precipitation regime the current rainfall totals reflects variable slope and aspect. Particularly strong rainfall comes to the Czech Republic from northwest. It was assessed that the raindrops fall on average by 5° angle (Kolejka and Kretek 1997, 2001). Due to these facts the distribution of precipitation was derived taking into account slope and aspect (Table 2).

As it was mentioned also precipitation model was created for geographical drought risk assessment. Recalculation of total rainfall from the period of precipitation coming before the heat period (drought period) was set is based on several assumptions, which are essential for the issue simplification. However, it does not need to be at the expense of reliability of systematic extrapolation of station data to the relief. The recalculation is based on data available from precipitation stations located in the area of interest or in its neighbour. Although we cannot even prove the occurrence of precipitation in the lowland areas at one altitude level (summer precipitation may have heterogeneous nature reflecting a significant proportion of convective precipitation) used simplifying assumptions that theoretically homogeneous precipitation field is interrupted by the relief with respect to altitude, slope and aspect during expected northwest direction of rainfall-bearing winds. Considering generalization and general usability of a model no other wind directions were took into account. Authors understand that wind coming from northwestern direction can be also changed by the terrain, but this issue is difficult to model. Linear regression was used to determine the dependence of total rainfall on altitude. Recalculation using slope and aspect characteristics required further simplification namely assumption that rainfall-bearing wind changes the direction of rain drops from vertical axis up to 5°. Using descriptive geometry calculations precipitation falling on a horizontal surface (100 % at an angle of 85°) were recalculated to combinations of eight aspect categories and four slope categories

Table 2 Percent correction for supposed distribution of rainfall (Kolejka and Kretek 1997, 2001)

	0°–3°	3.1°–8°	8.1°–15°	more than 15°
N	100	100.5	100.5	99
NE	100	100	100	100
E	100	99.5	98	94.5
SE	100	99	97.5	91.3
S	100	99.5	98	94.5
SW	100	100	100	100
W	100	100.5	100.5	99
NW	100	101	100	97.5

Fig. 4 Visualization of precipitation surface results of Hustopeče area

(0–3°—plain; 3, 1–8°—flat slope; 8, 1–15°—low slope; above 15°—steep slope). Current precipitation distribution represents the total rainfall recorded before the expected drought period mentioned during air temperature model calculation. Precipitation model calculation considered linear regression considering rainfall and elevation of stations located inside or near the area of interest (Hustopeče, Kobylí, Klobouky, Ždánice, Prušánky, Těšany a Dolní Věstonice). The resulting image of precipitation distribution documents in Fig. 4.

4 Fictitious Raster Surface Creation: The Case of Post-Industrial Landscape

The origins of many elements of the contemporary cultural landscape of the Kamenicko region situated in the Liberec Region in North Bohemia date back to the late 18th century, when the relatively harmonious landscape, covered in forests and pastures and subject to small-scale farming, began to be transformed industrial activities in the southern foothills of the Jizerské Mts. in the watershed region between the Lužická Nisa (Baltic Sea) and Jizera River (North Sea). The multi-functional post-industrial landscape Kamenicko is a legacy of the industrial revolution. There have been gradually formed numerous enterprises of glass, textile, wood, engineering and electrical industries on its territory. Many municipalities

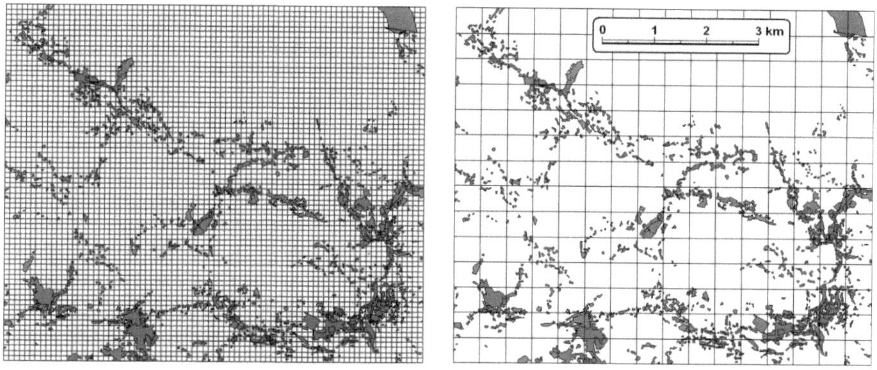

Fig. 5 Overlay of the Kamenicko region post-industrial sites with 100 × 100 and 500 × 500 m^2 grids

almost fused together now, have grown in the original forest landscape. A detailed land use mapping was carried out and it distinguished objects and areas of different age and origin. Two experiments were carried out to identify cores and peripheries of studied the post-industrial landscape. The 100 × 100 and 500 × 500 m square grids were applied for the area classification respecting the representation (share) of post-industrial areas and objects in any square (Fig. 5). The primary cartographic square results presentations were transformed using by both the IDW method and neighborhood statistics into the well understandable area maps.

Given that GIS software packages offer certain tools enabling efficient evaluation of the regular discrete networks (e.g. Focal statistics), it was possible to experiment with some methods, but also with different size of surface elements of geometric networks. To determine the proportion of areas of industrial heritage in post-industrial landscape regular grids of squares covering a map of the area at a scale of 1:10,000 were selected. The ratio of post-industrial site within the corresponding reference site was calculated using a well-established procedure in each of the squares. The representation of data followed the scheme of contemporary landscape typology (Kolejka et al. 2000), where the minimum percentage ratio indicates the number of words necessary for allocation of the contemporary landscape type to the given square (only based on the secondary landscape structure, regardless of other structures), so that at least one of the words of the denomination included the term "post-industrial" landscape. This scale enables a reliable denomination of a contemporary landscape type within the grid element. The key values for classification and typology include 0 %, up to 11 %, up to 19 %, up to 36 %, up to 91 % and over 91 % (Fig. 6) and come from following landscape sites typology:

(a) Monofunctional sites: a single form of land use constitutes a minimum of 91 % of the reference unit area.
(b) Bifunctional sites: two forms of land use jointly cover a minimum of 91 % of the reference unit area, whereby the ratio of the dominant land use form is at

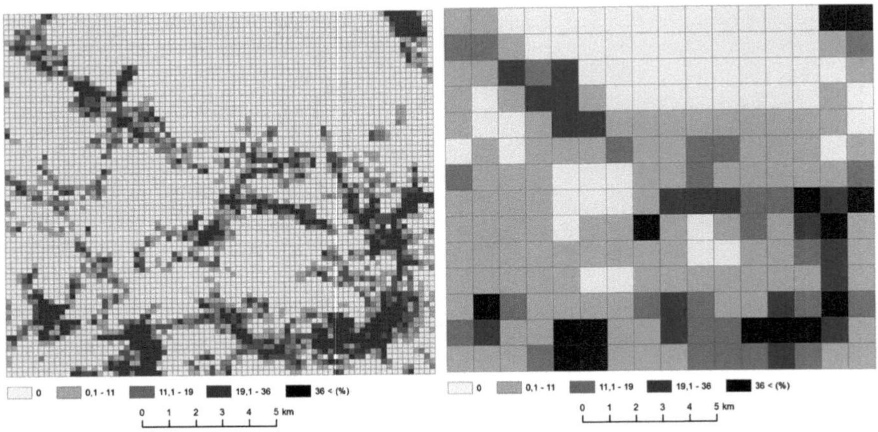

Fig. 6 Calculation of the percentage ratio of post-industrial sites within the 100 × 100 and 500 × 500 m^2 on the territory of the Kamenicko region

least 55 % (in total over 1/2) and the ratio of the second land use form is at least 2/3 of the dominant form, i.e. approx. 36 % (in total over 1/3).

(c) Trifunctional sites: three forms of land use jointly represent at least 91 % of the reference unit area, whereby the ratio of the dominant land use form is approx. 43 % (less than 1/2), the ratio of the second land-use form represents at least 2/3 of the first one, i.e. approx. 29 % (in total less than 1/3) and the ratio of the third dominant form is at least 2/3 of the second dominant form ratio, i.e. at least 19 % (in total less than 1/4).

(d) Polyfunctional sites: four land-use forms jointly constitute at least 91 % of the reference unit area, whereby the ratio of the most dominant represents at least 38 % (over 1/3), the ratio of the second dominant form represents 2/3 of the dominant form, i.e. approx. 25 % (1/4 and more), the ratio of the third dominant form is at least 2/3 of the preceding form ratio, i.e. approx. 17 % (less than 1/5) and the ratio of the last (fourth) dominant land-use form represents at least 2/3 of the preceding form ratio, i.e. approx. 11 % (over 1/10).

It is possible to use an interpolation method related to the center of squares, which transform discrete values to isolines. Geographical research uses this procedure very often. Very popular but not very accurate is Inverse Distance Weighted (IDW) method (Di Piazza et al. 2011) or much reliable NEIGHSTA (Neighborhood Statistics) (Zhang et al. 2007). IDW is based on fact that certain values of know pair of points are related to each other, but their similarity is inversely related to the distance between them (Lu and Wong 2008). Moreover, the distance parameter is applied uniformly without any empiric determination considering the distribution of data inside the study area. In spite that, many authors (e.g. Pig et al. 2004;

Lloyd 2005) used IDW because of its popularity and straightforward and non-computationally intensive processing. Finally, resulting values are very close to the nearest values and increase "bull eyes" creation. These almost concentric circles near isolated points make the surface too fictitious and inaccurate.

Neighborhood statistics uses tool creating output values for each cell location based on the location value and the values identified in a specified neighborhood. A moving window or search radius can be used. For our research Focal Statistics tool was applied. It is based on overlapping neighborhoods to calculate a specified statistic for the cells within a specified neighborhood around each input cell. This tool offers many optional parameters and it is also its disadvantage because user must set the size of moving window, its shape, type of statistic etc. The final result is represented by smoothed surface but how fictitious is a fictitious surface knows only a user. On the other hand delineating area of similarity of some attribute is due to using this technique very useful although a number of iteration with changed parameters must be done. A statistical validation using comparing final surface with for instance weighted density surface supported with calculated distance band is recommended.

Both methods were tested over industrial heritage data related to centers of reference squares considering grid of 100 × 100 and 500 × 500 m (Fig. 7a, b).

Optimal results have been achieved by Neighborhood statistics method (Focal statistic) using grid of 100 × 100 m with moving window concerning 9 × 9 pixels where each pixel contained proportion of postindustrial areas (Fig. 8—left). This method derived areas with graded representation of proportion of industrial heritage in Kamenice study area. As a result of comparison with different sized moving windows (Fig. 7a), delineated smoothed areas fit the best the basic distribution of input data. In other cases results were too fragmented (Fig. 7 a, moving window of 3 × 3 or 6 × 6 pixels) or too smoothed and generalized (Fig. 7a, moving window of 12 × 12 or 15 × 15 pixels). A clear disadvantage of this processing is fact that defined areas do not always represent graded representation of industrial heritage areas in continuous appearance. Almost similar results can be found at IDW outputs working with a grid of 500 × 500 m (Fig. 8—right). In this case visible disadvantage is evident in the occurrence of isolated circular shapes "bull eyes" in grid spatial distribution. After all both methods have similar results although certain differences are obvious. Also, very similar is the statistical analysis of the representation of different forms of use of industrial heritage areas in areas with proportion bigger than 36 %. A more detailed comparison leads to the conclusion that areas derived by Neighborhood statistics based on a grid of 100 × 100 m and using moving window of 9 × 9 pixels are better fitted over the areas of industrial heritage than areas obtained by the IDW method working with a grid of 500 × 500 m. Significant value shift in larger squares to its centers causes inaccuracy.

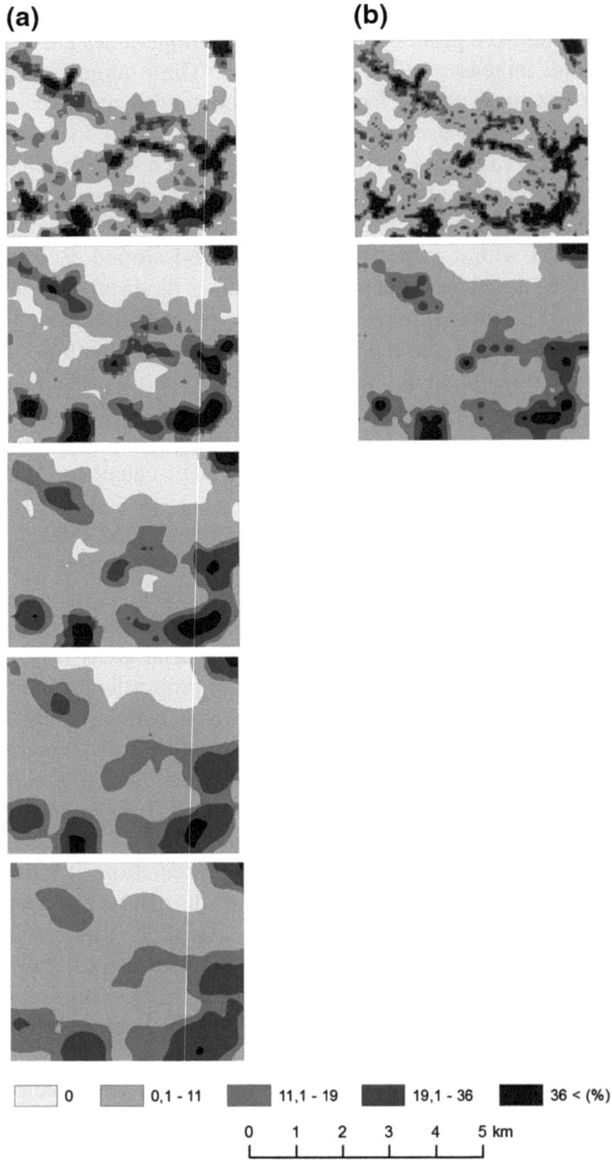

Fig. 7 Results of Neighborhood statistics method (Focal Statistics) over squares of 100 × 100 m (**a**—*left column*) using moving window of (in descent order) 3 × 3, 6 × 6, 9 × 9, 12 × 12 and 15 × 15 pixels; (**b**—*right column*) IDW outputs over grid of 100 × 100 m (*top*) and 500 × 500 m (*bottom*)

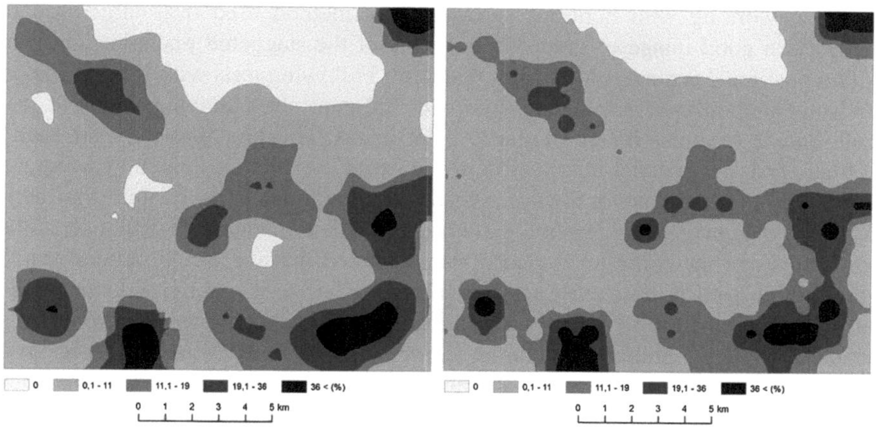

Fig. 8 Conversion outputs of data processing selected as optimal using the Neighborhood statistics method (Focal Statistics, *left*) and the IDW method (*right*) presenting the percentage ratio of post-industrial sites within original squares in isoline fields for the purposes of conducting a classification and typology of the Kamenicko post-industrial landscape

5 Conclusion

Modeling of spatial distribution of geographic phenomena gives in reality a picture of territorial distribution of values of modeled features. Regardless the terrain is the limiting factor the final picture of phenomena values is variously distant from the terrain composition. The landscape features closely bound with the terrain allow the reader of the picture quite easily identify the ties of mutual relationships. Such pictures—called real surface models—demonstrate a visible dependency of the phenomenon on the terrain. If the difference between modeled phenomena values and the terrain is big, it means outside the direct understanding capability of the reader, then the picture represents a fictitious surface. Modeling of air temperature and precipitation distribution proved that using algorithms considering statistical relations of studied phenomenon on terrain characteristics generates depending on the DEM resolution relatively real picture of natural characteristic (e.g. temperature, precipitation). Of course, we used well known dependence of temperature and rainfall on altitude and terrain characteristic (slope and aspect). In case of air temperature model the fact of the model accuracy to a real state is much acceptable then in case of precipitation model because the distribution of solar energy can be modeled using valid coefficient of relative insolation. The amount of precipitation is influenced by many variables (e.g. change of wind direction, difficult terrain morphology, actual synoptic situation over area etc.) that cannot be fully incorporated. As described above, the precipitation model considers dependence on altitude, collected data from neighbor stations and slope and aspect characteristic of the terrain in relation to an angle of falling rain drops influenced by these variables and coming from the prevailing direction (northwest).

When working with fictitious surfaces IDW method used over larger square generates a good image of graded appearance of the staggered phenomenon even with a positional error resulting from that shift. Following tests with Neighborhood statistics method over smaller squares with different sized moving windows can contribute to create an image similar to IDW results. This time, however, the fitting of final areas over spatial distribution of industrial heritage areas is more accurate, although, in turn, it cannot always get graded continuous areas of industrial heritage. GIS data processing technology aimed at the determination of gradation of the post-industrial landscape level clearly demonstrated that the Neighborhood statistics method itself is unsuitable for this purpose because it requires different sized moving window testing without having been previously known which moving window result is optimal. In contrast, combination with IDW having larger grid resolution approximately optimal spatial distribution of final areas can be obtained even with certain inaccuracy. When IDW offers over the grid of 500 × 500 m good spatial pattern, mentioned testing within Neighborhood statistics method will get similar spatial pattern giving more accurate spatial distribution of searched areas representing graded level of post-industrial landscape.

In both cases, reading real and fictive surfaces in 3D perspective makes the reader able to imagine a very high quality of phenomenon values spatial distribution. A correct reading of such pictures stays at the beginning of scientific research and the explanation of the phenomenon interaction with the environment. The importance of all kinds of surface models, especially those not directly linked with the terrain, is still not fully understood and accepted.

Acknowledgments Suggestions related to the methods of definition, classification, typology and evaluation of post-industrial landscapes were elaborated under the grant project "The Fate of Czech Post-Industrial Landscape" number IAA 300860903 supported by the Grant Agency of the Academy of Sciences of the Czech Republic and the grant project "The use of geoinformation technology for drought risk assessment" number 7AMB12GR017 supported by the Grant Agency of the Ministry of Education, Youth and Sports. This work was also supported by the program Safety and Security Research of Ministery of Interior, Czech Republic in the Disaster management support scenarios using geoinformation technologies project (VG20132015106).

References

Arrouays D, McBratney AB, Minasny B, Hempel JW, Heuvelink GBM, MacMillan RA, Hartemink AE, Lagacherie P, McKenzie NJ (2014) The GlobalSoilMap project specifications. In: Arrouays D, McKenzie N, Hempel J, Richer de Forges A, McBratney AB (eds.) GlobalSoilMap: basis of the global spatial soil information system. CRC Press, Taylor and Francis Group, London, p 494

DeGaetano AT, Belcher BN (2007) Spatial interpolation of daily maximum and minimum air temperature based on meteorological model analyses and independent observations. J Appl Meteorol Climatol 46:1981–1992

Di Piazza A, Lo Conti F, Noto LV, Viola F, La Loggia G (2011) Comparative analysis of different techniques for spatial interpolation of rainfall data to create a serially complete monthly time series of precipitation for Sicily, Italy. Int J Appl Earth Obs Geoinf 13(3):396–408

Goodale CL, Alber JD, Ollinger SV (1998) Mapping monthly precipitation, temperature and solar radiation for Ireland with polynomial regression and digital elevation model. Climatol Res 10:35–49

Jarvis CH, Stuart N (2001) Uncertainties in modelling with time series data: estimating the risk of crop pests throughout the year. Trans GIS 5:327–343

Johnson GL, Daly C, Taylor GH, Hanson CL (2000) Spatial variability and interpolation of stochastic weather simulation model parameters. J Appl Meteorol 39:778–796

Kolejka J, Kretek M (1997) Odhad rizika dopadů klimatických změn na vybraná chráněná území přírody Opavska pomocí GIS. Geografie, č. IX, PdF MU, Brno, pp 4–22

Kolejka J, Kretek M (2001) Klimatické změny a ochrana přírody. GIS jako účelný nástroj pro hodnocení rizik. Geoinformation 1:28–31

Kolejka J, Lipský Z, Pokorný J (2000) Ráz krajiny České republiky. GIS a DPZ pomáhají v jeho identifikaci a hodnocení, GEOinfo 7(2):24–28

Kurtzman D, Kadom R (1999) Mapping of temperature variables in Israel: a comparison of different interpolation methods. Climatol Res 13:33–43

Lloyd CD (2005) Assessing the effect of integrating elevation data into the estimation of monthly precipitation in Great Britain. J Hydrol 308:128–150

Lu GY, Wong DW (2008) An adaptive inverse-distance weighting spatial interpolation technique. Comput Geosci 34:1044–1055

Marquínez J, Lastra J, García P (2003) Estimation models for precipitation in mountainous regions: the use of GIS and multivariate analysis. J Hydrol 270(2–3):1–11

McKenzie NJ, Gessler PE, Ryan PJ, O'Connell DA (2000) The role of terrain analysis in soil mapping. In: Wilson JP, Gallant JC (eds) Terrain analysis: principles and applications. Wiley, New York, pp 245–265

Mitas L, Mitasova H (1988) General variational approach to the interpolation problem. Comput Math Appl 16(12):983–992

Openshaw S (1983) The modifable areal unit problem. Concepts and techniques in modern geography, vol 38. Geobooks, Norwich

Pig JL, Green CJ, Zartman RE, Bronson KF (2004) Exploring spatial dependence of cotton yield using global and local autocorrelation statistics. Field Crop Res 89(2–3):219–236

Smith RB (1979) The influence of mountains on the atmosphere. Adv Geophys 21:87–230

Smith MP, Zhu AX, Burt JE, Stiles C (2006) The effects of DEM resolution and neighborhood size on digital soil survey. Geoderma 137:58–69

Spreen WC (1947) A determination of the effect of topography upon precipitation. Trans Am Geophys Union 28:285–290

Vahalík P, Klimánek M (2013) Using GIS and maximum likelihood classification to model forest altitudinal zones. In: Implementation of DSS tools into forestry practice. Technical university in Zvolen, Zvolen pp 59–70

Vašků Z (1971) Metodika vyhodnocení potřeby odvodnění a závlah. Státní meliorační správa, Praha. In: Ambros Z (ed) Praktikum geobiocenologie. Ediční středisko MZLU v Brně, Brno, 2003, pp 1–98

Weisse AK, Bois P (2001) Topographic effects on statistical characteristics of heavy rainfall and mapping in the French Alps. J Appl Meteorol 40(4):720–740

Willmott CJ, Matsuura K (1995) Smart interpolation of annually averaged air temperature in the United States. J Appl Meteorol 34:2577–2586

Wotling DG, Bouvier C, Danloux J, Fritsch JM (2000) Regionalization of extreme precipitation distribution using the principal components of the topographical environment. J Hydrol 233:86–101

Zhang Ch, Jordan C, Higgins A (2007) Using neighborhood statistics and GIS to quantify and visualize spatial variation in geochemical variables: an example using Ni concentrations in the topsoils of Northern Ireland. Geoderma 137:466–476

Impact of GDAL JPEG 2000 Lossy Compression to a Digital Elevation Model

Jan Růžička and Kateřina Růžičková

Abstract Mobile platforms bring possibilities to use distributed services in several users' environments. One of the conditions of the mobile platform is to transfer as less as possible amount of data via network. When dealing with geodata in mobile environment it is always necessary to use compression algorithms to lower their size. Digital elevation models in a GRID form are usually compressed with LZW or DCT technique, but the best compression ratio is now available with wavelet algorithms. The compression can be loss or loss less. The paper describes results of tests of JPEG 2000 loss compression technique. We studied impact of JPEG 2000 loss compression on quality of DEM. We have tested several options of OpenJPEG library to find how these options can change resulting compressed DEM. The main part of the research was focused on pre-filtering of DEM before compression. We have find out a way how to minimise average error of loss compression when keeping the same compression ratio.

Keywords Compression · JPEG 2000 · Filtering · Digital elevation model

1 Introduction

The issue of influence of raster compression on data quality had been solved in many articles. For example Vatolin et al. (2005) deal with comparing several software tools for compression data according to JPEG 2000 format. As an introduction on compressing DEM can be given in thesis of Inanc (2008), witch provides overview of compressions techniques applied for DEM. Mittal et al. (2013)

J. Růžička · K. Růžičková (✉)
Institute of Geoinformatics, HGF, VSB-Technical University of Ostrava,
17. listopadu 15, 708 33 Ostrava, Czech Republic
e-mail: katerina.ruzickova@vsb.cz

J. Růžička
e-mail: jan.ruzicka@vsb.cz

K. Růžičková and T. Inspektor (eds.), *Surface Models for Geosciences*,
Lecture Notes in Geoinformation and Cartography,
DOI 10.1007/978-3-319-18407-4_17

studied, how data for DEM creation can be changed via compression of source satellite images. Gladkova and Grossberg (2006) and Gladkova et al. (2007) focused their research on analysing several compression techniques including JPEG 2000 to MODIS and hyperspectral data.

A lot of papers describe technical specification prepared by software producers. There are also results of several compression ratios and their impact to error in the resulting DEM published by Microimages, Inc. (2009). They analysed impact of loss JPEG 2000 compression to ASTER data, which is quite high.

Ben-Moshe et al. (2007) describes a new technique for simplification of DEM based on analyses of DEM before compression. It is named Image Compression Terrain Simplification (ICTS). The ITCS technique is compared with another techniques including JPEG 2000 compression technique. The paper shows results that ICTS gives similar or even better results for simplification of DEM than JPEG 2000. What is missing in the paper are parameters used for JPEG 2000 compression, mainly compression ratio.

2 Methods and Software

2.1 Methods

The research was focused on two basic ideas:

- How sensitive is the error in the compressed data to parameter BLOCKSIZE and can we use it to reduce the impact of the compression to the DEM quality.
- Can the mean filtering of the data before compression produce less error in the compressed data.

The comparison was based on the simple procedure:

- The original DEM was filtered by mean filter.
- The original DEM and the result of filtering were compressed by loss wavelet compression and stored to JPEG 2000 format (with different parameters described later).
- The compressed files were uncompressed and compared with original data using simple map algebra (difference = original_dem − compressed_dem).
- For the difference layers was calculated value that represents average error realised by data compression.

Used mean filter has the following values:

```
MATRIX 3
1 1 1
1 1 1
1 1 1
DIVISOR 9
```

The compression has been done in four ratios:

- 200:1
- 100:1
- 20:1
- 10:1

The BLOCKSIZE parameter has been specified in three values:

- 512
- 1024
- 2048

3 GDAL

The GDAL allow to use three different open source libraries for wavelet compression according to JPEG 2000 specification. Unfortunately the only one was used for our research purposes. The following table shows problems with each library (Table 1).

The used library was OpenJPEG. There were problems with several tested files, but most of the tested files were compressed correctly. This is probably a bug in the library and should be corrected in the future.

The OpenJPEG library was build from SVN repository to have a latest version of the library.

The basic command for compression to JPEG2000 with OpenJPEG library was:

gdal_translate -of "JP2OpenJPEG" -co "QUALITY = 1" -co "BLOCKSIZEX = 1024" -co "BLOCKSIZEY = 1024" demfilter1.tif demfilter1.tif.j2k

The parameter QUALITY has been set to values: 10, 5, 1, 0.5 to reach the specified compression ratios. The parameters BLOCKSIZE has been set to values: 512, 1024, 2048. Other parameters were used with default values, because they do not have an impact to a quality of a compression.

Table 1 Libraries for GDAL

Library	Problem
Libjasper	The compression ratio can be specified, but it has no effect between 10:1 and 500:1. The library specifies the ratio itself
Kakadu	Compilation process is quite complicated and has not been finished by author of the paper
OpenJPEG	Some of the tested files were compressed in the results with a lot of noise that was not possible to filter out. This is probably a bug in the library

4 Grass GIS

GRASS GIS was used for filtering the original DEM with mean filter, for calculating with map algebra and for counting average error.

The command for filtering the data was:

r.mfilter input = dem output = demfilter1 filter = filter1.txt

The command for map algebra was:

r.mapcalc 'dem_dem.tif.j2k.tif = dem-dem.tif.j2k.tif'

5 Data

The library was tested on five tiles from ASTER DEM version 2. Each tile has a resolution 3601 × 3601 pixels and spatial resolution about 0.01°. In the following table are listed basic characteristics for selected tiles.

From Table 2. is obvious that mainly the flat areas were selected. That selection was made by expectations that for the flat areas the filtering should produce the best results.

The data were available in GeoTIFF format without compression, with encoding of values using range of UInt16 domain. The size of each tile in original format was about 28 MB.

6 Results

The following five tables show the results of calculations. The results are discussed in the chapter discussion and conclusion. The results are based on average error that was counted as sum of errors for each individual pixel divided by number of pixels.

The average error can be used for several purposes, but for other several purposes can be important distribution of the error in the whole DEM and maximal error. The following two tables show distribution of error for compression ratio

Table 2 Libraries for GDAL

Tile	Minimal elevation	Maximal elevation
N23E026	409	1098
N33E081	4342	6375
N49E017	88	1163
N51E021	1	341
S24E125	295	477

100:1 for the tile N23E026 with BLOCKSIZE 1024. The maximum error for not filtered data was 47 m and for filtered data it was 53 m.

The following two tables show distribution of error for compression ratio 20:1 for the tile N49E017 with BLOCKSIZE 1024. The maximum error for not filtered data was 6 m and for filtered data it was 33 m.

7 Discussion and Conclusion

From Tables 3, 4, 5, 6 and 7 is obvious that the BLOCKSIZE 1024 gives the smallest error for all levels of compression and for all tested tiles. So there is not need for ASTER DEM data to use another than default BLOCKSIZE value that is 1024. For other discussion we used only results with BLOCKSIZE = 1024.

Table 3 Impact of compression to tile N23E026

Filtered	Compression ratio	Block size (pixels)	Average error (m)
No	200:1	512	5.8413434788
Yes	200:1	512	5.3532582706
No	200:1	1024	5.4606829955
Yes	200:1	1024	4.8807861465
No	200:1	2048	5.4813481336
Yes	200:1	2048	4.885607773
No	100:1	512	4.0456516406
Yes	100:1	512	3.1500869
No	100:1	1024	3.7438103258
Yes	100:1	1024	2.795019141
No	100:1	2048	3.7461034189
Yes	100:1	2048	2.7379252469
No	20:1	512	0.7462615101
Yes	20:1	512	0.8957982528
No	20:1	1024	0.4930430245
Yes	20:1	1024	0.6542141207
No	20:1	2048	0.609507017
Yes	20:1	2048	0.7624123356
No	10:1	512	0.3767458374
Yes	10:1	512	0.8122391255
No	10:1	1024	0.0963257221
Yes	10:1	1024	0.5737475651
No	10:1	2048	0.2021722344
Yes	10:1	2048	0.6926780112

Table 4 Impact of compression to tile N33E081

Filtered	Compression ratio	Block size (pixels)	Average error (m)
No	200:1	512	8.7217949348
Yes	200:1	512	8.6787087668
No	200:1	1024	8.3779604403
Yes	200:1	1024	8.3703773081
No	200:1	2048	8.3392846305
Yes	200:1	2048	8.3425146259
No	100:1	512	6.7825648727
Yes	100:1	512	6.8121760432
No	100:1	1024	6.6063102592
Yes	100:1	1024	6.6317583108
No	100:1	2048	6.6271325631
Yes	100:1	2048	6.6606450382
No	20:1	512	3.732998432
Yes	20:1	512	2.9196588377
No	20:1	1024	3.5608164784
Yes	20:1	1024	2.7426009669
No	20:1	2048	3.6206660944
Yes	20:1	2048	2.8138112458
No	10:1	512	1.7225701985
Yes	10:1	512	2.6346513022
No	10:1	1024	1.5241541332
Yes	10:1	1024	2.4730518174
No	10:1	2048	1.6048564374
Yes	10:1	2048	2.5526437047

Tables 3, 4, 5, 6 and 7 show that for compression ratio 200:1 the filtered data give smaller error than original data. The difference between errors is between 0.007 and 0.579 m.

For the compression ratio 100:1 are the results similar (except the tile with highest elevation). The difference between errors is between −0.025 and 0.948 m. The result for the tile N33E081 is close to zero, but what is more interesting is that the result for tile N23E026 is close to 1 m. For the tile N23E026 is the average error for not filtered data about 4 m and for filtered data 3 m. That is improve in 25 % and it could be interesting for several applications. When considering the distribution of the error we can see in the Tables 8 and 9 that the distribution is not so different.

For the compression ratio 20:1 are the results (except the tile N23E026, where are the filtered data with higher error) even better when comparing average error. The difference between errors is between −0.161 and 1.065 m. For two tiles

Table 5 Impact of compression to tile N49E017

Filtered	Compression ratio	Block size (pixels)	Average error (m)
No	200:1	512	7.2025538896
Yes	200:1	512	7.1445872552
No	200:1	1024	6.9104690365
Yes	200:1	1024	6.8661799104
No	200:1	2048	6.9288282028
Yes	200:1	2048	6.8633231643
No	100:1	512	6.1518125616
Yes	100:1	512	5.836271914
No	100:1	1024	5.9580645816
Yes	100:1	1024	5.6325006453
No	100:1	2048	6.0004222962
Yes	100:1	2048	5.6750086622
No	20:1	512	3.2997735595
Yes	20:1	512	2.2283458859
No	20:1	1024	3.0743997876
Yes	20:1	1024	2.0134329683
No	20:1	2048	3.1570105993
Yes	20:1	2048	2.1086771154
No	10:1	512	1.1898974189
Yes	10:1	512	2.0523082044
No	10:1	1024	0.9489432608
Yes	10:1	1024	1.8444519369
No	10:1	2048	1.0514626094
Yes	10:1	2048	1.9479006302

(N49E017 and S24E125) is the difference between errors about 1 m and that is improve in 33 % and it could be interesting for several applications. But we have to consider the distribution of the error and from Table 11 is obvious that the number of pixels with error higher than 3 m is enormous in comparison to the results of not filtered data described in the Table 10. Also the maximum error is 33 m for filtered data.

For the compression ratio 10:1 are all the results better for not filtered data.

The test should be done for more tiles of ASTER data and for another DEM data as well. Even from presented results we can conclude, that when are DEM data (from ASTER source) compressed into JPEG 2000 format with GDAL tool and OpenJPEG library that the user should consider filtering the data. When the compression ratio is from 20:1 to 200:1 then the filtering can improve the average error

Table 6 Impact of compression to tile N51E021

Filtered	Compression ratio	Block size (pixels)	Average error (m)
No	200:1	512	6.535689699
Yes	200:1	512	6.337613414
No	200:1	1024	6.3057875019
Yes	200:1	1024	6.0749517957
No	200:1	2048	6.3383421758
Yes	200:1	2048	6.1166968107
No	100:1	512	5.6986560939
Yes	100:1	512	5.0205892544
No	100:1	1024	5.4986851056
Yes	100:1	1024	4.7879202304
No	100:1	2048	5.5468173124
Yes	100:1	2048	4.8213145613
No	20:1	512	2.4868854119
Yes	20:1	512	1.8504729741
No	20:1	1024	2.2364937507
Yes	20:1	1024	1.6295754959
No	20:1	2048	2.3128838675
Yes	20:1	2048	1.7339232268
No	10:1	512	0.9868178954
Yes	10:1	512	1.7198964526
No	10:1	1024	0.7394989867
Yes	10:1	1024	1.5018379834
No	10:1	2048	0.8564899241
Yes	10:1	2048	1.6074981794

or left the average error at the same size. The improvement of the error can reach 33 % of the average error. When the compression is 100:1 then the distributions of the error for not filtered and filtered data are almost equal.

The final conclusion is that the filtering should be used mainly for ratio 100:1 for areas with small variance in elevation.

As described in Ben-Moshe et al. (2007) ICTS technique gives similar or even better results for simplification of DEM than JPEG 2000. What is missing in the paper are parameters used for JPEG 2000 compression, mainly compression ratio. We were not able to test ICTS technique yet, so it would be very interesting to do a research in that area. We would like to recommend to compare our technique with ICTS technique for DEM compression (simplification).

Table 7 Impact of compression to tile S24E125

Filtered	Compression ratio	Block size (pixels)	Average error (m)
No	200:1	512	6.2306892598
Yes	200:1	512	6.0936592253
No	200:1	1024	6.0593004612
Yes	200:1	1024	5.909516942
No	200:1	2048	6.11625323
Yes	200:1	2048	5.9609665185
No	100:1	512	5.6848097751
Yes	100:1	512	5.109245781
No	100:1	1024	5.492256116
Yes	100:1	1024	4.8706002167
No	100:1	2048	5.552220483
Yes	100:1	2048	4.9080530949
No	20:1	512	3.1804059334
Yes	20:1	512	2.0791826239
No	20:1	1024	2.9353852848
Yes	20:1	1024	1.8694861752
No	20:1	2048	3.0062762966
Yes	20:1	2048	1.9709853345
No	10:1	512	1.1675331477
Yes	10:1	512	1.9444637281
No	10:1	1024	0.9293747355
Yes	10:1	1024	1.7352174151
No	10:1	2048	1.0397344809
Yes	10:1	2048	1.8395679222

Table 8 Error distribution for not filtered data for the tile N23E026 with compression ratio 100:1

Error (m)	Number of pixels
0	1,479,984
1–5	8,258,059
6–25	3,228,018
25–40	1127
>40	13

Table 9 Error distribution for filtered data for the tile N23E026 with compression ratio 100:1

Error (m)	Number of pixels
0	1,934,438
1–5	9,192,064
6–25	1,839,507
25–40	1142
>40	50

Table 10 Error distribution for not filtered data for the tile N49E017 with compression ratio 20:1

Error (m)	Number of pixels
0	7,067,717
1	5,423,233
2–3	4,75847
>3	404

Table 11 Error distribution for filtered data for the tile N49E017 with compression ratio 20:1

Error (m)	Number of pixels
0	6,119,138
1	5,601,369
2–3	1,169,941
>3	76,753

Acknowledgments The ASTER GDEM data were obtained through the online Data Pool at the NASA portal http://asterweb.jpl.nasa.gov/gdem-wist.asp.

References

Ben-Moshe B, Serruya L, Shamir A (2007) Image compression terrain simplification. In: Proceedings of the 19th Canadian conference on computational geometry (CCCG2007), pp 125–128. Available at. http://cccg.ca/proceedings/2007/05b1full.pdf

Gladkova I, Gottipati S, Grossberg M (2007) A new lossless compression algorithm for satellite earth science multi-spectral imagers. In: Proceedings of the SPIE, compression, communications, and archiving III, 6683 (668307). Available at. http://glasslab.engr.ccny.cuny.edu/u/grossberg/pubs/imageralgSPIE_07.pdf. Aug 2007

Gladkova I, Grossberg M (2006) A lossless compression algorithm for hyperspectral data. In: Proceedings of the SPIE, satellite data compression, communications, and archiving II, 6300 (630001). Aug 2006

Inanc M (2008) Compressing terrain elevation datasets. Thesis submitted to the graduate faculty of Rensselaer Polytechnic Institute. New York, 62 p. Available at. http://www.ecse.rpiscrews.us/Homepages/wrf/theses/inanc-phd.pdf

MicroImages, Inc. (2009) JPEG2000 Compression for global ASTER DEM. Available at. http://www.microimages.com/documentation/TechGuides/75jp2AsterDEM.pdf

Mittal ML, Singh VK, Krishnan R (2013) Effect of data compression on DEM accuracy derived from satellite images. Docstoc, 6 p. Available at. http://www.docstoc.com/docs/160718730/effect-of-data-compression-on-dem-accuracy-derived-from-satellite

Vatolin D et al (2005) JPEG 2000 image codecs comparison. CS MSU Graphics and Media Lab. Available at. http://compression.ru/video/codec_comparison/jpeg2000_codecs_comparison_en.html

Design of Test Base for Determine Volume Accuracy

Václav Šafář

Abstract This paper describes how the first test base for Remotely Piloted Aerial Systems (RPAS) was built. Aerial photos was done with non-metric cameras Canon 700D. Using the software Agisoft PhotoScan were created orthophotos and Digital Surface Model (DSM). Ortophotos and DSM were calculated in four variants. Computing variants are different in number of ground control points (GCP) entered. Rectangular plane coordinates of check points were read from the orthophoto maps in all variants. Heights of these points were read from DSM in all variants too. The article analyzes the accuracy of check points in all variants. These experiences are nested into the design parameters of the new test base for using in opencast mining. Proposal for a new test base must accept the conditions laid by the Czech Mining Authority for determination of the surface, area and volume, which will comply with upcoming amendment regulation No 435/92.

Keywords RPAS · Non-metric cameras · DSM

1 Introduction

RPAS are used at present in many fields of human activity. In comparison with the common manned aircraft, in some cases the RPAS is more suitable. The RPAS advantage is based on its significantly low price, low price of the sensors, high operability, versatility and adaptability. In addition, the RPAS pilot/operator training is much cheaper. The RPAS is irreplaceable in photographing small areas with Ground Sample Distance (GSD) of 1–2 cm. However if we consider the area larger than tens square kilometres, the standard manned aircraft with digital photogrammetric cameras is more efficient. Therefore it is obvious, that when

V. Šafář (✉)
The Research Institute of Geodesy, Topography and Cartography,
Ústecká 98, 250 66 Zdiby, Czech Republic
e-mail: vaclav.safar@vugtk.cz

K. Růžičková and T. Inspektor (eds.), *Surface Models for Geosciences*,
Lecture Notes in Geoinformation and Cartography,
DOI 10.1007/978-3-319-18407-4_18

comparing the characteristics and parameters of various means, all possibilities, advantages and disadvantages of the particular project must be considered.

RPAS with non-metric cameras are using for mapping production now. A couple of years ago, it was very difficult to determine distortions in whole area of photo. Mainly to determine the distortion of some commercial objectives—it was almost impossible due to the computer capacity. Currently, the distortion can be calculated for the whole originally non-metric image, basically pixel-by-pixel. And the picture can be transformed to a metric one. Distortion must be calculated, that the interpolation of coordinates of the picture points located between the considered points has no significant influence on the accuracy of interpolated image coordinates of these points.

Agisoft PhotoScan have solutions for determining distortions and all other internal orientation parameters of a non-metric camera, the determination of external orientation parameters of non-metric images as compared in Bartoš et al. (2014), Marčiš (2013).

2 Building the First Test Calibration Base for RPAS

To find out the achievable 3D positioning accuracy of the RPAS systems and non-metric cameras, the Research Institute of Geodesy, Topography and Cartography (VUGTK) built (by courtesy of EUROVIA CS company) a test base with 101 control points on a non-used section of the D11 highway near town of Hradec Králové in September 2013. The points were marked with white colour directly on the road asphalt surface. The marks were in shape of a 25 cm circle with two central sectors (see Fig. 1).

The highway section is oriented within the x-axis of the national coordination system, i.e. the x-coordinates change but the y-ones remain the same. The base is located in two lanes of the half of the highway section in direction Hradec Králové—Praha. Marked are the points of the outside edge of the outside lane and of the inside edge of the inside service lane with step of 10 m on 100 m section. Furthermore the points with step of 20 m on 200 m section and the points with step of 50 m on 650 m section are marked. In addition, all central points of such created images (squares and rectangles) are marked. For the point layout see Fig. 2. All other figures of the calibration base are left-turned by 90° to better comply with the A4 format of this paper.

Fig. 1 Sample of ground control point

Fig. 2 Layout of ground control points

The points were surveyed twice. For the first by a terrestrial method—trigonometric coordinate determination with Leica TCA 2003 total station with respect to the national points of fundamental geodetic control. For the second by the independent GNSS technology with geometric levelling for the height determination. The national S-JTSK datum was used, regarding the GNSS technology—the S-JTSK (Křovák 2013, CR 2005 quasigeoid model) datum. The Baltic after Adjustment (Bpv) datum was used for the height determination. GNSS survey was performed twice, independently and with sufficient time interval. To determine the heights of all marked points, the very precise levelling between the 3090 point (elevation 235.066 m) and KH-055-9 point (elevation 230.875 m) was performed. The particular control points were determined by the precise levelling with respect to the points determined by the very precise levelling.

3 RPAS Test Flight

The calibration base was photographed in February 2013. The test sensing flight was intended for determination of the orthophoto positional accuracy and vertical accuracy of the digital surface model. Non-metric Canon 700D camera, size of the

Fig. 3 UAV Hexacopter G6, by courtesy of UpVision, Ltd

sensing chip CMOS 5184 × 3456 pixels (22.3 × 14.9 mm), pixel size on the chip 4.384 × 4.384 μm, with objective Canon EF 24 mm f/2.8 IS USM. The 6-rotor helicopter called UAV Hexacopter G6 was used as the carrying vehicle—see Fig. 3.

The test area was photographed with nominal GSD of 1 cm from average height above terrain of 68 m. The area was sensed in three 5-line blocks with mutual block overlap—see Fig. 4. The entire area is north-south oriented. Total of 1235 pictures was obtained with 80 % longitudinal overlap and 60 % transverse overlap. The camera was fixed with its shorter side in the flight direction. The images covered the area of ca 20 ha. Arrangement of flight stripes in the blocks in the relation to the distribution of GCP such that no rotation of the blocks. WGS84 coordinates of the image projection centres were recorded. The centres were later transformed into JTSK datum. The parameter file is one of the inputs for the photogrammetric processing with the Agisoft PhotoScan 1.0.0 build 1480 software (Agisoft Company Documents 2013). For need of the photogrammetric calculation, a total of 34 initial points were manually measured on the images. The average number of radius vectors for one initial point is 18.

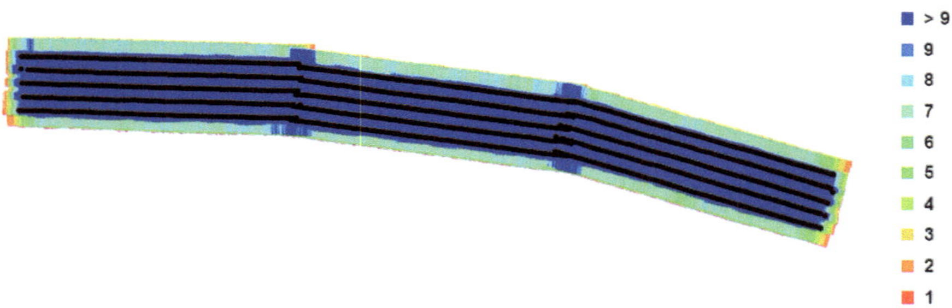

Fig. 4 Configuration of photo blocks and scale with the number beams of tie points

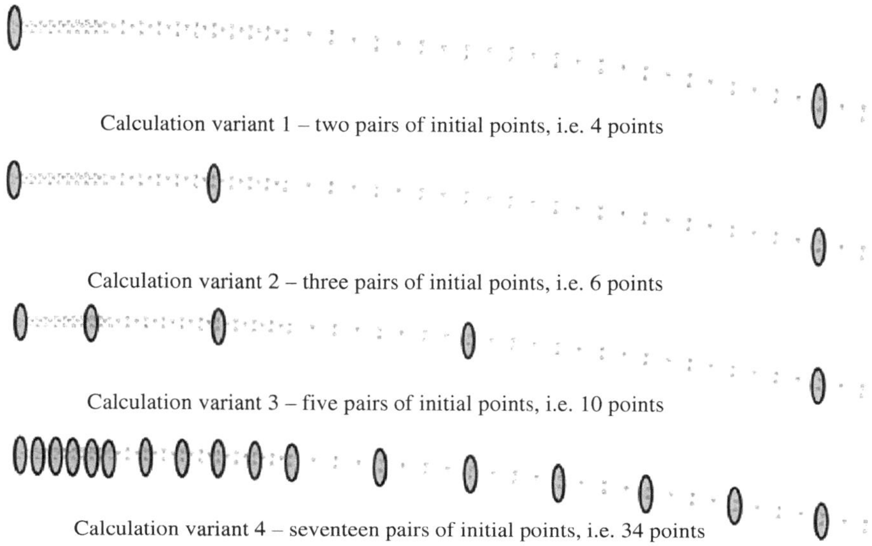

Fig. 5 Distribution of the ground control points in calculation variants

Gradually, four calculations with different number of used initial points were designed and performed. The first variant of the calculation was performed with four known initial points—two at the beginning and two at the end of the sensed area. The initial point layout of the second variant was the same as at the first one, however with two additional points inside the area. The third variant consisted of the second one with additional four points (altogether 10 points). The last variant included all 34 measured points. These points were used for calculation of residual errors on the control points in all calculation variants as the points without mean error to analyze the positional accuracy of created orthophoto and generated elevation model. The Agisoft PhotoScan 1.0.0 build 1480 was used (Agisoft Company Documents 2013). The control point coordinates were determined by the operator. He inserted on the centre of marked point in the orthophotomap a cell on the.dgn drawing with the point number, using Microstation V8i software. From the .dgn file, the numbered cells were subsequently exported to a .txt file and processed in Excel Microsoft. The point heights were determined from generated spatial model using the RoadMap software at the locations of points with zero elevation derived during identification of the marked points (Fig. 5).

Regarding the distribution of the initial points, the coordinate differences on the control points calculated in the variants 2, 3 and 4 are only irrelevantly dependent on the number of photogrammetric bases between the initial points entering the photogrammetric calculations (Table 1).

Table 1 RMSE of calculation variant

Variant description (Values of mean errors in (m) on the check points)		RMSExy		MAX		MIN		RMSEz	MAX	MIN
		Y	X	Y	X	Y	X	Z	Z	Z
Variant 1	Calculation—4 GCP input points, 96 check points	1.014	0.080	1.664	0.107	−0.361	−0.173	2.148	3.753	−0.906
Variant 2	Calculation—6 GCP input points, 94 check points	0.041	0.041	0.113	−0.020	−0.020	−0.107	0.038	0.120	−0.113
Variant 3	Calculation—10 GCP input points, 90 check points	0.015	0.015	0.040	0.036	−0.035	−0.042	0.039	0.143	−0.094
Variant 4	Calculation—33 GCP input points, 67 check points	0.012	0.011	0.032	0.025	−0.014	−0.026	0.035	0.115	−0.093

4 Proposals for Test Base Layout for Opencast Mining

In the Czech Republic, currently many mining companies consider using the RPAS devices and non-metric cameras for determination of drawn rock volume, e.g. Koliáš et al. (2014). Requirements for accuracy of the mapping and determination of the drawn rock volume can be satisfied. Systems are easy available on the market and many mapping applications and software for data processing have been developed.

On the basis of the experience with establishing the first RPAS base, determination of the relation between GSD and RMSE values in 3D position and also on the basis of experience with creating the methodology for determination of residual errors on initial and control points, it seems suitable to establish and permanently maintain a test and calibration base in area of Krušné hory foothills.

The ideal area could be between towns of Kadaň and Chomutov, particularly the area of the homogenizing disposal site of the Tušimice power plant (Figs. 6 and 7).

Fig. 6 Zone of brown coal opencast mining in Krušné hory foothills

Fig. 7 Area of homogenizing disposal site of Tušimice power plant

The RPAS calibration base should be at least of 600 × 200 × 20 m (length, height, wide) and consist of at least 48 signalized GCP spread with a regular horizontal and vertical distance. 14 of them will be entering points and 34 of them as check points, i.e. points, which will not participate in the photogrammetric processing. Before starting the works, a check sensing of the calibration base should be performed with given camera and given RPAS to assure the metrological link-up and subsequent determination of the final volume or volume decrease (mining wall) or increase (dump, pile, etc.). The base must be accessible via a public road and its description, parameters, pictures, and coordinates of initial and control points must be presented on the internet. The basic parameters for sensing the test base should be as follows. GSD 1–3.5 cm. Longitudinal overlap at least 75 %. Transverse overlap at least 50 %. Initial 14 points surveyed manually before the photogrammetric processing. Manual reading of horizontal coordinates from the orthophotomap and vertical coordinates from the digital terrain model. Final RMSE less than 1.5 × GSD in horizontal and 2.5 × GSD in vertical coordinates. If the above mentioned limits are not reached, the calculation should not be used for the volume determination. If the parameters are met, it is guaranteed, that the volume determination is with accuracy of 1 % and the error is random. These values will be further tested on known shapes (both natural and man-made) during 2015.

5 Conclusions and Recommendations

The first test base built by VUGTK is the first step of testing RPAS method for utilization in the surface mining. The tests of horizontal and vertical accuracy of photogrammetric methods, which use non-metric cameras and DPLS, represent a significant part of modern photogrammetry in the Czech Republic and are adequate alternative to the aerial sensing utilizing the manned aircraft and to both terrestrial and aerial laser scanning intended for mapping and determination of material volume in the surface mining with high accuracy. One of the main conclusions of the test described in this paper is determination of the number of initial points for number of photogrammetric bases. The test proved that to reach the horizontal and vertical accuracy of particular map points for volume determination with 1 % accuracy it is sufficient to establish one initial point for 100–120 image bases.

However this ratio can significantly change in case of using RTK methods on the RPAS board, if we know the input parameters of the centre coordinates of images made by non-metric cameras with XYZ accuracy of about 10 cm (in future also the parameters of camera rotation angles at exposition instant gained from an IMU device). In such case, only 6 initial points per the locality are sufficient. Precise parameters of the external orientation of images in the block [or at least their positions as in case of Topcon products—RPAS MaVinci (Topcon Company Documents 2014)] significantly decrease the block calculation time. Moreover, it is necessary to set the limit values for the base ration, i.e. the relation of distance between two neighbouring images to distance of the sensed object (flight elevation above the terrain), further the limit condition for the object observation angle in the image for correlation calculation, and the limit conditions and relations between the focal distance, size of the sensing sensor and size of the sensing element on the sensor. Further the relation of the limit (and needed) detail for the volume determination. It will be necessary to specify the relations between the texture type and size of used GSD with goal to eliminate, or to specify, locations unsuitable for creation of 3D model, where the insufficient texture of the object surface on the images does not enable the correlation calculation.

References

Agisoft Company Documents (2013) http://www.agisoft.ru/

Bartoš K, Pukanská K, Sabová J, Weiss G (2014) The application of open-source photogrammetric software for the purposes of cultural heritage documentation. GeoScience Engineering, VŠB-TU Ostrava. ISSN 1802-5420

Koliáš J, Blín J, Blín J (2014) The future of aerial mapping with autonomous drones. International conference of geodesy and mining survey 2014, XXI conference of society of mining surveyors and geodesists, Tábor, 13–15 Oct. ISBN 978-80-248-3599-0

Marčiš M (2013) Quality of 3D models generated by SFM technology. Slovak J Civil Eng 21(4):13–24. ISSN 1210-3896

Topcon Company Documents (2014) http://topcon-positioning.eu

New Automatic Accuracy Evaluation of Altimetry Data: DTM 5G Compared with ZABAGED® Altimetry

Jakub Šilhavý and Václav Čada

Abstract The paper tackles the issue of accuracy of digital terrain models, which is a long term research area at the University of West Bohemia in Pilsen. In 2014, a new generation of digital terrain models were completed in the Czech Republic. This fact initiated an impulse for new measures and tests in this field. The accuracy tests have been realized by comparing the Digital Terrain Models (DTM) of the Czech Republic—the 5th Generation (DTM 5G) and the previous generation of altimetry data from ZABAGED®. The test have been performed in two selected test fields in the Pilsen Region. This comparison was done already in previous research, but it was built on pre-released laser scanning data before the final filtration and gross error clearing. In order to confirm the results from the previous testing, the comparison has been repeated using the final dataset. The robust method of DTM accuracy evaluation developed at the University of West Bohemia in Pilsen and a highly automated implementation of the comparative method on 250 local test sites (total area of 85 km^2) have been applied in this project. The algorithm's implementation is based on ArcPy libraries provided by Esri. The key parts of the Python source code are described in detail and the whole algorithm is published as open-source. The tests proved that the pre-release dataset was good enough for accuracy testing. The results are similar with the previous research. It underlines the value of the new generation of altimetry data in the Czech Republic. Their accuracy is significantly higher than the previous generation in ZABAGED®.

Keywords Automated processing · ArcGIS tools · Gross error visualization · Pilsen region · Altimetry

J. Šilhavý (✉) · V. Čada
Geomatics Section of Department of Mathematics, Faculty of Applied Sciences, University of West Bohemia in Pilsen, Univerzitní 22, 306 14 Pilsen, Czech Republic
e-mail: jsilhavy@kma.zcu.cz

V. Čada
e-mail: cada@kma.zcu.cz

K. Růžičková and T. Inspektor (eds.), *Surface Models for Geosciences*, Lecture Notes in Geoinformation and Cartography,
DOI 10.1007/978-3-319-18407-4_19

1 Introduction

This work builds on the knowledge of previous works made at the University of West Bohemia in Pilsen (UWB). The quality and accuracy of state maps' DTMs were tested by Vyčichlová and Čada (2001), Šíma and Egrmajerová (2004). The original methodology for a robust DTM checking was presented by Fiala and Šíma (2006) and applied by Čada and Šilhavý (2013) when the product ZABAGED® altimetry—3D contour lines (furthermore referred to as the ZABAGED® altimetry) was tested.

The statistical methods are commonly used to asses DTM accuracy when the differences in z-value are calculated. Höhle and Höhle (2009) compares DTMs from airborne laser scanning (ALS) and digital photogrammetry with limited number of check points from ground surveying using GPS/RTK. Müller et al. (2014), Fiala and Šíma (2006) compare tested DTM against referenced DTM which leads to more values used in statistic computations. The mentioned authors do not automate the assessment process. Otherwise, this research comes with automatic tool for DTM accuracy evaluation based on methodology from Fiala and Šíma (2006).

Two generations of qualitatively different DTMs covering the Czech Republic were analysed. The previous generation is represented by the ZABAGED® altimetry and the new generation by the Digital Terrain Model (DTM) of the Czech Republic—5th Generation (furthermore referred to as DTM 5G).

This paper is bound to previous research by Čada and Šilhavý (2013), who used the pre-release ALS data after automatic classification of laser beam before the final filtration and gross error clearing (furthermore referred to as DTM from ALS). Nowadays, the new generation DTM 5G is available on the market.

The goals of this paper are to confirm the previous results achieved by Čada and Šilhavý (2013), make comparison on larger area and present the algorithm which is able to automatically compare two DTMs without subjective area delimitation according to terrain type and land cover.

2 Digital Terrain Models' Characteristics

The ZABAGED® altimetry was created during the years 1994–2000 by digitizing the analogue Base Map of the Czech Republic (Základní mapa ČR) 1:10,000. The altimetry has a form of 3D contour lines and covers the whole area of the Czech Republic. The basic 2-m contour interval is complemented with 1-m contours or a net of points in flat areas. In mountainous areas, the contour interval is 5 m. The altimetry accuracy is characterized by the root mean square error (mH) which depends on the terrain slope (mH = 1 m where the slope is 0°, mH = 1.5 m where the slope is up to 10° and mH = 2.5 m where the slope is up to 25°).

Between the years 2006 and 2008, the ZABAGED® altimetry was improved by using a stereo-photogrammetric processing focused on terrain edges (bank lines, earthworks, diggings, terrain steps etc.). The new product was named ZABAGED®—improved altimetry.

For the purpose of the models' comparison, the optimized DTM ZABAGED® was created from ZABAGED® improved altimetry using the ArcGIS Desktop (3D Analyst) environment. The optimization consists of altimetry elimination under specific objects (building blocks, lakes, rocky areas).

3 Digital Terrain Model of the Czech Republic—5th Generation (DTM 5G)

DTM 5G originated from airborne laser scanning (ALS) which was completed in 3 periods (the Central zone of the Czech Republic between March and October 2010, the Western zone between March and June 2011 and the Eastern zone until November 2013).

The LiteMapper 6800 system (IGI GmbH) was used for ALS with the RIEGL LMS-Q680 airborne laser scanner which was equipped by Global Positioning System (GPS) and Inertial Measurement Unit (IMU). The average flight heights of the ALS were 1200 and 1400 m above the terrain depending on the height of the vegetation. The scanning blocks were 10 km wide and maximally 60 km long.

The ALS data were georeferenced into the UTM/WGS 84-G873 working coordinate system and the Baltic height reference system for further processing. The automatic land cover and surface objects' data separation was made using the robust filtering method. The success rate of the automatic data separation was estimated to 30–40 % in the time of growing vegetation season and 90 % in the time outside the growing vegetation season (Brázdil 2012).

The errors caused by the automatic data separation were interactively eliminated during the next phase by manual check of the robust filtering method. This phase resulted to the ALS point cloud represented the terrain, vegetation and buildings. This dataset was used in the previous research (Čada and Šilhavý 2013) and is referred to it as DTM from ALS.

Several steps were applied to achieve the final product DTM 5G. An important step was the elimination of low vegetation in the areas covered by agriculture cultivated land (the areas of cultivated land were defined by the state agricultural land registry (LPIS database) or manually digitized by means of hillshade relief) or areas with temporary terrain roughness or areas without terrain breaks (ditches, terrain edges). In those areas, the points with lowest altitude were found in 5×5 m grid covering the area. Such identified points were marked as terrain representatives. In other areas the workflow was identical using 1×1 m grid.

In areas larger than 25 m^2 where no points were identified (water areas, areas under buildings, partly in dense vegetation), the terrain interpolation was made using 5 × 5 m grid.

The final step of DTM 5G creation consists of terrain representatives selection to achieve the same density of points over the area and smoothing the result surface when the maximal smoothing distance was 0.16 m. The height of every terrain representative point was corrected by the influence of the systematic error. This error was estimated to 0.0034 m based on measuring on comparative baselines (soccer fields, tennis courts, squares) in the Central zone. The accuracy of points on comparative baselines was characterized by root mean square coordinate and height error mXY = mH = 0.03 m when using the RTK method. The final point cloud in the UTM/WGS 84-G873 coordinate system was transformed to the DTM 5G product in the national S-JTSK coordinate system using local transformation formulae of VÚGTK.

The DTM 5G product is distributed in form of triangular irregular network (TIN). According to Brázdil (2012), the supposed RMSE of heights is 0.18 m in the terrain without continuous vegetation and buildings and 0.30 m in the terrain with dense vegetation.

4 Methodology

The Czech method of DTM checking presented by Fiala and Šíma (2006) was applied in the same manner as by Čada and Šilhavý (2013) in order to preserve the same research conditions.

The method is based on surface differences where the quality of models is determined by difference volume (V) related to difference area (S). The mean height error (mH) is estimated by empiric equation:

$$\mathrm{m_H^2} = \sigma_\mathrm{H}^2 + \mathrm{s_H^2} \tag{1}$$

$$\mathrm{m_H^2} = 1.25 * \mathrm{a_H^2} \tag{2}$$

$$a_H = \frac{\sum |V|}{S}\,(\mathrm{m}) \tag{3}$$

$$s_H = \frac{\sum V}{S}\,(\mathrm{m}) \tag{4}$$

More detailed description of this method can be found in Fiala and Šíma (2006), Čada and Šilhavý (2013).

5 Implementation

Fiala and Šíma (2006) implemented the robust method using the Atlas DTM software. The scripting was not used thus the manual workflow was required. In Čada and Šilhavý (2013), the semi-automated scripts using the ArcGIS software were presented. The scripts are extended to use a full automatic workflow (see Fig. 1).

By user defined processing area is automatically divided into 1 km^2 polygons. The program computes the comparison for each input polygon. It is also possible to provide a path to a directory with a set of shapefiles which includes customized areas to be processed. The statistics are computed for every shapefile separately. It allows defining classes for the computation. This capability can be used for example to separate areas with different slope or vegetation. The area of 1 km^2 is recommended setting for stable computing. More computing memory is required when this value would be increased.

The steps of automatic division are depicted in Fig. 2. The input processing extent is overlaid by an automatically generated grid of polygons, which are intersected by the input extent. It is hard to interpolate TIN analysis for small and narrow polygons, thus polygons smaller than 0.3 km^2 are deleted.

The tested and referenced DTMs in form of TIN representation are required as input to the surface comparison. The TIN representation is interpolated from source data using CreateTIN tool from the ArcGIS Toolbox. In large areas the interpolation of TIN could be demanding on memory and CPU. Therefore the TIN is created from source data for each polygon separately.

The mapping of source data to parameters for the TIN construction is made in the user configuration. The input data path, Z coordinate location and surface feature type are specified in parameters according to the CreateTIN tool. For example:

param_TIN(testedDTMDir + "ndexContours.shp", "Shape.Z", "softline")
param_TIN(referencedDTMDir + "kozel.gdb\\points_XYZ", "Shape.Z", "masspoints")

In this way, the input data can be either ALS point cloud or 3D features such as contours, spot heights and terrain edges.

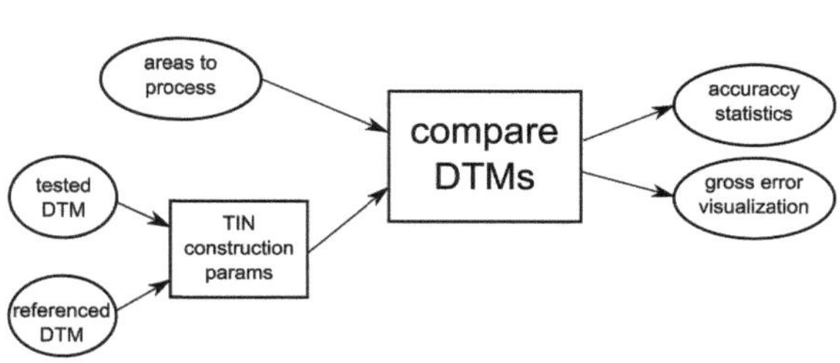

Fig. 1 Schema of algorithm to automatic altimetry accuracy evaluation

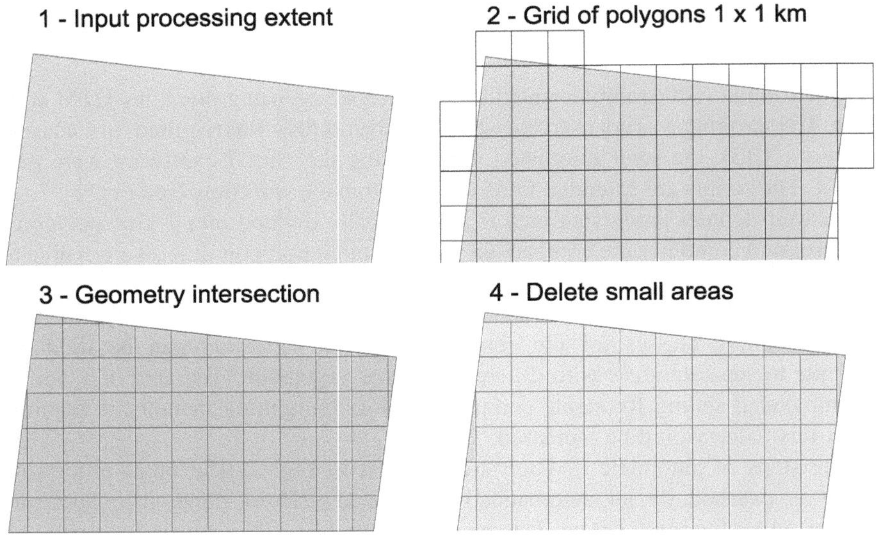

Fig. 2 Automatical creation of process areas

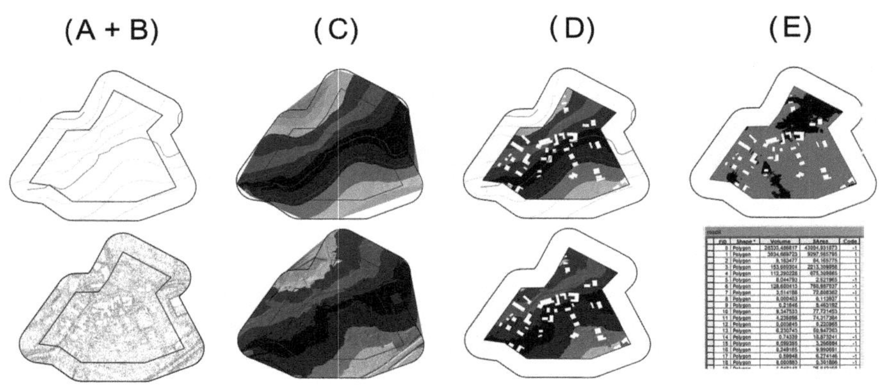

Fig. 3 Algorithm's workflow

The workflow of the algorithm has five key steps (Fig. 3).

A. Buffer creation around the input polygon
B. Clipping the source data using the buffer
C. TIN creation from clipped source data
D. Clipping TIN to the original extent and erasing areas excluded from the comparison
E. Computing the surface differences to obtain the statistics

In border areas of the input polygon the quality of TIN interpolation can be affected by missing data due to irregular shape. For this purpose the TIN is calculated inside the buffer zone around the input polygon. The size of the buffer zone is defined by the user as a parameter. It is recommended to estimate it from the characteristics of the source data (shape of the input polygons, the density of source data). In our case, the parameter was chosen 50 m due to average distance between two contour lines in the tested dataset.

After the TIN creation it is cropped to the original size of the input polygon. Moreover, the areas of buildings, lakes and rocks were erased from the TIN. The selection of areas to be excluded was made according to the genesis of DTM ZABAGED®. The Edit TIN tool was used with the parameter hard clip set to input polygon and parameters hard erase set to layers representing areas to be excluded.

In final step the geometric comparison between the tested and reference surfaces is made in order to get the volumetric differences. The Surface Difference tool from the ArcGIS Toolbox has two important outputs. The output raster contains height differences between surfaces interpolated for each cell. The output polygons contains all necessary information to compute accuracy statistics.

6 Results

The three datasets used in this research are shown in Fig. 4—(A) Plzeň, (B) Kozelsko, (C) Brdy. The small polygons depicted inside datasets A and B represent the original tested areas used in previous research Čada and Šilhavý (2013).

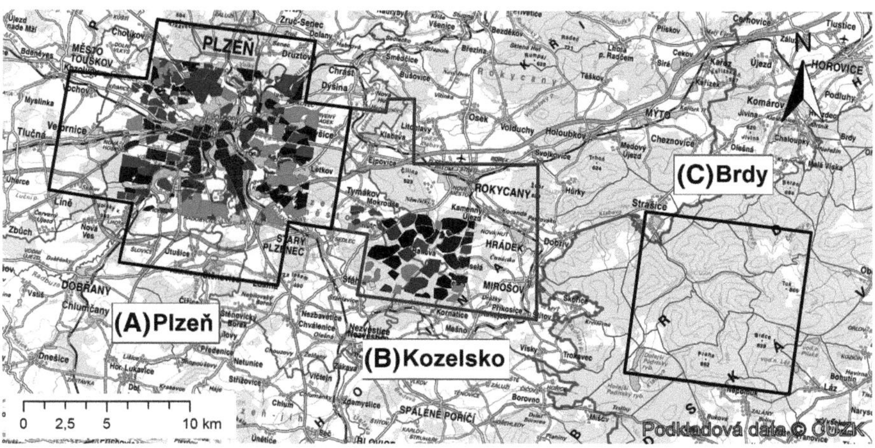

Fig. 4 Overview of tested areas

The results of three computations are presented in the form of tables and verbal description. First of all the results from the previous research by Čada and Šilhavý (2013) are summarized. After that the confirmation of these results using the new dataset is shown. Lastly, the modified technology is applied to all tested areas (Fig. 4).

7 Previous Research Summary

At the time of the research performed by Čada and Šilhavý (2013) the DTM 5G was not available in the final version for tested areas in the Pilsen Region. Thanks to the collaboration of the UWB with the Land Survey Office the DMR from ALS were obtained for testing.

The quality testing of the DTM ZABAGED® altimetry was made in two areas including Plzeň and Kozelsko within the total area of 85 km^2 (small polygons in Fig. 4). The results showed the quality problems in expressing morphological shapes of the relief and the occurrence of extensive areas with height errors exceeding three times the standard deviation. The maximum deviations reached value up to 30 m (see Fig. 5 in Čada and Šilhavý (2013) and example on the server http://home.zcu.cz/~jsilhavy/hrubeChybyDEM.html).

The final accuracy of DTM ZABAGED® was characterized by RMSE of heights mH = 0.86 m (see Table 1). This parameter was exceeded about 50 % in forested areas. The worst accuracy was recorded in such areas with extreme slope (more than 15°) where the height error reached about 130 % [see Tables 5 and 6 in Čada and Šilhavý (2013)].

The systematic error was −0.23 m. This means that DTM ZABAGED® is systematically up the reference model DTM from ALS.

8 Influence of Difference Between DTM from ALS and DTM 5G

The discussion about results of the research of Čada and Šilhavý (2013) pointed out that they might be affected by the fact, that the DTM from ALS was used instead of the final DTM 5G product. Thanks to the Land Survey Office, the product DTM 5G was obtained in order to repeat the analysis of the previous research with this new data. The results' summary from Čada and Šilhavý (2013) is shown in Table 1 whereas full results from the new computation in the same areas are shown in Table 2 and Table 3.

The new analysis confirms the original conclusion made by Čada and Šilhavý (2013) that the local gross errors in results are not caused by using DTM from ALS. Therefore this dataset can be used for the accuracy evaluation of DTM ZABAGED®.

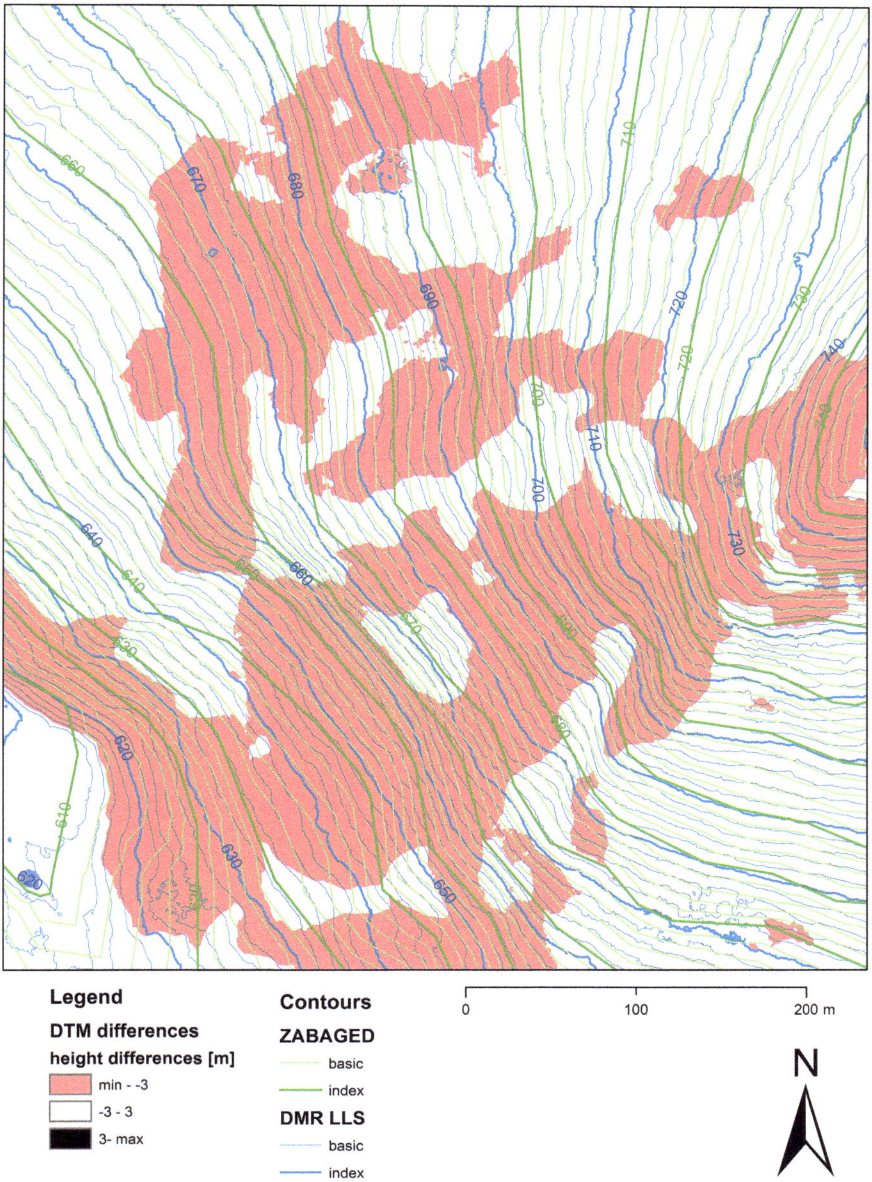

Fig. 5 Gross error occurence in the area Brdy

Table 1 The summary of results from research work of Čada and Šilhavý (2013)

Area	aH (m)	sH (m)	σH (m)	mH (m)	∑|V| (m^3)	∑V (m^3)	S (km^2)
Plzeň DTM ALS	0.76	−0.16	0.94	0.95	16,828,420	−3,474,831	22.1
Kozel DTM ALS	0.66	−0.25	0.78	0.82	38,637,139	−14,839,135	58.8
Total	0.69	−0.23	0.83	0.86	55,465,559	−18,313,966	81.0

Table 2 The summary of altimetry accuracy in the area Kozelsko

Type of relief	Land cover	aH (m)	sH (m)	σH (m)	mH (m)	∑|V| (m^3)	∑V (m^3)	S (km^2)
Planar	Forest	0.43	−0.17	0.51	0.53	2,039,534	−794,902	4.8
	Field	0.41	0.01	0.52	0.52	629,675	7882	1.5
	Urban area	0.52	−0.20	0.62	0.65	138,032	−52,194	0.3
Slope (up to 15°)	Forest	0.97	−0.26	1.18	1.21	7,628,252	−2,039,891	7.9
	Field	0.52	−0.11	0.64	0.65	1,865,534	−398,590	3.6
	Urban area	0.72	−0.15	0.89	0.90	318,501	−66,753	0.4
Slope (more than 15°)	Forest	1.23	−0.18	1.52	1.53	3,833,945	−567,178	3.1
	Field	0.90	−0.21	1.11	1.13	216,443	−51,135	0.2
	Urban area	0.76	−0.16	0.94	0.95	115,358	−23,923	0.2
Total Kozelsko		0.76	−0.18	0.94	0.95	16,785,275	−3,986,685	22.0

Table 3 The summary of altimetry accuracy in the area Pilsen

Type of relief	Land cover	aH [m]	sH [m]	σH [m]	mH [m]	∑|V| [m^3]	∑V [m^3]	S [km^2]
Planar	Forest	0.61	−0.36	0.67	0.76	2,767,086	−1,641,262	4.5
	Field	0.34	−0.06	0.43	0.43	1,716,521	−289,563	5.0
	Urban area	0.50	−0.23	0.57	0.62	4,083,105	−1,931,586	8.2
Slope (up to 15°)	Forest	0.75	−0.41	0.85	0.94	5,284,510	−2,861,804	7.0
	Field	0.46	−0.24	0.52	0.57	6,667,702	−3,521,908	14.5
	Urban area	0.65	−0.40	0.71	0.82	4,775,731	−2,926,711	7.3
Slope (more than 15°)	Forest	1.61	−0.18	2.01	2.01	7,739,914	−862,065	4.8
	Field	0.62	−0.28	0.72	0.78	1,992,980	−902,975	3.2
	Urban area	0.96	−0.31	1.16	1.21	3,724,348	−1,209,864	3.9
Total Plzeň		0.66	−0.28	0.78	0.83	38,751,898	−16,147,737	58.4
Total both areas		0.69	−0.25	0.83	0.86	55,537,173	−20,134,422	80.4

Table 4 The summary of altimetry accuracy in the areas Pilsen, Kozelsko and Brdy

Area	aH (m)	sH (m)	σH (m)	mH (m)	$\Sigma\|V\|$ (m^3)	ΣV (m^3)	S (km^2)
Pilsen (DTM 5G)	0.61	−0.18	0.74	0.76	114,809,262	−32,913,682	187.8
Kozelsko (DTM 5G)	0.66	−0.14	0.81	0.82	72,989,259	−16,008,122	111.3
Brdy (DTM ALS)	1.12	0.05	1.40	1.40	111,279,613	4,584,583	99.5
Total	0.75	−0.11	0.93	0.94	299,078,133	−44,337,221	398.5

9 Evaluating of ZABAGED® Altimetry Using Modified Technology

The subjective delineation and classification of areas according to the type of relief and land cover was time consuming in the current method. Moreover, not the whole area is taken into the comparison due to spaces between areas. This reasons led to modify the algorithm to automatically creating of tested areas covering the whole study area. The results using the modified algorithm are summarized in Table 4 for the areas Kozelsko and Plzeň (see Fig. 4a, b).

In both areas 340 automatically generated polygons with the total area of 318 km^2 were tested. The RMSE of heights was found 0.76 m for the Pilsen Region and 0.82 for the Kozelsko region. Better results compared with the previous 0.86 m (see Table 1) might be influenced by larger area with more flat terrain and agriculture land cover.

The modified algorithm did not allow the automatic classification according to the relief type and land cover but it only served for the height accuracy comparison. The automatic classification of the tested areas could extend the algorithm in further research.

10 Extending Test Areas to Brdy

It was easy to extend the comparison to the new area using fully automatic method. The data from the research of Vichrová and Čada (2012) in the area of Brdy (see Fig. 4c) was used to compare 107 new areas in total area of 100 km^2.

RMSE of heights 1.4 m (see Table 4) was influenced by the terrain type where the forested areas with steep terrain are dominated. The accuracy of such a type of terrain was found from 1.21 to 1.53 m in similar terrain in Kozelsko (see Table 2). The gross errors occurence is depicted in Fig. 5.

11 Conclusion

The analyses clearly demonstrated the need to test the ZABAGED® altimetry before using it in GIS analyses when the hight accuracy better than 1 m is required.

The previous study of Čada and Šilhavý (2013) aimed at getting the results dependent on the relief type and land cover. New results using the DTM 5G confirmed that using the DTM from ALS has no negative impact to accuracy of altimetry comparison when the robust method of DTM checking is used.

DTM 5G is an important source for improving accuracy of ZABAGED®. The altimetry is not only an object to be improved. The planimetric component of this database will be also influenced by changes in altimetry (rivers, roads, railroads, walls etc.).

The evaluation of height accuracy of ZABAGED® is very important for user applications which still use DTM ZABAGED® as an input for analyses (flood modelling, maps as source for insurance contracts, protecting agricultural land from erosion where the land consolidation is designed, creating analytical documentation for local and regional space planning).

The presented tools are able to evaluate the ZABAGED® altimetry and effectively visualize areas with the gross errors. The algorithms are available on the website http://home.zcu.cz/~jsilhavy/DTM.

References

Brázdil K (2012) Technical report DTM 5G. Zeměměřický úřad a Vojenský geografický a hydrometerologický úřad. http://geoportal.cuzk.cz/Dokumenty/TECHNICKA_ZPRAVA_DMR_5G.pdf. Accessed 13 Dec 2014

Čada V, Šilhavý J (2013) Comparison of accuracy of the ZABAGED® altimetry—3D contour lines product with data gained by aerial laser scanning of the entire territory of the Czech Republic. Geodetický a kartografický obzor 59/101(8):173–181. ISSN 0016-7096

Fiala R, Šíma J (2006) The Czech method of DTM checking. EuroSDR Official Publication, vol 51, pp 87–94

Höhle J, Höhle M (2009) Accuracy assessment of digital elevation models by means of robust statistical methods. ISPRS J Photogram Remote Sens 64(4):398–406. http://dx.doi.org/10.1016/j.isprsjprs.2009.02.003. ISSN 0924-2716

Müller J, Gärtner-Roer I, Thee P, Ginzler C (2014) Accuracy assessment of airborne photogrammetrically derived high-resolution digital elevation models in a high mountain environment. ISPRS J Photogram Remote Sens 98:58–69. http://dx.doi.org/10.1016/j.isprsjprs.2014.09.015. ISSN 0924-2716

Šíma J, Egrmajerová L (2004) Accuracy verification of DTM ZABAGED®. Geodetický a kartografický obzor 50/92(11):213–231. ISSN 0016-7096

Vichrová M, Čada V (2012) Digital terrain model of the second military survey—part of the military training area Brdy. Geoinformatics 2011(6):346–353. ISSN 1802–2669

Vyčichlová V, Čada V (2001) The quality assessment of state map 1:5000. Kartografické listy 9:79–92. ISSN 1336-5274

Hydraulic Erosion Modeling on a Triangular Mesh

Věra Skorkovská, Ivana Kolingerová and Bedrich Benes

Abstract Although hydraulic erosion modeling on a GIS terrain models has been addressed by a body of previous work, it still remains an open problem. In GIS, raster representation and triangular irregular networks (TIN) are the most commonly used surface models, because they are simple and offer implicit topological information. However, these data structures do not allow the simulation of erosion on concave terrain features, such as caves or overhangs. Other methods, more commonly used in the computational fluid dynamics, use volumetric data representation. They are able to model the 3D features, but they usually have high memory requirements and are computationally demanding. We propose a novel solution to the hydraulic erosion modeling problem that uses a triangular mesh data structure. Our framework allows for adaptive changes of the mesh resolution according to the local complexity of the terrain, which leads to lower memory requirements when compared to the volumetric approaches. Our data structure also supports the visualization of the concave 3D features, allowing the simulation and visualization of erosion on terrain elements such as tunnels or caves.

Keywords Hydraulic erosion · Erosion · Terrain modeling · Smoothed particle hydrodynamics · Triangular mesh

V. Skorkovská (✉) · I. Kolingerová
NTIS—New Technologies for Information Society, University of West Bohemia, Univerzitní 8, 306 14 Plzeň, Czech Republic
e-mail: vskorkov@ntis.zcu.cz

I. Kolingerová
e-mail: kolinger@kiv.zcu.cz

V. Skorkovská · I. Kolingerová
Department of Computer Science and Engineering, Faculty of Applied Sciences, University of West Bohemia, Univerzitní 8, 306 14 Plzeň, Czech Republic

B. Benes
Department of Computer Graphics Technology, Knoy Hall of Technology, 401 N. Grant Street, West Lafayette, IN 47907-2021, USA
e-mail: bbenes@purdue.edu

K. Růžičková and T. Inspektor (eds.), *Surface Models for Geosciences*, Lecture Notes in Geoinformation and Cartography,
DOI 10.1007/978-3-319-18407-4_20

1 Introduction

The problem of hydraulic erosion simulation is a common open problem in GIS. The modern GIS systems, such as ArcGIS, mostly use raster data structure (regular heightfields) to provide such simulations. Regular heightfields are sufficient for most surface analysis done in GIS, but cannot be used for modeling of concave features, such as tunnels, caves or overhangs. However, these features often result from the hydraulic erosion.

The existing hydraulic erosion methods in GIS are often based on the raster data structure. The use of this structure allows for the creation of fast and interactive solutions at the cost of being constricted to only 2.5D non-adaptive models. Triangular irregular network (TIN) data structure brings the option to adaptively change the resolution of the terrain based on the local complexity of the surface, however, the restriction to 2.5D models still applies. A survey on hydrologic modeling in GIS can be found, e.g., in DeVantier and Feldman (1993) or Aksoy and Kavvas (2005).

Other approaches are capable of creating fully 3D models using a volumetric representation of the terrain, but these methods are generally very memory consuming and far from being interactive. The fully 3D hydraulic erosion simulation is more common in the field of computational fluid dynamics. An essential part of the erosion modeling is the simulation of fluid dynamics. The fluid dynamics can be described by the Navier-Stokes equations for incompressible Newtonian fluids (Acheson 1990) and its solutions can be divided into two main categories: Eulerian and Lagrangian approaches (Bridson 2008).

The Eulerian approaches divide the volume of the scene and track the fluid properties at fixed points. As the fluid moves through the scene, the tracked properties change according to the fluid state at the given position. Usually a uniform grid is used to simplify the necessary computation, however, the uniform space division leads to higher memory requirements. This approach gives accurate results, but it is very computationally expensive.

Benes et al. (2006) represent the scene as a regular grid, where each voxel is classified as either FULL (a voxel full of water, can contain dissolved material), EMPTY (an empty voxel containing only the air) or MAT (a voxel containing material). The authors define the processes that lead to the change of the state of the voxel. The method is capable of simulating fully 3D effects, however, it is not capable of running with real time response. Wojtan et al. (2007) simulate the corrosion and erosion of solid objects. The surface is stored as a level set. The erosion is then simulated by advecting the level set inward, the deposition by advecting it outward.

The Lagrangian approach methods represent the fluid using particles. The fluid properties are tracked at the particles as they move through the scene. This representation simplifies the fluid simulation as some of the conditions of the Navier-Stokes equations are automatically satisfied. The methods are generally faster, but less accurate than the Eulerian methods.

Smoothed-particle hydrodynamics (SPH) (Gingold and Monaghan 1977) is a particle-based approximative numeric solution to Navier-Stokes equations. The properties of the particles are calculated over the smoothing radius, a distance, over which the particles are influenced by each other. The contributions of each particle are weighted by their distance and density according to the smoothing kernel function used.

SPH is very often used in fluid simulation (Adams et al. 2007; Kipfer and Westermann 2006), but to the best of our knowledge, Krištof et al. (2009) were the first to combine the SPH fluid simulation with erosion. The SPH particles flow through the scene, eroding the terrain and transporting the sediment. The diffusion of the sediment is defined by a donor-acceptor scheme between the SPH particles. The terrain is stored as a raster, limiting the use of this method to 2.5D effects. Fully 3D terrain features, such as caves or overhangs, cannot be simulated with this approach.

In this paper we propose a novel solution to the hydraulic erosion modeling problem. Our method combines the use of the SPH for the fluid simulation and the surface triangular meshes for the representation of eroded objects. The triangular mesh can be seen as a compromise between the TINs and the volumetric representations. It preserves the adaptability feature of the TIN models, but also allows the creation of models containing caves or overhangs. However, inner cavities cannot be modeled using this approach. The main advantage of simulating the fluid using the SPH method is that the particles are located only in the affected regions, limiting the computation to these areas.

2 Data and Methods

In this paper we propose a novel solution to the hydraulic erosion modeling problem. We simulate the fluid with the Smoothed-particle hydrodynamics (SPH) particles by using the approach of Krištof et al. (2009). The terrain is represented as a surface triangular mesh in order to simulate the erosion of elements such as tunnels or caves. An example is shown in Fig. 1. The model of a river bed has a higher resolution in the areas, where the erosion will be applied (Fig. 1a).

2.1 Data Structure

A surface triangular mesh is used to represent the terrain data. The mesh can either be derived from a raster data obtained, e.g., from an aerial scan, or it can be computed on a given set of points by a planar triangulation, or it can be prepared using a modeling software. Usually the approaches can be combined to produce the desired input terrain data: the base mesh is derived from the scanned data and further details, such as caves or tunnels, are added using a modeling software.

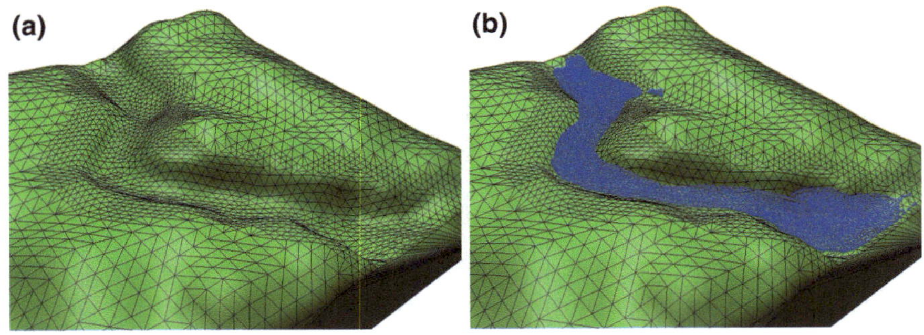

Fig. 1 An example of the triangular mesh model in erosion simulation. **a** The original model has higher resolution in the erosion areas. **b** The fluid represented by the SPH particles

3 Method Overview

Every iteration of the proposed method consists of the following steps:

1. The properties of the SPH particles, such as pressure, speed and direction, are calculated.
2. The fluid-terrain interaction is computed.
3. Each particle moves in the calculated direction.
4. The erosion/deposition sediment exchange is calculated for the particles colliding with the terrain.
5. The triangular mesh is updated according to the amount of sediment eroded/deposited at the vertices.

3.1 SPH Simulation

We simulate the fluid using the Lagrangian SPH method (Gingold and Monaghan 1977). The main advantage of the method lies in its adaptability—the particles are restricted to the areas containing the fluid. The computation is then performed only in the regions where the fluid is present.

The SPH defines the way the properties of the particles, such as pressure, speed and direction, are computed. The pressure is calculated based on the density of the particles and determines the speed and direction of the movement of the particle.

3.2 Fluid-Terrain Interaction

The motion of a particle is defined by the particle properties, by the gravity, and by the surrounding terrain. The fluid-terrain interaction is a vital part of the erosion simulation. A particle is influenced by all the mesh faces within the reach of its radius. To speed up the search for the nearby faces, a regular spatial subdivision is used. Using this auxiliary data structure, we do not have to search for the faces in the whole mesh, but only in the cells that fall into the radius of the particle. Each face f in the selected cells is checked for an interaction with the particle p as follows:

- The line l is constructed that goes through the position of the particle with the direction of the normal vector of the face f.
- If the line l intersects the face f, the orthogonal distance d between the face f and the particle p is calculated.
- If the distance d is smaller than the radius of the particle, the face f influences the motion of the particle p.

The face f contributes to the force affecting the particle with a force calculated using the penalty-force method (Amada 2006):

$$f = k^s dn + k^d (v \cdot n)n, \tag{1}$$

where k^s is the penalty force stiffness, k^d is the damping coefficient, d is the penetrated distance, v is the velocity of the particle, and n is the normal vector of the face.

3.3 Erosion and Deposition

When the particle collides with the terrain, we simulate the erosion of the surface or the deposition of the sediment, depending on the velocity and the direction of the particle. We calculate the erosion rate using a method similar to the one presented in the work of Wojtan et al. (2007). The erosion rate is calculated using the equation by Partheniades (1965):

$$\varepsilon = k(\tau - \tau_c)^a, \tag{2}$$

where ε is the erosion rate, k is the erosion constant, τ is the shear stress, τ_c is the critical shear stress, and a is a constant usually set to 1. The critical shear stress τ_c is a threshold value that has to be exceeded for the erosion to take place. The shear stress τ is a force applied to the solid boundary and is defined via a power-law model (Wojtan et al. 2007):

$$\tau = K\theta^m, \tag{3}$$

where $K = 1$ is a constant, θ is the shear rate and m is a constant determined by the material. The shear rate θ can be approximated from the fluid velocity relative to the surface:

$$\theta = \frac{v_{rel}}{d}, \tag{4}$$

where v_{rel} is the relative fluid velocity and d is the orthogonal distance between the particle p and the face f.

If the shear stress τ exceeds the value of the critical shear stress τ_c, the erosion rate ε has a positive value, and erosion takes place. The amount of the eroded sediment is uniformly divided among the vertices of the face f. Each vertex has a parameter assigned to store the amount of eroded (deposited) sediment, the contributions from individual particles are summed up and the mesh is only updated once in each iteration of the algorithm. The eroded sediment is carried by the particle until the conditions for the deposition are fulfilled.

The deposition occurs when the shear stress τ does not exceed the value of the critical shear stress τ_c. The erosion rate ε is negative and the sediment carried by the particle is deposited onto the vertices of the face f.

3.4 Mesh Modification

Each terrain vertex keeps track of the amount of sediment that was removed or added to it during the erosion/deposition step. Eventually, the mesh needs to be updated to reflect this change of volume.

Each affected vertex needs to be moved in the direction of the normal vector according to the amount of associated sediment. However, we cannot use the amount of sediment to directly determine the vertex displacement. Each vertex belongs to several faces and these faces can generally differ greatly in size and shape. The same vertex displacement could thus lead to different volume changes in two parts of mesh with distinct resolution.

As we need to control the volume changes regardless of the local resolution of the scene, we have to calculate the vertex displacement based on the exact volume change. The total volume change caused by the displacement of a vertex is a sum of contributions of all the affected faces. Figure 2a shows an example of sediment deposition onto a vertex. The deposition caused the vertex to move from the position D to the position A, with the vertex displacement a. The volume change for one of the faces equals to the volume of the area bounded by the original face and the face created by displacing the vertex, this area always forms a tetrahedron (Fig. 2b).

The volume of a tetrahedron V_{i} can be calculated as follows:

$$V_i = \frac{|a \cdot (b \times c)|}{6}, \tag{5}$$

where a, b and c are the edges of the tetrahedron sharing the vertex D. Using Eq. (5), we can get the total volume change V for the vertex displacement $|a|$ as

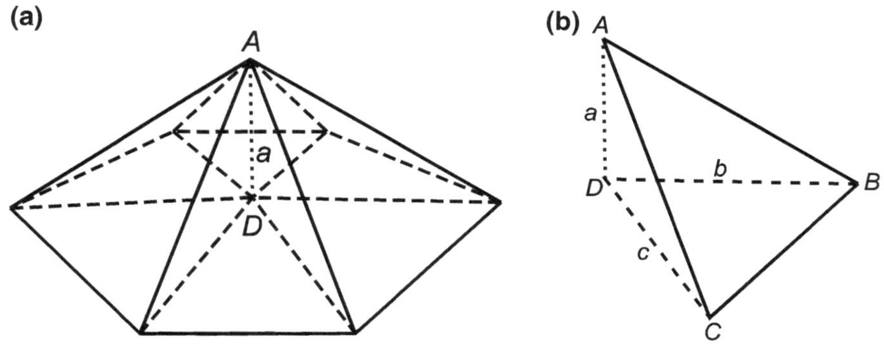

Fig. 2 The volume change caused by the displacement of a single vertex (**a**) and the volume change of one of the affected faces (**b**)

$$V = \sum_{0}^{n-1} V_i, \tag{6}$$

where n is the number of faces sharing the displaced vertex. It can be derived that for $a' = ax$, where x is a constant, the total volume $V' = Vx$.

To compute the correct vertex displacement a', we first calculate the volume change V for the vertex displacement $|a| = 1$ in the direction of the normal vector. The vertex displacement a needs to be adjusted so that the volume change is equal to the amount of eroded/deposited sediment s:

$$a' = a\frac{s}{V}, \tag{7}$$

where a' is the required vertex displacement.

3.5 Mesh Inconsistency

The erosion can cause an inconsistency in the mesh. If a part of the mesh is heavily eroded, it is possible for two parts of the mesh to overlap, creating a topology inconsistency. A simplified 2D example of such a case is shown in Fig. 3. Figure 3a shows a simple valid scene, the red point represents the vertex, where the erosion will be applied. The eroded scene in Fig. 3b shows an example of a created topology inconsistency. While erosion can cause creation of holes or breaking the mesh into pieces, deposition can cause unwanted mesh merging.

The repair of topology inconsistencies in the mesh is a well-known problem and many methods exist that address it, e.g., cf Attene et al. (2013). The problem is difficult due to the numerical inaccuracy of floating point arithmetic, which can

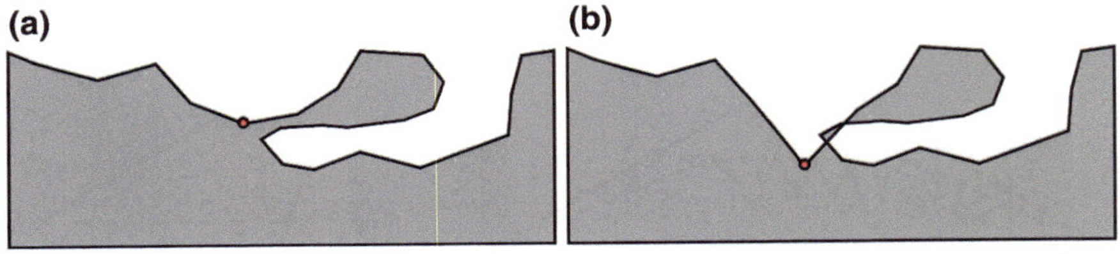

Fig. 3 An example of a mesh inconsistency. **a** The original scene. **b** The eroded scene with an inconsistency

cause the algorithms to collapse. As the inconsistencies are created very rarely during the erosion simulation, we do not repair them. We detect the inconsistency when it is created, and stop the simulation.

4 Discussion and Results

We have developed our framework in C++ and run it on an Intel Core 2 Duo clocked at 2 GHz. We use the *Fluids v.2* library (Hoetzlein 2009) for SPH fluid simulation that provides the necessary functionality and computes the particle properties during the simulation. The Navier-Stokes equations are scale-sensitive and the actual simulation scale of the fluid in the used implementation is approximately up to tens cm^3. As the terrain represents a large-scale scene, this inconsistency of scales can lead to less realistic behavior of the fluid.

We have verified the proposed algorithm by simulating several different erosion scenarios. Erosion is a long term process and a comparison with real-world data would be necessary to verify our algorithm. We rely the correctness on the computational fluid dynamics and we verified our solution visually.

We created a simple scenery of a lake being filled by water to validate the erosion process and to compare our solution to the results of the reference method by Krištof et al. (2009). The scene is shown in Fig. 4. The lake represents a local

Fig. 4 An example of an erosion simulation of an overfilling lake

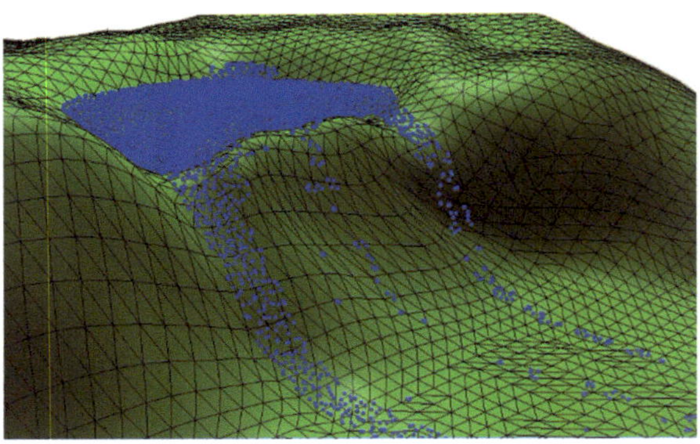

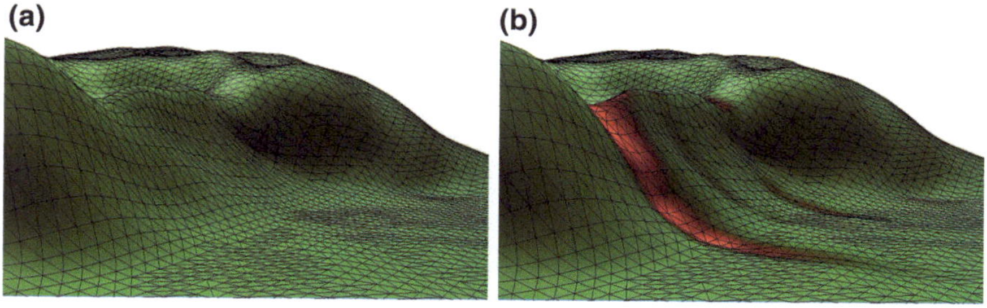

Fig. 5 An example of an erosion simulation. **a** The original terrain. **b** The eroded terrain; *red* color highlights the eroded regions

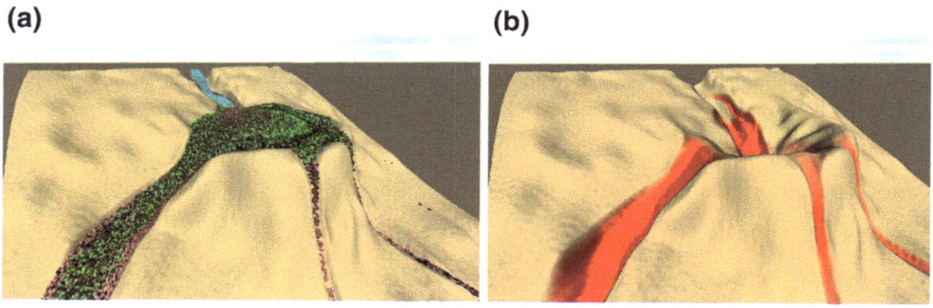

Fig. 6 The results of the reference method (Krištof et al. 2009). **a** The fluid eroding the terrain. **b** The eroded terrain; *red* color highlights the eroded regions

depression, which is being filled with water particles. The lake eventually fills up and the fluid flows over the edge and erodes the underlying terrain surface. We use a simple particle-based visualization for the fluid, as we are interested in the resulting eroded terrain itself, not in the realistic image of the water.

Figure 5 captures the alteration of the terrain during the simulation. Figure 5a shows the original terrain before the erosion was applied. The final eroded scene is depicted in Fig. 5b with the eroded regions highlighted in red color. The results of the reference method are shown in Fig. 6.

Unlike the reference method, our method can also simulate the fully 3D scenes with concave features. We modeled the demonstration scene captured in Fig. 7a. The scene consists of a block of material with a tube-like cavity. The water runs through it and erodes its surface (Fig. 7b). The fluid has the largest momentum before it collides with the mesh for the first time, hence the erosion is the strongest at that point. The fluid flow is affected by the changed surface of the mesh (Fig. 7b). The full simulation of both presented examples can be seen in Skorkovská et al. (2014).

Computational requirements were measured for the presented simulations. The lake scene (Fig. 4), consists of 12,528 particles and 9,068 faces, we measured an execution time of one iteration of approximately 2.9 s. Each iteration of the concave

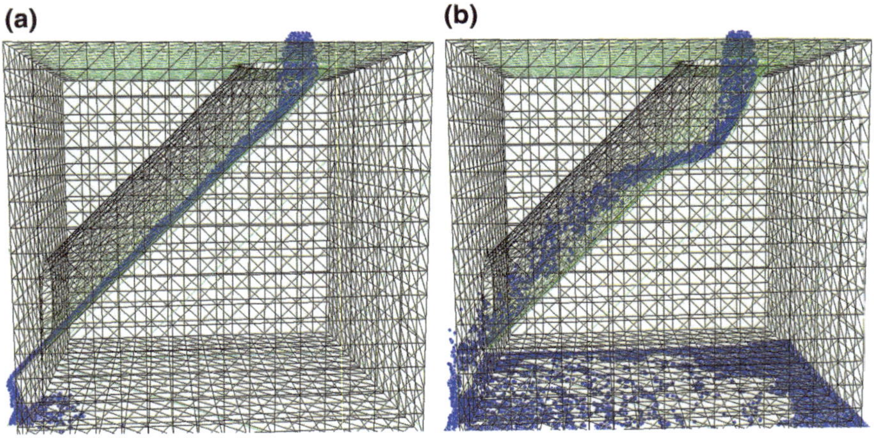

Fig. 7 Erosion simulation of concave features. **a** The original scene. **b** The water erodes the mesh, leading to change of the flow of the fluid

scene simulation (Fig. 6), consisting of 12,528 particles and 4,706 faces, took approximately 0.7 s. The goal of our implementation was to verify the applicability of our approach, without the emphasis on the speed of the simulation.

5 Conclusion

A novel solution to the hydraulic erosion simulation problem was proposed in this paper. Our solution combines the terrain represented as a triangular mesh with a fluid simulated as a particle system. This approach is capable of simulating fully 3D scenes containing features such as caves or overhangs with lower memory requirements than the existing volumetric solutions.

The triangular mesh representation has an advantage over other approaches in its adaptability. The resolution of the scene can be adjusted according to the shape and complexity of the terrain. The resolution can be increased in the regions that require more attention and preserved in other parts of the mesh. However, the use of the triangular meshes brings new challenges during the erosion simulation. The erosion can cause a creation of an inconsistency in the mesh, which would cause an incorrect behavior of the simulation. These inconsistencies occur very rarely and we address them by stopping the simulation to prevent the collapse of the simulation.

There are many possible avenues to improve the proposed algorithm. The problem of mesh inconsistency has to be solved in order to be able to simulate erosion of more complex scenes. The implementation of multiple materials could be the next step to achieve more realistic results. Last but not least, the implementation could be optimized to speed up the simulation.

Acknowledgments We would like to thank to Jan Ježek (University of West Bohemia, Pilsen) for his valuable advice and comments. This work has been supported by the European Regional Development Fund (ERDF)—project NTIS (New Technologies for Information Society), European Centre of Excellence, CZ.1.05/1.1.00/0.2.0090 and by the project SGS-2013-029—Advanced Computing and Information Systems.

References

Acheson DJ (1990) The Navier-Stokes equations. In: Elementary fluid dynamics. Clarendon press, Oxford, pp. 201–220. ISBN 978-0-19-859679-0

Adams B, Pauly M, Keiser R, Guibas LJ (2007) Adaptively sampled particle fluids. ACM Trans Graph 26(3):48:1–48:7

Aksoy H, Kavvas ML (2005) A review of hillslope and watershed scale erosion and sediment transport models. Catena 64(2):247–271

Amada T (2006) Real-time particle-based fluid simulation with rigid body interaction. In: Dickheiser M (ed) Game programming gems 6. Charles River Media, Rockland, pp 189–205

Attene M, Campen M, Kobbelt L (2013) Polygon mesh repairing: an application perspective. ACM Comput Surv 45(2):15:1–15:33

Benes B, Těšínský V, Hornyš J, Bhatia SK (2006) Hydraulic erosion. Comput Animation Virtual Worlds 17(2):99–108

Bridson R (2008) The equations of fluids. In: Fluid simulation for computer graphics, Ak Peters series. Taylor & Francis, Boca Raton, pp. 3–16. ISBN 978-1-56881-326-4

DeVantier BA, Feldman AD (1993) Review of GIS applications in hydrologic modeling. J Water Resour Plann Manage 119(2):246–261

Gingold RA, Monaghan JJ (1977) Smoothed particle hydrodynamics: theory and application to non-spherical stars. Mon Not R Astron Soc 181(3):375–389

Hoetzlein R (2009) Fluids v.2—a fast, open source, fluid simulator. http://www.rchoetzlein.com/eng/graphics/fluids.htm. Accessed 15 Oct 2014

Kipfer P, Westermann R (2006) Realistic and interactive simulation of rivers. In: Proceedings of graphics interface 2006. Canadian Information Processing Society, Toronto, ON, pp. 41–48

Krištof P, Benes B, Křivánek J, Šťava O (2009) Hydraulic erosion using smoothed particle hydrodynamics. Comput Graph Forum 28(2):219–228

Partheniades E (1965) Erosion and deposition of cohesive soils. J Hydraul Div Am Soc Agric Eng 91:105–139

Skorkovská V, Kolingerová I, Benes B (2014) Video example. http://home.zcu.cz/~vskorkov/Hydraulic_Erosion/. Accessed 15 Oct 2014

Wojtan C, Carlson M, Mucha PJ, Turk G (2007) Animating corrosion and erosion. In: Proceedings of the third Eurographics conference on natural phenomena. Eurographics Association, pp. 15–22

Registration of Optical and Radar Satellite Images Using Local Features and Non-rigid Geometric Transformations

Mohamed Tahoun, Aboul Ella Hassanien and Ralf Reulke

Abstract Optical and radar sensors provide us with different physical properties which require to be combined for different remote sensing applications like change detection. In this paper, a feature-based method for registering RapidEye and Pléiades optical images with TerraSAR-X radar images is presented. Different feature detectors/descriptors have been tested during the keypoint extraction and matching processes. In the extraction phase, a keypoint filtering process is proposed in order to control the number of extracted keypoints from both optical and radar images for a real time processing on high resolution satellite images. A speckle noise removal is performed on TerraSAR-X radar images as a pre-processing step in order to reduce the number of false detections. After getting the matches between the input images (list of tie points), a full registration process is followed by testing and applying different non-rigid image transformation models (e.g. affine and polynomial) in order to warp the sensed image on the reference one. The registered version is obtained after resampling both images using interpolation methods. The results showed a good performance for some detectors/descriptors like SURF, GFTT and Harris on optical images and SIFT on radar images. There is also a varying performance of the tested warping and resampling methods in addition to a full registration framework using SIFT and global homography transformation with linear blending. The tested optical and radar images are in different spatial resolutions and covering the area of Berlin Brandenburg airport in Germany.

M. Tahoun (✉) · R. Reulke
Department of Computer Science, Humboldt University in Berlin,
Rudower Chaussee 25, 12489 Berlin, Germany
e-mail: tahoun@informatik.hu-berlin.de

R. Reulke
e-mail: reulke@informatik.hu-berlin.de

A.E. Hassanien (✉)
Department of Information Technology, FCI, Cairo University,
Ahmed Zewail St. 5, Giza, Cairo 12613, Egypt
e-mail: aboitcairo@cu.edu.eg

K. Růžičková and T. Inspektor (eds.), *Surface Models for Geosciences*,
Lecture Notes in Geoinformation and Cartography,
DOI 10.1007/978-3-319-18407-4_21

Keywords Satellite image registration · Local features · SIFT · SURF · GIS Ostrava 2015

1 Introduction

Matching and registration of satellite images are basic tasks when processing remote sensing data taken from same/different sensors or at different times or from different viewpoints (Juan and Gwon 2009). Some achievements regarding the two processes have been remarked but there is still a need for an accurate registration scheme on high resolution satellite images with different modalities (Hong and Schowengerdt 2005; Khan et al. 2011). On the other hand, image registration is a well-known task in computer vision and image processing fields where two or more images are aligned by determining a geometric transformation between the source and the reference images. Image registration approaches can be generally classified into: feature-based (e.g. Points or edges) and intensity-based (directly uses the intensity values) or mixed methods. A survey of image registration techniques is found in Deshmukh and Bhosle (2011).

A local feature is considered as an image pattern which differs from its direct neighbourhoods and is usually represented by points, edges, corners or by other features (Bouchiha and Besbes 2013). Based on such representations, many feature detectors have been proposed, for example, Harris and FAST detectors work based on corner detection while SIFT and SURF work based on points or blobs. The last two detectors have been recently used in many applications and recorded a good performance against rotation, scaling and blurring (Juan and Gwon 2009). Corners usually represent a point in which the directions of two edges have a clear change while blobs are known as region of interest. Generally, the main steps of image registration process are: feature identification, feature matching, spatial transformation, and finally the resampling step. The feature identification includes the detection of the required features like keypoints or ground control points (GCPs) from the input images. Once the features are extracted from images, the matching process starts by comparing the feature descriptors of the extracted keypoints and then finding the set of correspondence between these descriptors. Mean square differences like Euclidean distance or root mean square (RMS) error, and mutual information are examples of similarity functions or metrics. This step is followed by finding a geometrical transformation that aligns the source image on the reference one using the set of similarity points or matches. Different methods are existing for performing such transformations and they are classified as rigid and non-rigid image transformations. In rigid transformations, the shape and the size of the objects do not change while they do change in non-rigid transformations like affine and similarity transformations. Based on the transformation model, the resampling techniques like linear interpolation are used to resize and then composite the final registered image. Good features extracted from images are supposed to be robust against rotation, scaling, illumination changes or invariant against any geometric changes. An evaluation of local features on satellite images is introduced in Tahoun et al.(2014).

Image registration has many applications on remote sensing data where images are taken from the same or different sensors like change detection and fusion. In this work, we focus on registering multi-sensor data where images are taken from different sensors (optical and radar) using non-rigid geometric transformations. The paper is organized as follows: the next section will give more details about the feature detection step. After this, the whole registration scheme is presented in the following section. Finally, the detailed experiments and results are discussed followed by some findings and conclusions.

2 Feature Detection

Image registration starts by detecting and extracting features from the input images. Feature detectors are used to extract features from the sensed and the reference images. Good detectors are designed to be invariant to different transformations that could be performed on images like scaling, transition and rotation or being robust against noise. This section highlights the tested detectors in our experiments.

Scale invariant feature transform (SIFT) has been presented by Lowe in the year 2004 (Lowe 2004). It has four major steps including: scale-space extrema detection, keypoint localization, orientation assignment and finally building the keypoint descriptor as in Fig. 1a. In the first step, the points of interest are identified by scanning both the location and the scale of the image. The difference of Gaussian (DoG) is used to perform this step and then the candidates of the points are localized to sub-pixel accuracy. Then the orientation is assigned to each keypoint in local image gradient directions to obtain invariance to rotation. In the last step, a 128-keypoint descriptor is built and ready for the matching process. SIFT gives good performance but still have some limitations against strong illumination changes and big rotation angles.

Speeded-up robust features (SURF) is a local invariant interest point or blob detector (Bay et al. 2006). It is partly inspired by the SIFT descriptor and is also used in static scene matching and retrieval. It is invariant to most of the image transformations like scale and illumination changes in addition to small changes in viewpoint (Schmidt et al. 2010). It uses integral images or an intermediate representation for the image and contains the sum of gray scale pixel values of image. Then a Hessian-based interest point localization is obtained using Laplacian of Gaussian of the image (Fig. 1b). Both SIFT and SURF are scale- and rotation-invariant detectors and descriptors.

Beside SIFT and SURF, other recently used and tested detectors include: (1) Features from accelerated segment test (FAST): is a corner detection method (Rosten and Drummond 2006; Schmidt et al. 2010). Its importance lies in its computational efficiency as it is faster than many famous feature extraction methods

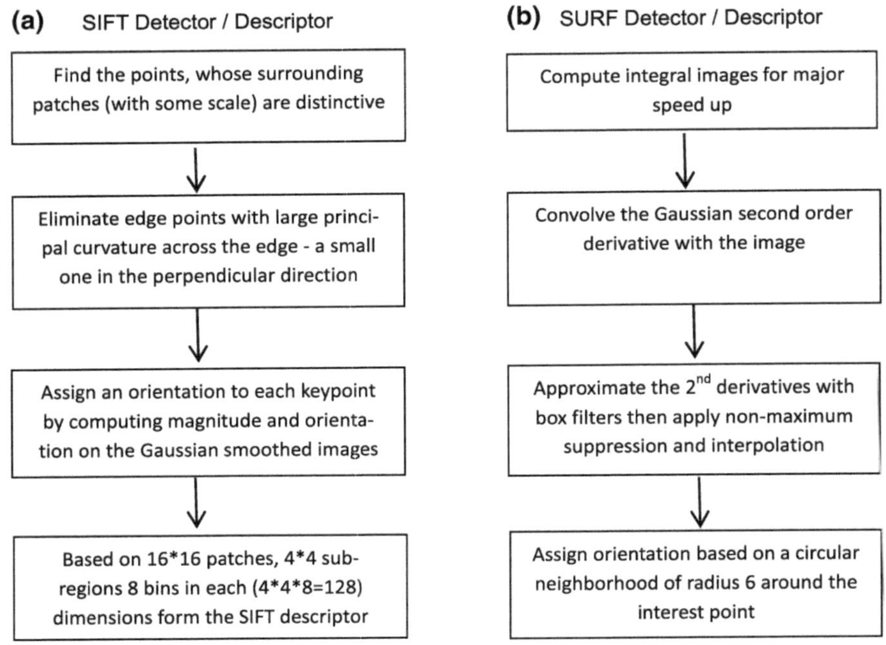

Fig. 1 The steps for building **a** SIFT detector/descriptor and **b** SURF detector/descriptor

(e.g. Difference of Gaussian (DoG) and Harris). It uses a Bresenham circle of radius 3 to find out whether a selected point is a corner or not? Each pixel in the circle is given a number from 1 to 16 clockwise. If a set of contiguous pixels inside the circle is brighter or darker than the candidate pixel then it is classified as a corner. (2) Oriented BRIEF-binary robust independent elementary features (ORB): is a local feature detector based on binary strings (Calonder et al. 2010). It depends on a relatively small number of intensity difference tests to represent a patch of the image as a binary string. The construction and matching of this local feature is fast and performs well as long as invariant to large in-place rotations is not required. ORB is basically a fusion of FAST keypoint detector and BRIEF descriptor with many modifications to enhance the performance. First it uses FAST to find keypoints, then apply Harris corner measure to find top N points among them. (3) Harris: is a combined corner and edge detector based on the local auto-correlation function (Harris and Stephens 1988; Schmidt et al. 2010). It does not depend on rotation or shift or affine change of intensity. It extends the principle of Moravec's corner detector (Moravec 1980) by considering the local auto-correlation energy. Corners are usually good features to match especially with viewpoint changes. (4) Good features to track (GFTT): extracts the most prominent corners in the image as described in Shi and Tomasi (1994) where the corner quality measure at each pixel is calculated. Then a non-maximum suppression is applied. The corners with a quality less than a certain threshold are rejected and the remaining corners are sorted by the quality measure in a descending order. Finally, each corner for which there is a stronger corner at a distance less than a threshold is thrown away. The procedure for the extraction and matching processes using such feature detectors/descriptors could be summarized as in Tahoun et al. (2014): finding distinctive keypoints, taking a region around each keypoint in an invariant manner (e.g. scale or affine invariant), extracting and normalize the region content, computing a descriptor for the normalized region and finally, matching the obtained local descriptors in order to get the list of tie points between the input images. In our experiments, we have also tested five different descriptors including: SURF, SIFT, binary robust independent elementary features (BRIEF), Oriented BRIEF (ORB), and binary robust invariant scalable keypoints (BRISK). Good descriptors should be distinctive and have a good performance against viewing changes.

3 Image Registration Framework

The registration framework starts by applying a speckle noise removal on the TerraSAR-X radar image (Fig. 2). Such step increases the quality of SAR and active radar images against any possible degradation. This includes applying different adaptive filters like Lee or Frost (available in GIS software like ENVI with the possibility to change or use different noise models (multiplicative or additive)). The next step is to detect and extract the keypoints (KPs) or the ground control points (GCPs) from both optical and radar images. For high resolution satellite images, the number of detected KPs are usually huge and need to be optimized in order to exclude the false GCPs candidates. For this reason, a keypoint filtering process is performed if the number of the detected keypoints exceeds a user-defined threshold for reducing the running time and the hardware requirements of such high resolution satellite images (the initial threshold value is set to 500 KPs or more in the case of optical and radar images). The detailed procedure for extracting and matching the key-points from the input images is summarized as follows:

1. Extract the keypoints from the input images and compute their descriptors (KP1 and KP2 if there are two input images). If the number of the detected keypoints exceeds a predefined number, then the detected keypoints are filtered by sorting them according to their response. A keypoint response strength is measured according to its cornerness.
2. Once the keypoint descriptors are built, the similarity measures or metrics can be used to measure how the two sets of KPs are similar. In this step, the nearest neighbor is considered as the keypoint with minimum Euclidean distance for the invariant descriptor. Different image similarity measurements have been tested in the experiments. The Minkowski form distance is defined based on the Lp norm as in (1):

$$D_p(S,R) = (\sum_{i=0}^{N-1} (S_i - R_i)^p)^{(1/p)} \quad (1)$$

where $D_p(S,R)$ is the distance between the two feature vectors $S = S_1, S_2, \ldots\ldots S_{N-1}, R = R_1, R_2, \ldots\ldots R_{N-1}$ representing the descriptors of the extracted keypoints from the input images. Euclidean distance (p = 2) recorded the most stable results compared to Manhattan distance (p = 1) distance and fast library for approximate nearest neighbors (FLANN) matchers). Furthermore, the root mean square error (RMSE) is used to measure the difference between the predicted values and the observed values from RapidEye/Pléiades and TerraSAR-X as base and warped images and vice versa. The RMSE is formulated as in (2):

$$RMSE = \sqrt{\frac{\sum_{1}^{n} (x_i - \bar{x}_i)^2}{n}} \quad (2)$$

where x_i and $\bar{x}_i$ are the predicted and the observed values (Fig. 2).

3. The outlier removal step is done using RANSAC to exclude the inconsistent matches and help getting the list of tie points representing the actual matches between the two input images (set of inliers). RANSAC estimates the good

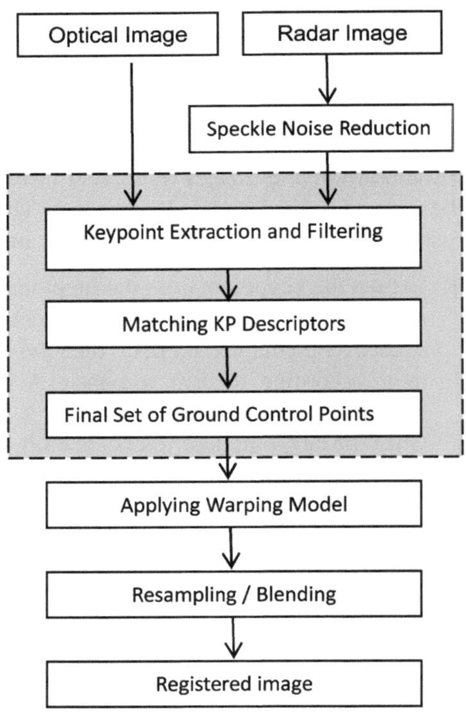

Fig. 2 The general framework of the experiments including the whole registration process between optical and radar images

points from the available set of data which containing a number of outliers via a random selection of some points and calculating the matrix. This process is repeated over the set of available points.

To evaluate the tested detectors/descriptors, the homography is computed between the filtered keypoints *H12* (from first group of keypoints to the second group). Then, the number of overlaps between *KP1* and the transformed keypoints from the second image is calculated *KP2′* (using inverse of *H1*). This number of overlaps is called the Corresponding Count $C(I_1,I_2)$. Then, by dividing $C(I_1,I_2)$ by the mean of the number of detected keypoints (d_1 and d_2), we get the Repeatability (Re) (Juan and Gwon 2009; Tahoun et al. 2014) as in (3):

$$\text{Re}_{1,2} = \frac{C(I_1, I_2)}{mean(d_1, d_2)} \tag{3}$$

The repeatability of a detector measures how well it determines the corresponding scene regions in the input images. It depends on the number of overlaps between the sets of KPs of the sensed and the reference images.

After getting the set of inliers, the warping phase starts using different non-rigid geometric transformations including affine and polynomial transformation models. We have also used a global homography transformation model as a warping method with a linear blending resampling technique. The tested transformation and resampling methods include:

1. Affine transformation: is a combination between linear transformation and translation. It's a simple transformation method and also known as rotation, scaling, and translation (RST). In affine transformations, the relation between two corresponding points (x,y) and (X,Y) could be formulated as in (4):

$$\begin{aligned} x &= m_1 + m_2X + m_3Y \\ y &= n_1 + n_2X + n_3Y \end{aligned} \tag{4}$$

where m_{1-3} and n_{1-3} are the transformation coefficients. Although it is faster than other transformations but it has some limitations and less accuracy than the polynomial model for example.
2. Polynomial transformation: is tested with different degrees depending on the number of GCPs. The first order polynomial warping is similar to (4) but includes *XY* interaction term which allows image shear as in (5):

$$\begin{aligned} x &= m_1 + m_2X + m_3Y + m_4XY \\ y &= n_1 + n_2X + n_3Y + n_4XY \end{aligned} \tag{5}$$

3. Triangulation warping: A triangulation of set of points $P = \{p_1, \ldots, p_n\}$ is a maximal planar subdivision with vertex set P. Delaunay triangulation warping fits triangles to the irregularly spaced GCPs and interpolates values to the output grid.

4. Global homography transformation: an input image (S) is transformed to a new image (N) by applying the following transformation as in (6):

$$N(x',y') = s(\frac{h_{11}x + h_{12}y + h_{13}}{h_{31}x + h_{32}y + h_{33}}, \frac{h_{21}x + h_{22}y + h_{23}}{h_{31}x + h_{32}y + h_{33}}) \quad (6)$$

where h is a 3×3 transformation matrix. The homography matrix values are randomly generated within a uniform distribution.

In order to perform the resampling step between the sensed and the reference images, the following methods have been tested: nearest neighbor (NN), linear/bilinear, and cubic interpolation. The nearest neighbor method uses the nearest pixel without any interpolation to create the warped image while the bilinear one performs a linear interpolation using four pixels to resample the warped image. For example, for any four points, we can get the values of a, b, c, and d by solving a linear system of equations including 4 variables and 4 equations as in (7):

$$\begin{aligned} &I(x,y) = ax + by + cxy + d = 0, \\ &Four\text{–}points : I(x,y), I(x+1,y), I(x,y+1), I(x+1,y+1) \end{aligned} \quad (7)$$

Then, to get x' between x and $x + 1$, for any y' between y and $y + 1$, the sum of squared differences (SSD) as in (8) is used to find $I(x', y')$:

$$\arg\min_{x',y'} \sum (W(x,y) - I(x+x', y+y'))^2 \quad (8)$$

where $W(x, y)$ is shifted by (x', y'). This procedure could be applied to the whole image or to a small displacement for instance. In the cubic interpolation, 16 pixels (4×4) are used to approximate the sinc function using cubic polynomials to resample the image. It takes usually longer time compared to the linear or the bilinear methods so it is used when the speed is not the issue. It's smoother and less in distortion than other methods. In short, the registration process depends on the quality of the tie points and on the type of the transformation model used in the warping phase in addition to the resampling method.

4 Experimental Results

The experiments have been run on satellite images with different spatial resolution including RapidEye and Pléiades optical images and TerraSAR-X radar images (covering the area of Berlin Brandenburg airport area in Germany (Table 1). RapidEye images have 5 spectral bands with ground resolution 5 m while Pléiades images have only 4 spectral bands with higher ground resolution (0.5 m) and finally TerraSAR-X band with 2.75 m ground resolution. The hardware and software

Table 1 The tested images from RapidEye, Pléiades and TerraSAR-X satellite data

Optical/radar images	RapidEye		Pléiades		TerraSAR-X
Original dimension	5000 × 5000		37,197 × 30,258		25,468 × 45,065
Ground resolution	5 m—5 spectral bands		0.5 m—4 spectral bands		2.75 m
Bands	Blue	440–510 nm	Blue	430–550 nm	Active X-band microwave
	Green	520–590 nm	Green	490–610 nm	
	Red	630–685 nm	Red	600–720 nm	
	Red E.	690–730 nm	NIR	750–950 nm	
	NIR	760–850 nm			

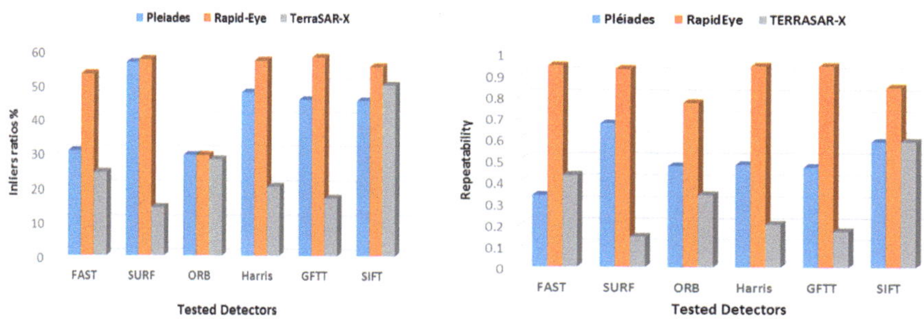

Fig. 3 The general inliers ratios (%)—*left*—and the repeatability—*right*—of the tested detectors on Pléiades, RapidEye and TerraSAR-X satellite images

specifications include: PC with Intel-core ™2 quad CPU 2.66 GHz and 12 GB RAM using Windows 7/2008 server (64 bit), Visual Studio-10 and OpenCV 2.49 in addition to MTLAB (Release 14) for Windows.

Different experiments and tests have been done in order to optimize and enhance the entire registration process. A comparison among the used feature detectors: FAST, SURF, ORB, Harris, GFTT and SIFT is presented in Fig. 3. SURF, Harris and GFTT detectors give a good performance on RapidEye and Pléiades optical images in terms of the inliers ratios and repeatability values while SIFT detector gives the best performance on TerraSAR-X radar images compared to other detectors. The best repeatability values on optical images are obtained using SURF while its performance is low on radar images. On the other hand, ORB has a better inliers ratios and repeatability values than SURF and Harris on radar images.

Furthermore, different binary-visual and vector-based descriptors have been also tested. The list includes: SIFT, SURF, BRIEF, ORB and BRISK as in Table 2. The results done on RapidEye and TerraSAR-X have shown that SIFT descriptor takes longer time to be built compared to other descriptors. Also, the matching time of SIFT descriptor using Euclidean distance is also greater than other descriptors but it still giving the most robust and stable performance (an example in Fig. 4).

Table 2 The tested descriptors and their performance on RapidEye and TerraSAR-X images

Item/Descript.		BRIEF	BRISK	ORB	SIFT	SURF
# of detected keypoints (KPs)	RapidEye	# of detected KPs: 271,747, filtered later to n KPs				
	TerraSAR-X	# of detected KPs: 518,470, filtered later to n KPs				
KPs detection time (in seconds)	RapidEye	13.947	13.639	13.940	11.510	15.515
	TerraSAR-X	21.846	18.248	18.578	18.453	22.768
KP filtering time (in seconds)	RapidEye	0.035	0.86	0.025	0.028	0.034
	TerraSAR-X	0.058	0.050	0.060	0.052	0.060
Descriptor-time (in seconds)	RapidEye	0.183	0.213	0.372	4.415	0.540
	TerraSAR-X	0.151	0.169	0.498	4.566	0.367
Matching time using Euclidean distance (in seconds)		0.074	0.133	0.113	0.090	0.068

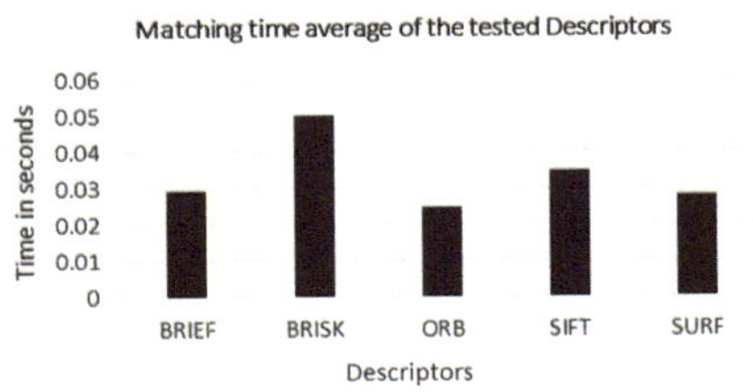

Fig. 4 A comparison among the matching time of the tested descriptors using SURF detector

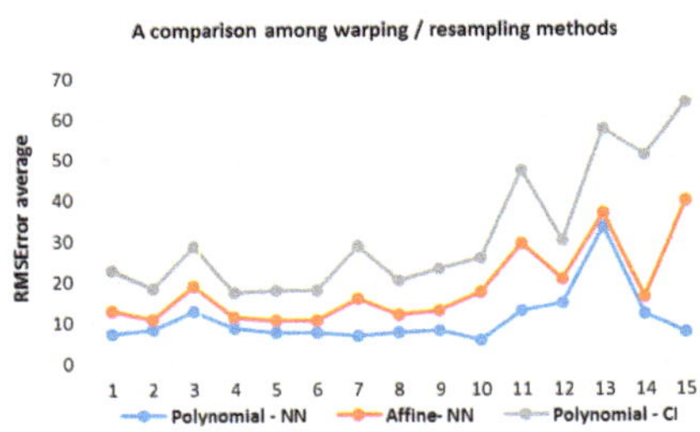

Fig. 5 A comparison among different non-rigid transformations and resampling models

On the other hand, Fig. 5, depicts a comparison among different transformation models including: polynomial and affine transformations with the resampling methods: NN and cubic interpolation on RapidEye and TerraSAR-X registration process. The polynomial transformation with nearest neighbor method recorded a better results compared to the other combinations. Figure 6 shows the original TerraSAR-X radar image before and after applying the speckle noise reduction filter. This pre-processing step enhances the registration process by reducing the expected false detections. A final test is done using SIFT detector and global

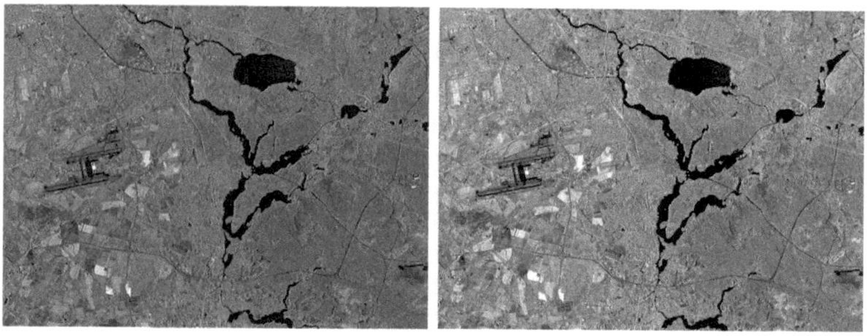

Fig. 6 An original TerraSAR-X image (*left*) and after the applying speckle noise removal using Lee adaptive filter (*right*)

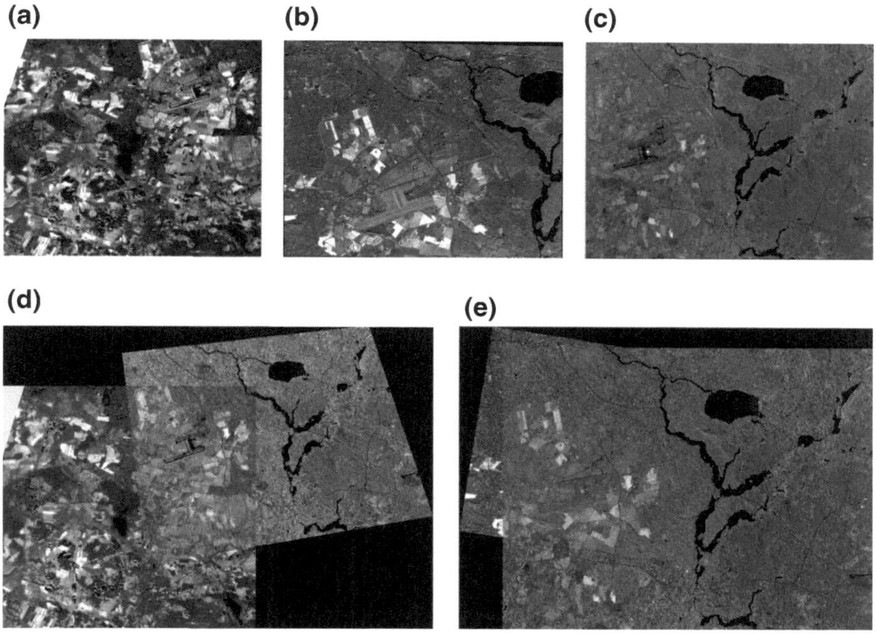

Fig. 7 The tested optical and radar images in (a), (b), and (c) while (d) and (e) show two examples of optical and radar registration using SIFT and global homography. **a** Rapid-Eye optical image band#3, date: 24.09.2010. **b** Pléiades optical image band#4, date: 12.10.2012. **c** TerraSAR-X radar image, date: 25.07.2010. **d** RapidEye–TerraSAR-X registered image. **e** Pléiades–TerraSAR-X registered image

homography transformation between the optical (RapidEye/Pléiades) and TerraSAR-X radar images (Fig. 7a, b and c).

In Fig. 7, a full registration (with mosaicking) is done between RapidEye and Pléiades optical and TerraSAR-X radar images using SIFT and global homography transformation. The same procedure for the detection and the matching steps is performed followed by the outlier removal using RANSAC. Global homography transformation and linear blending resampling are used to warp the TerraSAR-X image on the RapidEye as in Fig. 7d while the Pléiades image is warped on the TerraSAR-X image in Fig. 7e.

5 Conclusions

Matching and registration of multi-sensor satellite images are mandatory steps for many remote sensing applications. This work presents an enhanced scheme for registering RapidEye/Pléiades optical and TerraSAR-X radar images using local feature detectors and non-rigid geometric transformations. First, a speckle noise removal is performed on TerraSAR-X images as a pre-processing step for reducing any possible degradation. The scheme includes a keypoint filtering process that controls the ground control points and leads to reducing the running time and increasing the accuracy as well. RANSAC is used as an outlier removal for excluding the false matches and get an enhanced tie points' list between the sensed and the reference images. The results showed a good performance of SIFT detector on TerraSAR-X radar images while other detectors like SURF and Harris have a good performance on RapidEye/Pléiades optical images. A comparison among different non-rigid geometric transformation methods is also presented with a good performance for polynomial transformation with nearest neighbor resampling method. A full registration framework between the tested Rapid-Eye/Pléiades optical and TerraSAR-X radar images is done using SIFT detector and global homography transformation and linear blending. We aim to develop a generalized and more robust scheme for registering both optical and radar images for supporting multi-modal remote sensing applications.

References

Bay H, Goolhy LV, Tuytelaars T (2006) SURF: speeded-up robust features. In: Proceedings of the ninth european conference on computer vision, pp 404–417

Bouchiha R, Besbes K (2013) Automatic remote-sensing image registration using SURF. Int J Comput Theor Eng 5(1):88–92

Calonder M, Lepetit V, Strecha C, Fua P (2010) BRIEF: binary robust independent elementary features. In: 11th European conference on computer vision (ECCV), pp 778–792

Deshmukh M, Bhosle U (2011) A survey of image registration. Int J Image Process (IJIP) 5:245–269

Harris C, Stephens M (1988) A combined corner and edge detector. In: Proceedings of the 4th alvey vision conference, pp 147–151

Hong TD, Schowengerdt RA (2005) A robust technique for precise registration of radar and optical satellite images. Photogr Eng Remote Sens 71(5):585–594

Juan L, Gwon O (2009) A comparison of SIFT, PCA-SIFT and SURF. Int J Image Process (IJIP) 3(4):143–152

Khan NY, McCane B, Wyvill G (2011) SIFT and SURF performance evaluation against various image deformations on benchmark dataset. In: Bradley AP, Jackway PT (eds) DICTA, pp 501–506

Lowe DG (2004) Distinctive image features from scale-invariant keypoints. Int J Comput Vision 60(2):91–110

Moravec H (1980) Obstacle avoidance and navigation in the real world by a seeing robot rover. Tech Report CMU-RI-TR-3 Carnegie-Mellon University, Robotics Institute

Rosten E, Drummond T (2006) Machine learning for high speed corner detection. In: The 9th european conference on computer vision 1, pp 430–443

Schmidt A, Kraft M, Kasinski AJ (2010) An evaluation of image feature detectors and descriptors for robot navigation. In: ICCVG (2)'10, pp 251–259

Shi J, Tomasi C (1994) Good features to track. In: Proceedings of the IEEE conference on computer vision and pattern recognition, pp 593–600

Tahoun M, Shabayek A, Hassanien A, Reulke R (2014) An evaluation of local features on satellite images. In: Proceedings of the 37th international conference on telecommunications and signal processing (TSP), pp 695–700

Experimental Comparison of Sequential and Global Modelling Methods of Subsurface Horizon Sequence

Ladislav Vizi and Mária Benčoková

Abstract The objective of the presented paper is to compare two spatial modelling approaches for the boundary surfaces of a geological subsurface structure, where the exact depth locations of the modelled horizons are known only at sampled locations in the borehole intersections. The paper aims at two approaches: (1) traditional sequential modelling of each horizon or the layer thicknesses separately and (2) global approach based on multivariate modelling of each zone thicknesses simultaneously to keep the relationships between the internal zones and the total formation thickness of a subsurface structure. The experimental comparison of the sequential and global methods is performed on a simulated artificial model of the geological subsurface structure, which is used as a true reality, completely known throughout the estimated area. The model is used to compare obtained results of the different estimation approaches. The estimations are performed using an isotopic set of the depth and zone thickness variables, derived from the simulated geological structure.

Keywords Multivariate geostatistics · Cokriging · Horizons · Zones · Isochore maps · Cumulative thickness

1 Introduction and Basic Terms

From the beginning, geologists create maps of the subsurface geological structures and objects that are hidden in the earth's crust. Maps and map making is therefore an integral part of the work geologist. The final geological map represents a

L. Vizi (✉) · M. Benčoková
Institute of Geoscience, Faculty of BERG, Technical University of Košice, Park Komenského 15, 042 00 Košice, Slovak Republic
e-mail: ladislav.vizi@tuke.sk

M. Benčoková
e-mail: bencokova.maria@gmail.com

K. Růžičková and T. Inspektor (eds.), *Surface Models for Geosciences*, Lecture Notes in Geoinformation and Cartography,
DOI 10.1007/978-3-319-18407-4_22

numerical model of a mapped phenomenon, such as thickness or depth of the studied structure (Yarus and Chamber 2006). For instance, in petroleum geology, it is important to be able to map the surface of sedimentary bodies to provide a description of the structure and better understanding of the precise succession of sedimentary sequences and the deposition mechanisms (Chichi et al. 2000).

The presented paper deals with methods of spatial modelling of a sequence of horizons of a *subsurface structure*, which defines a geological object in the earth's crust. The aim of this paper is to describe and to compare the methods for modelling of a sequence of boundary horizons of a subsurface structure.

A subsurface *horizon* represents a two-dimensional surface, which is the interface of different layers of rock or contact between two successive lithostratigraphic units. In the presented paper, the different horizons will be denoted as "H_i" for the *i*-th horizon. Denotation H_{i+1} will represent the upper horizon in a sequence and, vice versa, H_{i-1} will be a lower horizon in the sequence.

Two consecutive horizons define a rock *zone*, which may represent, for example, a specific group of layer or a geological formation. In this paper, the particular zone will be denoted as "Z_i". This zone is vertically bounded by horizons H_i on the base and H_{i+1} on the top.

The vertical difference between the uppermost horizon (top of the last zone) and the lowermost horizon (base of the first zone) defines a *total thickness* of a particular formation (denoted as "F"), consisting of the sequence of "*L*" successive zones. Formation thickness represents the sum of the thicknesses of all zones.

The individual terms, described above, are shown in Fig. 1, which is a schematic vertical section consisting of five formations of quasi-homogeneous zones.

The main emphasis on the modelling of the horizons of the subsurface structures is given to the preservation of hidden geometric correlation between the geological units, which is based on two important assumptions:

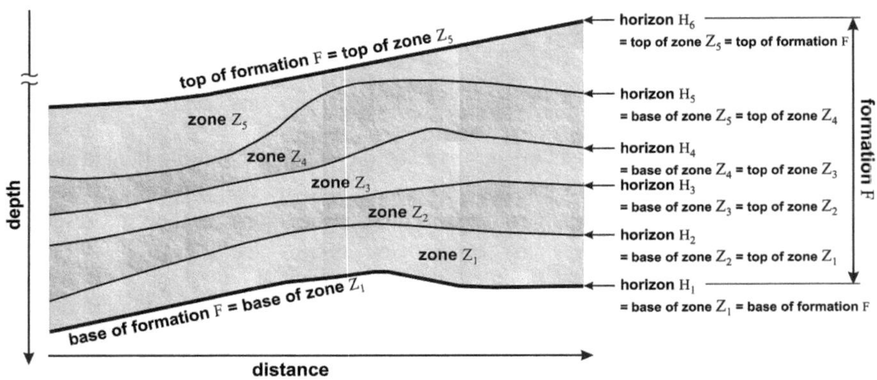

Fig. 1 A schematic vertical cross section explaining the basic terms used in the paper

1. The thickness of any zone within a specific formation must be equal to or greater than zero at each point of the studied domain. This means that the resulting *isochores map*, i.e. a map of vertical thickness, of the zone must be strictly positive (non-negative).
2. The sum of thicknesses of all zones within a particular formation must be equal to the total thickness of this formation.

The paper introduces a fictive numerical model of a subsurface structure representing an anticline fold with three zones and four horizons. This numerical model represents a true reality and it is used to compare the results of application of traditional sequential methods of horizon modelling and new global methods based on multivariate modelling (Sancevero et al. 2008). It must be noted that due to the unification process of modelling and comparison of the results, geostatistical methods will be used not only for global methods but also for traditional sequential ones.

2 Methods

The common feature of geometric modelling in geology is a relation among the certain number of layers that constitute a formation. It calls for a multivariate approach of spatial modelling to keep geometrical relations of sequence of the layers in the formation. As mentioned in Sect. 1, application of methods will be based on geostatistical modelling. By definition (Matheron 1962 in Journel and Huijbregts 1978), "*Geostatistics is the application of the formalism of random function to the reconnaissance and estimation of natural phenomena.*" Geostatistics is based on the theory of a *random function* $Z(\mathbf{x})$ (function, which takes unknown shape during the experiment, Ventcel'ová 1973). Its one realisation Prof. Matheron called *regionalised variable*, which is a function of space coordinates $\mathbf{x}$ (Matheron 1963, 1970, 1971).

Geostatistical methods of estimation are model-based. That means they call for a model of the spatial variability-*semivariogram* (shortly *variogram*). Variogram is a measure of one half the mean square differences between the values of $z(\mathbf{x})$ and $z(\mathbf{x} + \mathbf{h})$ in two locations $\mathbf{x}$ and $\mathbf{x} + \mathbf{h}$ separated by a vector $\mathbf{h}$ (Olea 1991). A *linear model of regionalisation*, composed of basic authorized structures of variability, is fitted to the experimental variogram (Goovaerts 1997). A natural extension of univariate modelling is multivariate approach, which extends the concept of a single-variable regionalisation to several regionalised variables that are spatially inter-correlated (Dowd 2004). Multivariate model of regionalization is called *linear model of coregionalisation*. Modelling of coregionalization of variability of N regionalised variables calls for inferring $N(N + 1)/2$ direct and cross variogram models (Goovaerts 1997).

A family of geostatistical interpolation methods based on variogram model is called *kriging*. Kriging is estimation techniques, based on a weighted linear combination of data and variogram model, which provides unbiased estimation with minimal variance of the errors (Isaaks and Srivastava 1989). A major advantage of kriging over other the interpolation methods is the possibility to mutually krige more than one variable, so called *cokriging*.

Traditional methods of subsurface horizons modelling are based on a *sequential approach* of the estimation process, which consists in an individual processing of each surface (Sancevero et al. 2008). The result of this classical modelling approach is achieved through an estimation process of individual depth or thickness variables "per-partes". The next step consists in the individual processing of the obtained depth or thickness estimations from a reference surface to get the final stratigraphic surfaces of the layers. Frequently observed disadvantage of this process is that the modelled surfaces cross each other. That leads to the negative thicknesses in some regions. It calls for artificial post-processing to eliminate these clearly unwanted results. In case of modelling of the thickness variables separately may happen that the sum of thickness estimations mismatches the total structure thickness. In this case, a volume correction must be applied.

For this work, geostatistical modelling was chosen due to the possibility of a mutual modelling. The objective is the application of an alternative way of modelling called *global approach* (Sancevero et al. 2008). The global approach is based on multivariate modelling of a sequence of the layers. Nevertheless, there are many situations where it is not always possible to quantify adequately the correlation between the thicknesses of overlain beds, even though geologically and geometrically it must be present (Dowd 1983). The global approach consists in modelling of the internal horizons of the structure, where the layers, zones or units of a structure are modelled simultaneously under consideration of the inherent relationships among the layers. The global approach used in this study is based on the *cumulative thicknesses* of all the layers from a reference surface on the top of a formation to the base of the formation as follows:

$$\begin{cases} Z_1^{\mathrm{cth}}(\mathbf{x}_\alpha) = \sum_{i=1}^{1} Z_i^{\mathrm{th}}(\mathbf{x}_\alpha) = Z_1^{\mathrm{th}}(\mathbf{x}_\alpha), \\ \vdots \\ Z_j^{\mathrm{cth}}(\mathbf{x}_\alpha) = \sum_{i=1}^{j} Z_i^{\mathrm{th}}(\mathbf{x}_\alpha), \\ \vdots \\ Z_L^{\mathrm{cth}}(\mathbf{x}_\alpha) = \sum_{i=1}^{L} Z_i^{\mathrm{th}}(\mathbf{x}_\alpha) = F^{\mathrm{th}}(\mathbf{x}_\alpha), \end{cases} \tag{1}$$

where "th" denotes the thickness variables and "cth" denotes the cumulative thickness variables (Fig. 2).

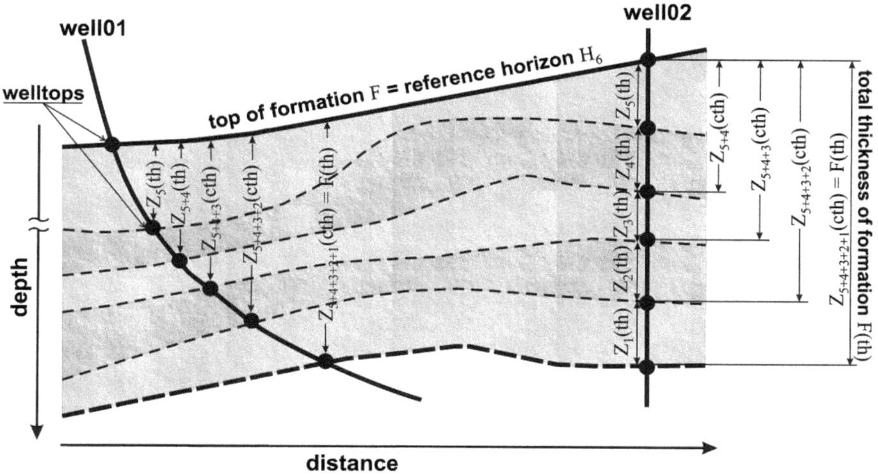

Fig. 2 The principle of deducing of the zone thicknesses and cumulative thicknesses within the formation based on the known reference surface and welltops. Dashed lines represent horizons H_i being modelled

The multivariate geostatistical methods are used for estimations of the cumulative thicknesses and a total thickness of the formation. The total thickness of the formation is used as secondary (auxiliary) variable, which controls the modelling results using a collocated cokriging method (Xu et al. 1992; Goovaerts 1997; Wackernagel 2003), and it is fully known between the top and base boundary surfaces. Individual thicknesses of the respective layers are obtained from the total thickness of the formation and then the intermediate surfaces are derived from a reference surface.

Modelling of the *cumulative proportions* of the thicknesses of each zone to the total thickness of the formation is modified method proposed by Prof. Dowd (Dowd 1983). If we denote by index "p" the proportion of the thickness of a zone, then the thickness proportion of zone Z_i at a sample position $\mathbf{x}_\alpha$ is defined by:

$$Z_i^{\mathrm{p}}(\mathbf{x}_\alpha) = \frac{Z_i^{\mathrm{th}}(\mathbf{x}_\alpha)}{F^{\mathrm{th}}(\mathbf{x}_\alpha)}\,. \tag{2}$$

The proportions of the zone thicknesses to the total thickness of the formation are new regionalised variables that are often better correlated and structured than the thicknesses alone. The zone proportions can get values ranking from 0 (absence of zone) to 1 (presence only one zone). The thickness proportions are cumulated in the same way as the thicknesses (Fig. 2). If we denote by index "cp" a cumulative proportion, then the new cumulative variables are obtained by replacing the index "cth" for "cp" in the expression (1) where the following relation must be held:

$$Z_L^{\mathrm{cp}}(\mathbf{x}_\alpha) = \sum_{i=1}^{L} Z_i^{\mathrm{p}}(\mathbf{x}_\alpha) = 1 \because Z_L^{\mathrm{cth}}(\mathbf{x}_\alpha) = \sum_{i=1}^{L} Z_i^{\mathrm{th}}(\mathbf{x}_\alpha) = \mathrm{F}^{\mathrm{th}}(\mathbf{x}_\alpha). \tag{3}$$

In the presented case study, both methods (cumulative thicknesses and cumulative proportions) give very similar results, therefore only the results of the first method will be given in detail in this paper.

3 Description of a Training Model and Input Data

Presented experimental comparison of different methods of modelling required a training, or known, numerical model. The model represents a completely known "reality" of stratigraphy of a subsurface structure. In practise, of course, this reality is known only in the well positions, or completely but indirectly by a seismic survey.

A fictional three-dimensional model of a subsurface structure (Fig. 3a) was created for the application of the sequential and global modelling approach and for final comparison of the obtained results. A final model could represent an anticline fold with an eroded uppermost part (Fig. 3b).

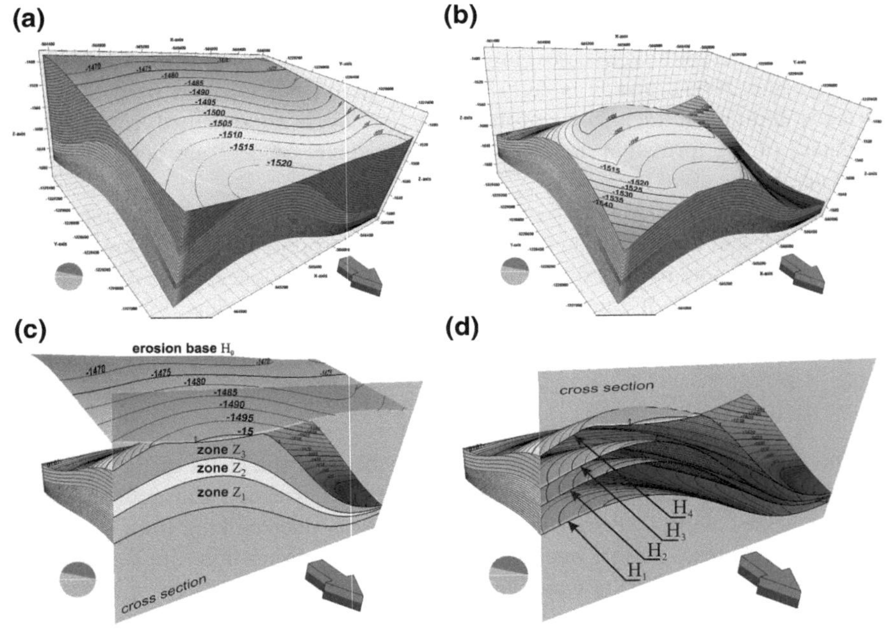

Fig. 3 Input 3D model of the subsurface geological structure representing an anticline fold, which is composed of three zones and four boundary horizons. Internal horizons H_3 and H_2 are the objects of interest

The first step was to simulate the Gaussian structure of variability on a regular two-dimensional grid with area 2500 × 2100 m with a mesh dimension 50 × 50 m. The Gaussian structure of variability was simulated to obtain a smooth surface. A set of realisations of a two-dimensional distribution of points was simulated using non-conditional Turning Bands method (Journel and Huijbregts 1978). One realisation of simulation that represented a dome-shape surface was selected. The selected surface included 2100 values with the standardized normal distribution. The values were transformed using simple algebra into the real space to represent the elevation values of the surface. The final surface represented one horizon of a subsurface structure model with an anticline shape.

The next step was to derive of the sequence of the horizons of the subsurface structure. The sequence was based on the surface that was derived in the first step. Next three horizons were subsequently derived using mathematical operations and generator of random numbers to ensure an absence of a functional relation between the four horizon values.

The four horizons, labelled from the bottom as H_1, H_2, H_3, H_4 (Fig. 3d), constitute three zones Z_1, Z_2, Z_3 (Fig. 3c). Note that the individual horizons constitute the boundary surfaces between the zones in the following sequence: the horizon H_1 is the base surface of the zone Z_1 and the horizon H_2 is top of zone Z_1, which is also the base surface of the zone Z_2, etc. All three zones form the formation "F". The top of the structure, where the thickness of the zone Z_3 decreases, is eroded by a base horizon H_0 of an upper formation. This ensures that there is no spatial relation between the thickness of the zone Z_3 and thicknesses of the zones Z_2 and Z_1.

The final model is a very simplified and smoothed version of reality. For the purpose of demonstrating the application of the various modelling methods and comparison of their pros and cons, however, is sufficient. The isochore maps of individual zones and the total formation were calculated as a difference of two successive horizons constitute bottom and top boundary surfaces (Fig. 4a–d). Table 1 gives the input statistical characteristics of the zone thicknesses on 2100 nodes $\mathbf{x}_o$ of the grid.

The final step was to obtain a discrete set of point data representing depth information (*welltops*) of individual horizons in the positions of the well path. In order to maintain constant information base for each method of modelling, the welltops were derived under the assumption of a vertical course of the wells through the structure. The grid nodes of the depth values of horizon H_0 were converted to a set of 2100 points (Fig. 5a, left). In total 104 points of 2100 nodes were filtered out (approximately 5 % of the 2100 simulated points). 104 sampling points were randomly, but eccentrically located within a particular grid cell. This ensured that the depth data available in positions $\mathbf{x}_\alpha$ do not share the same positions on the grid nodes $\mathbf{x}_o$ where the unknown values will be estimated (Fig. 5a, right). Finally, the 104 depth values of the individual horizons H_i from the grid nodes were interpolated into the nearest sample positions $\mathbf{x}_\alpha$ from the four surrounding node values using a bilinear interpolation. This resulted in an isotopic data configuration (Wackernagel 2003) of the depth measurements for each horizon, representing true vertical depth (TVD) of the horizons in 104 well positions (Fig. 5b).

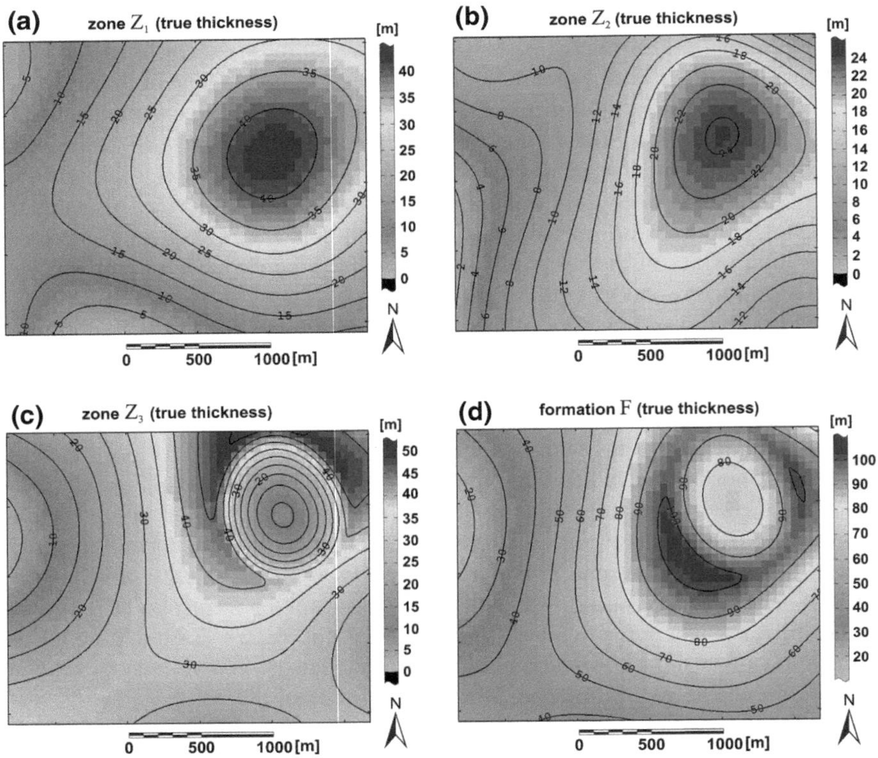

Fig. 4 The isochore maps of the true thicknesses of the individual zones Z_i (**a–c**) and the thickness map of the total structure formation F (**d**)

Table 1 Basic statistical characteristics of the true zone thicknesses and the formation thickness

Thickness	Minimum (m)	Maximum (m)	Mean (m)	Std. deviation (m)
Zone 3 [Z_3(th)]	3.42	50.99	27.64	10.10
Zone 2 [Z_2(th)]	1.54	24.28	14.06	5.51
Zone 1 [Z_1(th)]	1.53	42.74	22.79	11.22
Formation [F(th)]	16.39	106.61	64.49	22.97

Subsequently, the true vertical thicknesses (TVT) from the welltops were calculated for each individual zone. Table 2 gives the input basic statistical characteristics of the zone thicknesses in 104 well positions.

Compare to the Table 1, we see that the input data have a smaller range, but the higher variability except for the formation thickness F(th). The higher average values for the thicknesses of the zones Z_1 and Z_2 are caused by the preferential sampling of the top of subsurface structure, which is the most common case.

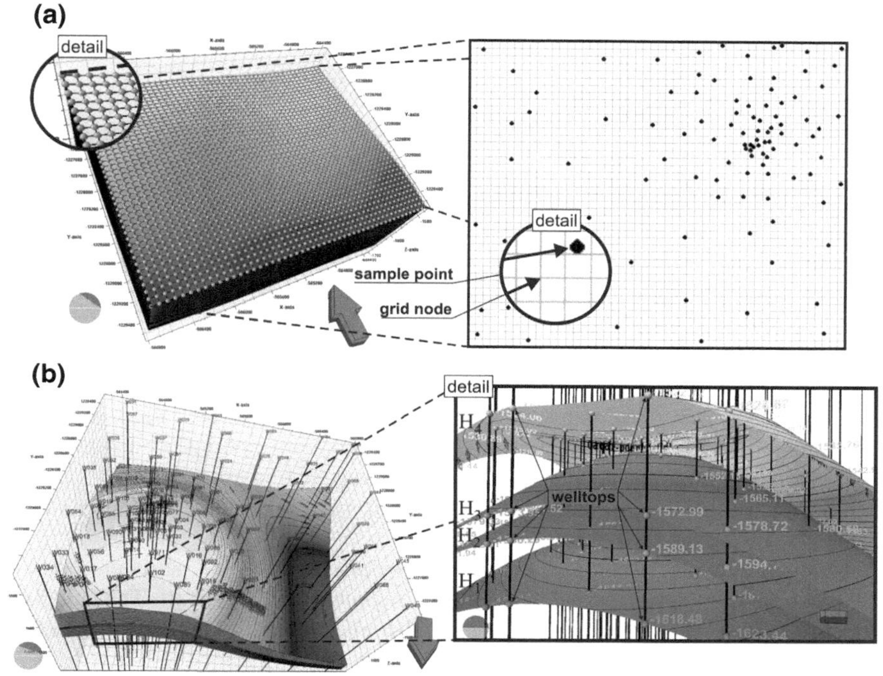

Fig. 5 Principle of conversion of the grid nodes to the set of points and selection of the sample points (**a**). Conversion of the sample points to the vertical lines, mimicking the survey borehole intersections of the subsurface structure, where the depth information of the zones was interpolated from the four surrounding node values (**a**, *left*) using a bilinear interpolation to obtain of the welltops data (**b**)

Table 2 Basic statistical characteristics of the sample zone and formation thicknesses in 104 wells

Thickness	Minimum (m)	Maximum (m)	Mean (m)	Std. deviation (m)
Zone 3 [Z_3(th)]	5.27	50.27	26.36	12.83
Zone 2 [Z_2(th)]	2.38	24.28	17.84	6.01
Zone 1 [Z_1(th)]	3.66	42.67	30.93	11.72
Formation [F(th)]	17.74	106.57	75.13	21.09

The paper will focus on modelling of internal horizons H_2 and H_3 that form of the boundary of the zone Z_2 (production zone or reservoir). We will assume that horizons H_1 and H_4 are completely known from seismic survey, but due to low seismic resolution internal horizons H_2 and H_3 could not be interpreted and they are known only in the 104 survey borehole positions $\mathbf{x}_\alpha$. Since the horizons H_1 and H_4 are known, we will assume that the total thickness of the formation F is also completely known on the estimated grid.

Table 3 Proportion of selected samples within the search neighbourhood used for estimation of values $z(\mathbf{x}_o)$

Number of samples	12	11	10	9	8	7	6	5	4
Proportion (%)	30.0	15.5	14.8	13.9	12.1	7.2	5.1	1.1	0.4

4 Estimation Neighbourhood

In practise, only a limited subset of sample data $z(\mathbf{x}_\alpha)$ is used in the estimation process. Typically, only n_o data closest to the location $\mathbf{x}_o$ are selected within a given search *estimation neighbourhood* centred on the location $\mathbf{x}_o$ being estimated (Goovaerts 1997). Several search strategies were tested due to highly irregular and preferential sampling pattern. Finally, unknown values of the 2100 grid nodes $\mathbf{x}_o$ were estimated using a moving neighbourhood with an isotropic radius of 650 m. The neighbourhood was split into 6 angular sectors with optimal 2 sample values per sector and minimal distance between two selected samples 250 m. Minimal number of samples in the neighbourhood circle used for estimation of unknown value $z(\mathbf{x}_o)$ were set to 4.

Minimal average distance of the samples selected within the neighbourhood to the location $\mathbf{x}_o$ is approximately 260 m and maximal one is 600 m. Table 3 gives details on the decreasing number of selected samples toward to the area boundaries. Due to the isotopic data configuration, the same neighbourhood was used for all estimation method.

5 Application of Sequential Methods

Figure 6 shows direct variogram models for the depth variables of the internal horizons H_2 and H_3. Due to evident non-stationary behaviour of the depth data, the quadratic trend for both variables was filtered out and the simple Gaussian structure was fitted on the residues. The variogam models are very similar because of significant correlation between the H_2 and H_3 depth variables. The variograms show very nice fitting up to distance 600 m what is the maximal average distance of samples within the search radius and grid nodes $\mathbf{x}_o$ where the unknown values are estimated. The variogam models were used to estimate the horizons H_2 and H_3 on the grid using non-stationary universal kriging method (Chiles and Delfiner 1999).

Figure 7 shows the isochore maps derived from the H_2 and H_3 horizon estimations. The obtained maps are not as smooth as the true ones because of higher variability of the input data compared to the true variability (see Tables 1 and 2). The formation isochore map is (and must be) the same as showed in Fig. 4d because the horizons H_1 and H_4 are known. What is more important is that we can see some black pixels on the south part of the isochore map of the zone Z_1 (Fig. 7a). In accordance to the map legend, it indicates the presence of some negative thickness values. It is due to crossing of the known horizon H_1 and estimated horizon H_2,

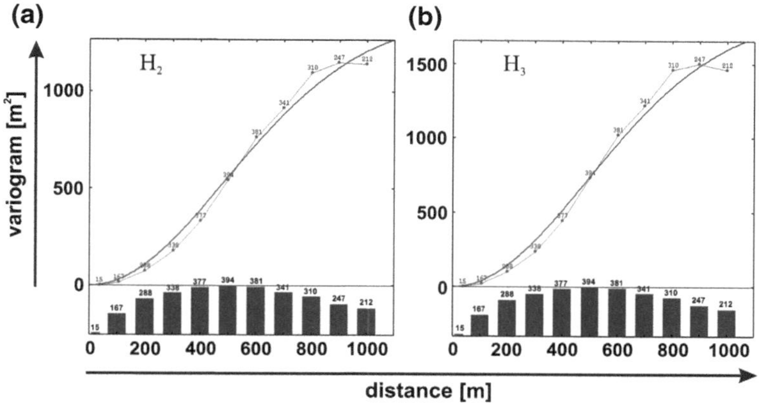

Fig. 6 Direct variogram models of the depth variables of the internal horizons H_2 and H_3

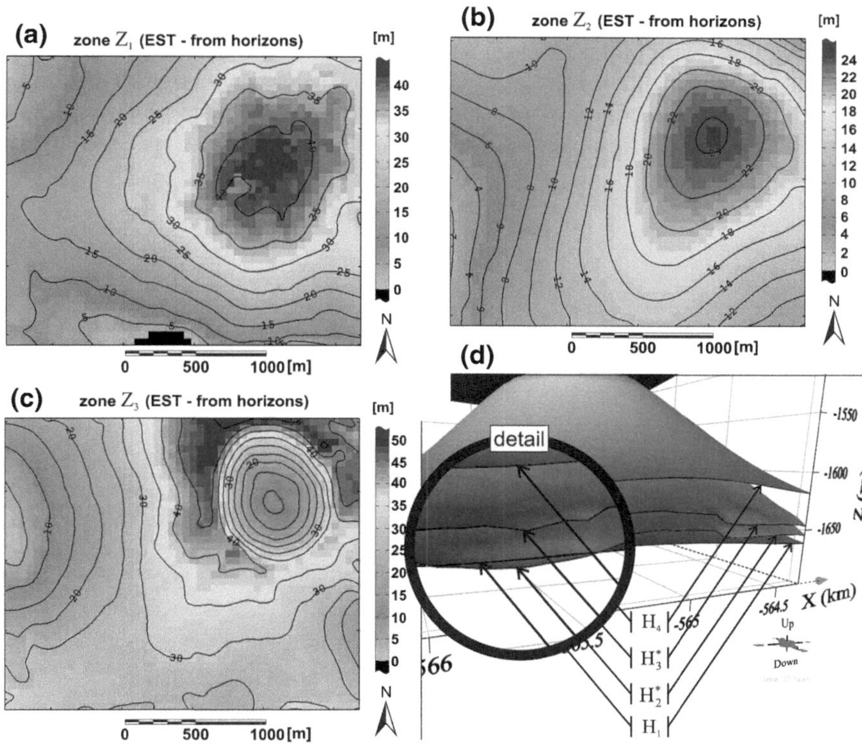

Fig. 7 The zone thicknesses derived from the depth of the internal horizons H_2 and H_3 estimations. Picture **d** shows the area where the estimation of the upper horizon H_2 crosses the true lower horizon H_1

Table 4 Basic statistical characteristics of the zone thicknesses derived from the horizon estimations

Thickness	Minimum (m)	Maximum (m)	Mean (m)	Std. deviation (m)
Zone 3 $[Z_3^*(th)]$	1.48	51.32	27.82	12.22
Zone 2 $[Z_2^*(th)]$	2.03	24.30	14.02	5.50
Zone 1 $[Z_1^*(th)]$	−3.56	43.14	22.60	11.37

what is shown in Fig. 7d. The result of the modelling therefore cannot be accepted without an artificial correction.

Table 4 gives the basic statistical characteristics of the zone thicknesses from the univariate horizon estimations. We can see the range of the zone Z_3 thickness is markedly overestimated at the expense of the thickness range of the zone Z_2. This overestimation goes further for the thickness range of the zone Z_1 where the negative values appear as it is demonstrated in Fig. 7a.

Figure 8 shows the variogram models for all three zone thicknesses. Due to the non-stationary behaviour of the thicknesses of the zones Z_1 and Z_2, the quadratic trend was removed from the input data. Stationarity of the zone Z_3 is caused by the erosion on the top of the structure, as stated previously.

The isochore maps of the zones Z_1 and Z_2 were obtained using universal kriging and the isochore map for the zone Z_3 was obtained using ordinary kriging (Armstrong 1998). Table 5 gives the basic statistical characteristics of the zone thicknesses from the univariate thickness estimations. The minimal and maximal values are slightly

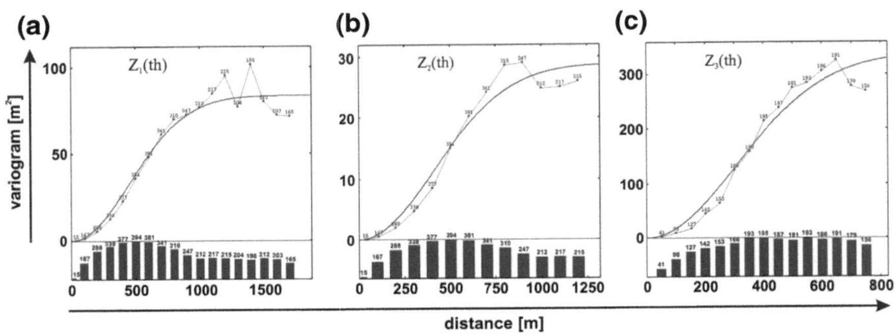

Fig. 8 Direct variogram models of the thickness variables of the zones Z_1, Z_2 and Z_3

Table 5 Basic statistical characteristics of the zone and formation thickness estimations

Thickness	Minimum (m)	Maximum (m)	Mean (m)	Std. deviation (m)
Zone 3 $[Z_3^*(th)]$	3.92	53.21	27.71	10.20
Zone 2 $[Z_2^*(th)]$	1.96	24.30	14.04	5.53
Zone 1 $[Z_1^*(th)]$	1.68	42.71	22.76	11.30
Formation [F(th)]	16.49	109.72	64.51	23.31

overestimated. The isochore maps show the similar “shaky” features as the ones showed in Fig. 7. More important is a comparison of the formation thickness derived from all zone thickness estimations to the true formation thickness.

In the Fig. 9 we see that the derived thickness of the formation does not match the true one. Differences ranking from −9.69 to +9.65 m with average value −0.02 m and standard deviation 1.51 m. In this case, the mean value is very low and shows only negligible underestimation of the total thickness within the studied area, but locally the differences are very high. Therefore, the result of the univariate zone thickness modelling cannot be accepted without a volume correction.

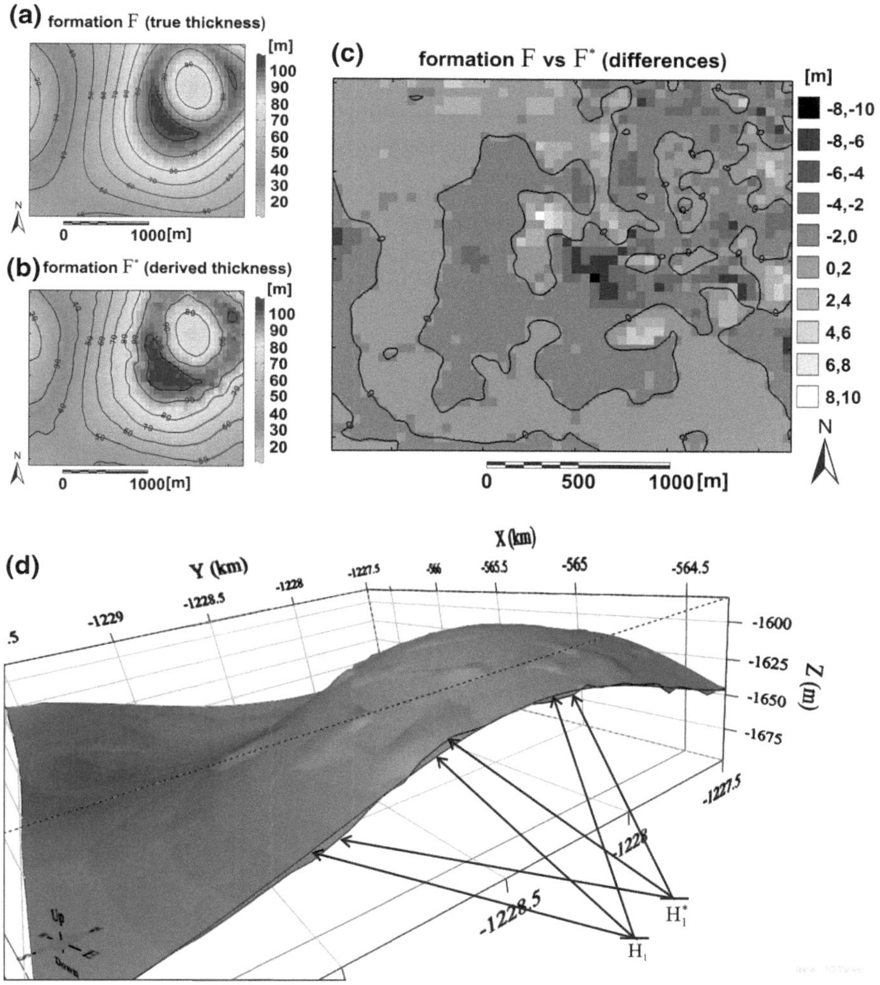

Fig. 9 The map of differences (**c**) of the true formation thickness (**a**) and the formation thickness obtained by adding together the thickness estimations of individual zones separately (**b**). Figure **d** shows differences between true horizon H_1 and the one derived from the zone thickness estimations

6 Application of Global Approach

For the global approach application, the thicknesses of the zones Z_3 and Z_2 were added together and a new cumulated thickness variable Z_{3+2} was created. The reasons to cumulate these two zones are as follows:

1. The zone Z_3 shows stationary behaviour unlike the zones Z_2 and Z_1. The stationarity of zone Z_3 is due to erosion of the top of the structure, as mentioned previously. Because the proportion of the Z_3 zone is much higher than the thinnest zone Z_2, the new cumulated variable also shows stationary behaviour.
2. Although there is a negligible correlation between zones Z_3 and Z_2 (−0.14), there is very strong correlation between the thickness of the zone Z_3 and the cumulated thickness Z_{3+2} (+0.89).
3. The presence of a well structured experimental cross-variogram between the thickness of the zone Z_3 and the formation thickness F guarantees the existence of a structured experimental cross-variogram between the cumulated thickness Z_{3+2} and the formation thickness F.

Figure 10 shows a complete multivariate model of coregionalisation of the thickness of zone Z_3, cumulative thickness of zones Z_3 and Z_2, denoted as Z_{3+2}, and total formation thickness F. The first row in the Fig. 10 shows the direct variogram

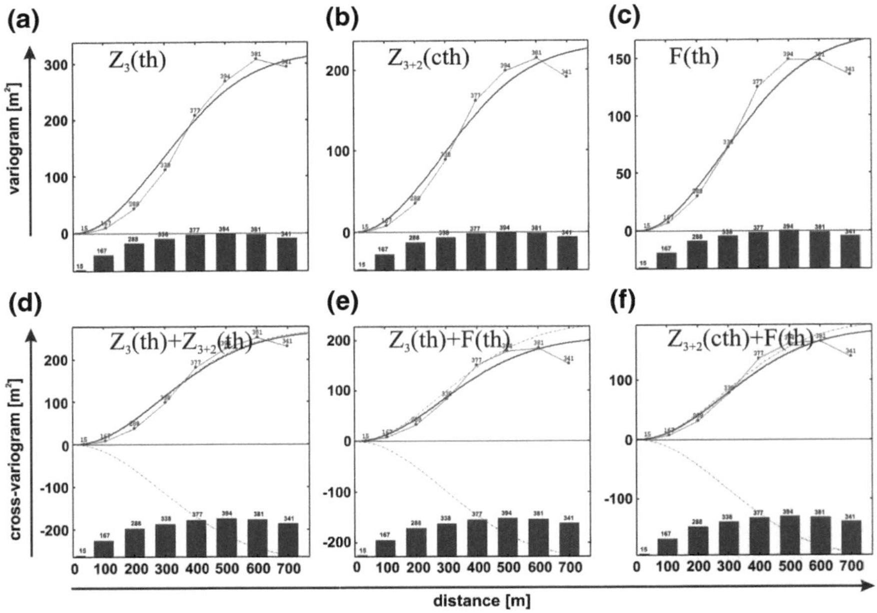

Fig. 10 A complete coregionalisation model of variability of the thickness of zone Z_3, the cumulative thickness of the zones Z_3 and Z_2 and the formation thickness. The first row shows direct variogram models and the second one shows respective cross-variograms between pairs of variables

models for the thickness variables and the second row shows the appropriate cross-variogram models for the pairs of variables. In fact, the direct variogram of the zone Z_3 thickness is the same as shown in Fig. 8c and its parameters were used as a reference to derive the coregionalisation model.

The isochore maps shown in Fig. 11 were obtained using multivariate cokriging of the thickness variable of zone Z_3, cumulative thickness variable Z_{3+2} and the formation thickness F. Since the formation thickness F is known not only in the sample positions $\mathbf{x}_\alpha$, but also in each of 2100 grid nodes $\mathbf{x}_o$, where the unknown values of the zone thicknesses are estimated, a collocated cokriging method was performed using the formation thickness F as an auxiliary variable. This ensures that the estimated zone thicknesses (in summation) respect the total thickness of the formation F. The isochore map of the zone Z_3 was cokriged directly together with the cumulated thickness Z_{3+2}. The isochore map of the zone Z_2 was obtained by subtracting the isochore map of zone Z_3 from the estimation of the cumulative

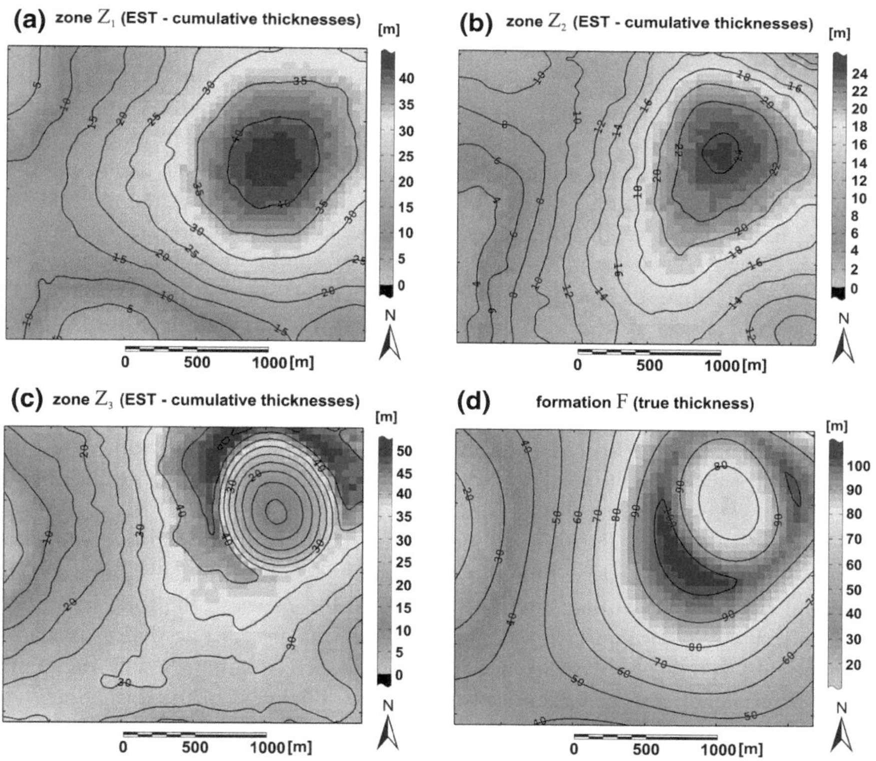

Fig. 11 The zone isochore maps derived from the multivariate estimation using global approach. Since the total formation thickness was used as an auxiliary collocated variable, the sum of the zone thickness estimations (**a–c**) is equal to the true formation thickness (**d**)

Table 6 Basic statistical characteristics of the zone thickness estimations cokriged from the cumulative thickness and the formation thickness

Thickness	Minimum (m)	Maximum (m)	Mean (m)	Std. deviation (m)
Zone 3 $[Z_3^*(\mathrm{th})]$	3.16	50.98	27.78	10.09
Zone 2 $[Z_2^*(\mathrm{th})]$	2.38	24.03	14.00	5.54
Zone 1 $[Z_1^*(\mathrm{th})]$	2.88	42.76	22.71	11.31

thickness Z_{3+2}. Finally, the isochore map of the zone Z_1 was derived by subtracting the cumulative thickness Z_{3+2} from the formation thickness F.

Table 6 gives the basic statistical characteristics of the zone thicknesses derived from the multivariate thickness estimations. The minimal value of the zone Z_3 is slightly underestimated and the ones for the zones Z_2 and Z_1 are overestimated. The maximal values are good reproduces as well as the mean and standard deviation values.

7 Conclusion

The paper demonstrates the application of two different approaches of estimating of the internal horizons of the subsurface geological structure—sequential and global.

The sequential methods are based on the univariate modelling of the internal horizons directly or direct modelling of the zone thicknesses. The first method has resulted in the negative thickness values for the lowermost zone Z_1 because of crossing of the true horizon H_1 by modelled internal horizon H_2. In fact, the negative values have appeared only in an extrapolation area, but it is expected that this situation may occur in some interpolation areas if a small number of the sparse data are used for estimation on a very fine grid, what if the most frequent case in practise. The second method has not guaranteed that the sum of the zone thickness estimations matches the total formation thickness. That means the univariate sequential methods call for an additional post-processing of the obtained results: elimination of the negative thickness values or replacing them by zero values in the first case or a volume correction in the second one.

The global approach is based on the multivariate modelling of the cumulative thickness of the zones and the total formation thickness. The alternative approach is to model the cumulative thickness proportions of the zones to the total formation thickness. The global approach has proved to be very useful one that produced a geometrically reliable geological model of the subsurface structure with maintaining the consistency between the zones and the total thickness of the formation. The multivariate methods, described and applied in the paper, provide an efficient way of estimating the isochore maps and deriving the internal horizons to build a subsurface structure model.

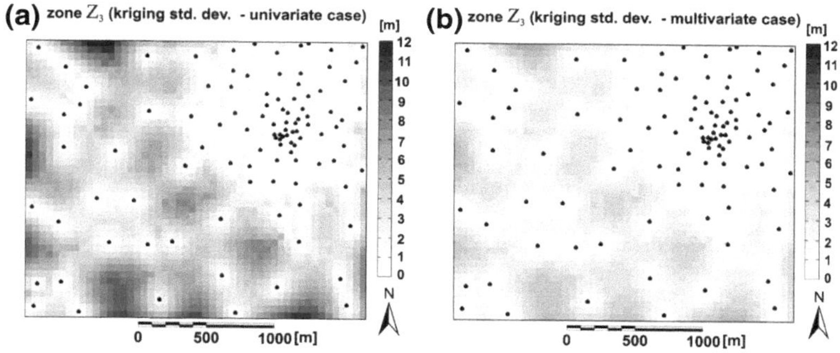

Fig. 12 A comparison of the kriging estimation variance (**a**) and cokriging estimation variance (**b**) for the thickness estimation of the zone Z_3

Another benefit of using the multivariate methods is decreasing of the estimation variance. Figure 12 shows a comparison of the kriging standard deviation map (a) and cokriging one (b) for the zone Z_3 estimation where we can observe almost 60 % decrease of the estimation variance for cokriging what could be advantageous for the global multivariate approaches.

It must be noted that the presented experimental comparison of the sequential and global approaches in this paper is limited to the one simulated structure of the anticline shape what has served as a true reality for the comparison of the modelling results. It would be interesting to compare the different approaches on more simulated structures of other shapes, with more complicated structural features, as for instance complex structural framework or anisotropy of a spatial variability.

Acknowledgments The authors are thankful to Nafta, a.s. Company for providing funding for Isatis software licencing. Funding for this study was provided by Slovak Scientific Grant Agency VEGA 1/0828/14.

References

Armstrong M (1998) Basic linear geostatistics. Springer, Berlin

Chichi H, Galli A, Ravenne Ch, Tesson M, de Marsily Ch (2000) Estimating the depth of stratigraphic units from marine seismic profiles using nonstationary geostatistics. Nat Resour Res 9(1):77–95

Chiles J P, Delfiner P (1999) Geostatistics: modelling spatial uncertainty. Wiley, New York

Dowd PA (1983) Conditional simulation of interrelated beds in an oil deposit. In: Verly G, Journel AG, Marechal A, David M (eds) Geostatistics for natural resources characterization. Reidel, Dordrecht

Dowd PA (2004) MINE5270 Multivariate geostatistics. Notes on MSc. In: Mineral resources and environmental geostatistics. University of Leeds, UK

Goovaerts P (1997) Geostatistics for natural resources evaluation. Oxford University Press, Inc., London

Isaaks EH, Srivastava RM (1989) An introduction to applied geostatistics. Oxford University Press, Inc., London

Journel AG, Huijbregts CJ (1978) Mining geostatistics. Academic Press, London

Matheron G (1962) Traité de Géostatistique Appliquée. Technip, Paris

Matheron G (1963) Principles of geostatistics. Econ Geology 58:1246–1266

Matheron G (1970) Random functions and their application in geology. In: Merriam DF (ed) Geostatistics—a colloquium. Plenum Press, New York

Matheron G (1971) The theory of regionalized variables and its application. Les Cahiers du Centre de Morphologie Mathématique. École Nationale Supérieure des Mines Paris. Fontainebleau, Paris

Olea RA (1991) Geostatistical glossary and multilingual dictionary. Oxford University Press, New York

Sancevero SS, Remacre AZ, Deraisme J, Souza O (2008) Volumetric characterization of resevoir using the multilayer approach. GEOSTAT 2008. Santiago de Chile

Ventcel'ová JS (1973) Teória pravdepodobnosti. Alfa, Bratislava

Wackernagel H (2003) Multivariate geostatistics, 3rd edn. Springer, Berlin

Xu W, Tran TT, Srivastava RM, Journel AG (1992) Integrating seismic data in reservoir modeling: the collocated cokriging alternative. In: 67th Annual Technical Conference and Exhibition of the Society of Petroleum Engineers held in Washington, DC. SPE 24742. Society of Petroleum Engineers, Washington DC, pp. 833–842, October 4–7

Yarus JM, Chamber RL (2006) Practical geostatistics—an armchair overview for petroleum reservoir engineers. SPE 103357. Distinguished Author Series, Society of Petroleum Engineers, pp. 78–87

Implementation of Poisson Faulting Algorithm with Circular Fault Function for Hardness Data Synthesis

Korneliusz K. Warszawski and Sławomir S. Nikiel

Abstract Natural erosion processes depend often upon terrain internal structure. Modelling such phenomena is very important in GIS-related (Geographic Information System) and computer graphics areas. Virtual environments takes advantage on improved realism of the images while GIS-based models benefits from plausible simulation output for urban development, landscape design or military disaster relief training. In cases, where terrain structure data cannot be derived from real geological records we can use physically-inspired models and use synthetic environments for simulation of wide class of terrains. This paper proposes a technique for generation of ground structure models related to their hardness. The method is based on the modified Poisson Faulting algorithm, that was originally used by Mandelbrot and Voss to generate fBs (fractional Brownian surfaces). The model defines hardness data to be simulated with regard to different geological materials and can be used to broaden variety of terrain models based on DEM (Digital Elevation Model) or height-field representation. We discusses the proposition of extending fractal generated terrain with hardness records. We use circle-based fault function for the Poisson Faulting algorithm. The results visually resemble natural objects observed in cross-section of soil and are highly similar to cross-section in open pit-mines. In contrast to a classic fractal parameter-driven models our proposition is a data-driven model for terrain synthesis, opening possibilities for dynamic simulations.

Keywords Terrain hardness model · Hardness-field · Terrain erosion modelling · Virtual environment

K.K. Warszawski (✉)
Faculty of Electrical Engineering, Computer Science and Telecommunications, University of Zielona Góra, ul. Szafrana 2, 65-246 Zielona Góra, Poland
e-mail: k.warszawski@weit.uz.zgora.pl

S.S. Nikiel
Faculty of Economics and Management, University of Zielona Góra, ul. Podgórna 50, 65-246 Zielona Góra, Poland
e-mail: s.nikiel@wez.uz.zgora.pl

K. Růžičková and T. Inspektor (eds.), *Surface Models for Geosciences*, Lecture Notes in Geoinformation and Cartography,
DOI 10.1007/978-3-319-18407-4_23

1 Introduction

Artificial terrains are used in areas as diverse as military simulation (e.g. Virtual Battlespace 3 from Bohemia Interactive Simulations), film and video game industry. The most straightforward approach to obtain digital terrain models is to use real elevation records, but when such an information is unavailable or the cost of its retrieval is prohibitive for the investment, a number of chaos- and fractal-driven algorithms can be successfully applied (Barnsley 1993; Crilly et al. 1993; Yokoya et al. 1989). It is possible then, to model visually convincing terrains and whole planets with chaotic models (based on fractal geometry). Typically, the background model is based on the fBs (fractional Brownian surface) (Mandelbrot 1982; Peitgen and Saupe 1988). Extending the fBm (fractional Brownian motion) with additional parameter defines fractal dimension of the underlying path structure. Fractal dimension is related to roughness or irregularity of the curve/surface. Two-dimensional array of fBs elevations over a polygonal mesh can be good approximation to a terrain (Yokoya et al. 1989; Musgrave et al. 1989).

The FFT-based (Fast Fourier Transform) digital modelling graphics applications provide better performance than those exploiting the classical cosine transform (Dixon 1994). Even more efficient terrain generating algorithms were proposed to solve that problem. The random MPD (Mid-Point Displacement) method was developed (Fournier et al. 1982) to quickly approximate fBs models for artificial landscapes. In order to obtain more realistically looking terrains the random-cuts, Poisson Faulting algorithm and the spherical subdivision method were proposed (Musgrave et al. 1989). Usually, a Gaussian distribution of random parameters is necessary to finely approximate the fBs model. It is very likely to observe the artefacts in the simplest terrain models generated with a straightforward fractal modelling methods. Most of them can be defined as visible stages of the grid, known as creases (Nikiel and Kirby 2000). It is possible to generalise the method of recursive mid-point subdivision to polygons of order 3–6 and 8 (Dixon 1994; Dixon et al. 2009). Subdivisions based on hexagons, were recommended by Mandelbrot (Dixon 1994). It is also possible to create an artificial terrain based on the IFS (Iterated Function Systems) interpolation techniques (Nikiel and Kirby 2000). If we extend the IFS from two- to three-dimensions, we can create the IFS surface patches, that resemble some cliff formations.

Computer Graphics models are based on surface representation with none or little information on internal structure (typically, the sub-surface texture models are used to mimic what is covered by the surface). In general, classic fractal parameter-driven terrain modelling can be described as sequential process (see Fig. 1). Fractal models, however, suffer from some visual artefacts. Strong sense of artificiality in fractal surfaces is caused mostly by the lack of reference to gravity and material deposition in the modelling methods. They are purely geometrical tools.

All the methods, referenced above, lead to artificial terrains that are highly symmetric. When reversed vertically, there is no visual difference in obtained surfaces. Fractal surfaces can be further processed, introducing weathering (smoothing), rivers

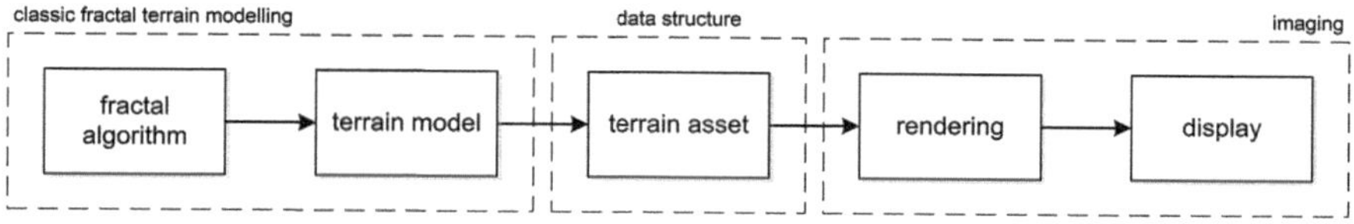

Fig. 1 Classic fractal terrain modelling and imaging process structure

(erosion) and other effects to remove or to conceal visual artefacts (Nagashima 1997). More advanced methods introduce basic knowledge about material of terrain like rocks, sand and mud (Benes 2007; Kristof et al. 2009; Bezin et al. 2011). Good knowledge about material susceptibility for erosion forces is a key element in realistic modelling of land deformation in geomorphological application (Bezin et al. 2014). When records of internal terrain structure is unavailable it can be synthesized with the method presented in the following sections.

2 Hardness Synthesis

The idea of hardness model is bound to particle-driven terrain synthesis model, where the controlled but stochastic stream of particles modify the terrain structure (Warszawski and Nikiel 2014a). In our preferred approach, data-driven modelling is a dynamic process (see Fig. 2).

The terrain erosion processes in natural phenomena are not purely geometric. Their influence depend on intensity of erosion forces (wind, rain, etc.) and resistance of the ground structure. The resistance is the point of interest in our research

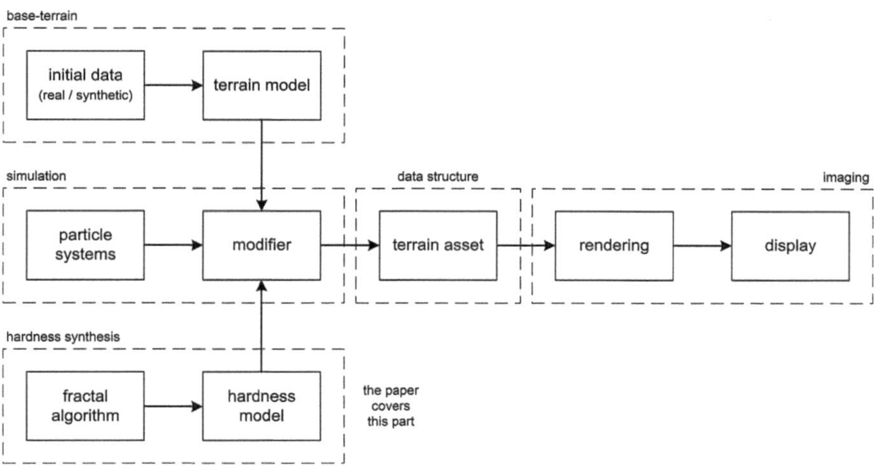

Fig. 2 The paper preferred terrain modelling and imaging process structure

work on hardness model for artificial terrain modelling. We suppose, that it may be helpful in erosion simulations in virtual environments and GIS-based projects.

2.1 General Structure

In our proposition, the terrain model is composed from two layers. The first one is a classic height-field. An additional, second layer stores information about the hardness of materials from which a given landscape will be constructed. This layer is called the hardness-field. We do not limit materials only to minerals. In our technique, one of them can be a water deposit or even an empty space-caverns (when the hardness is equal to zero). The mathematical model of this layer is described as a discrete function D in R^2 space, which is organized as *m-by-n* matrix, where $m, n \in N$. Individual records and dimensions are corresponding directly to the related height-field layer. The original hardness data generation technique (Warszawski et al. 2012) uses step-based fault function and provides models with sharp and irregular edges usually seen in rocky subsoil. The alternative method, proposed in this paper results in smoother and more regular surface comparing to the original one. It can be used for simulations of terrain erosion in milder processes.

The hardness synthesis for R^2 space models starts from fBs generation over the hardness-field layer. In this research, we used a modified Poisson Faulting method. The main idea of this variation is to replace the original step-based fault function with the circle equation and impose it on the layer. The general procedure is depicted in Fig. 3.

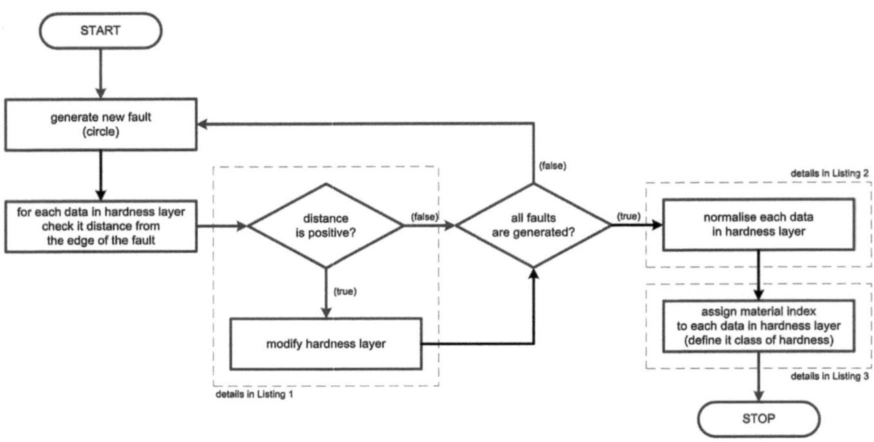

Fig. 3 General structure of the hardness synthesis algorithm with circular fault function

2.2 The Circular Fault Algorithm

Let $P_0 = (x_0, y_0)$ be a point with coordinates existing in a set defined with indexes of the model hardness layer *D*. Let *r* be a radius. Then P_0 and r define a circle on the layer *D* which is a border of the fault area. When the fault area is set, it is possible to check relations between the area and the cells of the model.

Let $P_{i,j} = (i, j)$ be a position of a cell in R^2 space where its coordinates are related to the indexes of the model layer D. Then the relation between the area and the cell can be described as the distance $S_{i,j}$ between them or referred to as the distance factor. The sign of its value also determines if cells are located inside or outside of the fault area. The calculation is shown in Eq. 1 and can be constructed as transformed equation of a distance from a point to the edge of the circle in R^2 space.

$$S_{i,j} = r^2 - \left((i - x_0)^2 + (j + y_0)^2\right) \tag{1}$$

The values stored in cells of the model hardness layer depend on the distance factor. Let $d_{i,j}$ be the current data stored in a cell from the model layer *D*. Let Δh be a difference between the heights on both sides of the fault circle. Then new cell data $d'_{i,j}$ is increased by the estimated fault height value, as is shown in Eq. 2 and Fig. 4 (left).

$$d'_{i,j} = d_{i,j} + \Delta h \tag{2}$$

The distance factor $S_{i,j}$ has positive value inside the circle fault area and reaches maximum at its centre. When it becomes negative outside the circle, it has no effect on the fault height value Δh. The Eq. 3 shows the dependencies. In our simulation, there is limited height of fault into range $\{\Delta h \in R \mid 0 \leq \Delta h < 1\}$, as is show in Fig. 4 (right).

$$\Delta h = \begin{cases} \sqrt{S_{i,j}/r^2}, & \text{if } S_{i,j} > 0 \\ 0, & \text{otherwise} \end{cases} \tag{3}$$

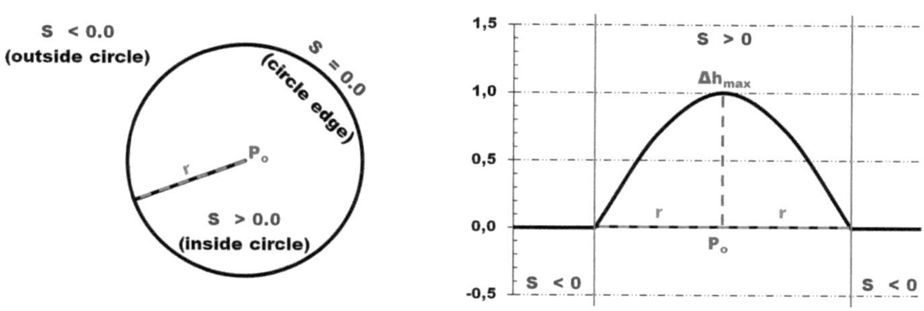

Fig. 4 Idea of poisson faulting method with circular fault function

2.3 Material Classification

The total number of geological materials used to construct the hardness layer is equal to the class of hardness of modelled terrain. This value refers to the model records indexing geological material and can be assigned for data $d'_{i,j}$ normalised to the interval $\{d'_{i,j} \in R^2 | 0 \leq d'_{i,j} < 1\}$. The Eq. 4 presents typical normalisation equation which can be done for each un-normalised cells $d_{i,j}$ with regard to *min* and *max* values of the records existed in the hardness layer with extra assumption that *min* < *max*.

$$d'_{i,j} = \frac{d_{i,j} - \min}{\max - \min} \tag{4}$$

Final modifications of the hardness-field layer assign all normalised cells to one of the geological materials. Its maximum is referred to the class of hardness of terrain model, so the value also informs us how many different materials are covered by the hardness model.

Let $d_{i,j}$ be the normalised record stored in a cell of the hardness layer D. Let λ be related to the class of hardness. Then new value of the cell $d'_{i,j}$ is equal to the floor of multiplication of cell data and the class of hardness. The restriction is done, if the initial record is equal to 1.0, to eliminate the possibility of exceeding the bounds of the hardness class model. This restriction limits all artefact records to the material with the highest index (see Eq. 5).

$$d'_{i,j} = \begin{cases} \lfloor d_{i,j} * \lambda \rfloor, & \text{if } d_{i,j} < 1 \\ \lambda - 1, & \text{if } d_{i,j} = 1 \end{cases} \tag{5}$$

Each stored indexes can be related to the Friedrich Mohs scale of mineral hardness and normalised again into (0, 1) range. The final hardness records $d''_{i,j}$ can be also generated with the Eq. 6 which automatically provides valid range for stored records.

$$d''_{i,j} = \frac{d'_{i,j}}{\lambda - 1} \tag{6}$$

The output of these calculations is the hardness-field layer of the terrain model. An example of the method progress is shown in Fig. 5.

2.4 Implementation

The processing of the data structure is ideal for parallelization and implementation in nVidia CUDA environment. This 'parallelism' of the proposed technique results in near real-time synthesis of the hardness model. The followed listings of the main method phases are shown as implementation examples.

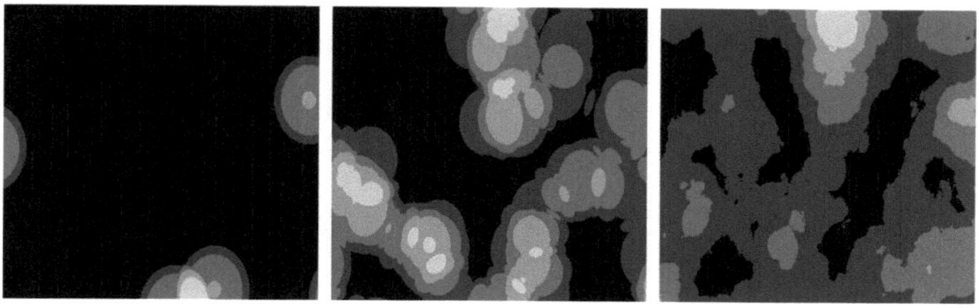

Fig. 5 Example of the method progress (*left*—after 5 steps; *middle*—after 50 steps; *right*—after 500 steps)

Listing 1. CUDA kernel for Poisson Faulting with circular fault function:

```
__global__ static void Faulting(float p0_x, float p0_y,      // position of fault center
                                float radius,                // radius of fault area
                                int cols, int rows,          // resolution of layer
                                float* dev_data)             // layer data on CUDA device
{
    int index = blockIdx.x * blockDim.x + threadIdx.x;       // cell index
    if (index >= cols * rows) return;                        // index cannot exceed layer size

    float px = static_cast<float>(index % cols);             // cell coordinates
    float py = static_cast<float>(index / cols);             // in 2D space

    if (   px > p0_x + radius || px < p0_x - radius          // execute only if cell
        || py > p0_y + radius || py < p0_y - radius) return; // is near circle center

    float sq_radius = powf(radius, 2.0f);                    // squared radius

    float distance = sq_radius - (powf(p1_x - p0_x, 2.0f)    // distance from fault circle
                              + powf(p1_y - p0_y, 2.0f));    // to this cell

    if (distance > 0)                                        // execute only if distance is positive
    {
        *(dev_data + index) += sqrtf(distance / sq_radius);  // increase value stored in cell
    }
}
```

Listing 2. CUDA kernel for layer data normalization:

```
__global__ static void Normalize(float min, float max,       // extreme values of layer
                                 int cols, int rows,         // resolution of layer
                                 float* dev_data)            // layer data on CUDA device
{
    int index = blockIdx.x * blockDim.x + threadIdx.x;       // cell index
    if (index >= cols * rows) return;                        // index cannot exceed layer size

    *(dev_data + index) = (*(dev_data + index) - min)        // normalize value stored in cell
                        / (max - min);                       // into 0.0 to 1.0 range
}
```

Listing 3. CUDA kernel for layer data hardenization:

```
__global__ static void Hardenize(int value,                  // class of hardness
                                 int cols, int rows,          // resolution of layer
                                 float* dev_data)             // layer data on CUDA device

{
    int index = blockIdx.x * blockDim.x + threadIdx.x;        // cell index
    if (index >= cols * rows) return;                         // index cannot exceed layer size

    if (*(dev_data + index) < 1.0f)                           // hardenize value stored in cell
    {
        *(dev_data + index) = static_cast<int>                // for each cell
                              (*(dev_data + index) * value); // with value less than 1.0
    }
    else                                                      // otherwise
    {
        *(dev_data + index) = static_cast<float>(value - 1); // eliminate artefacts
    }
}
```

3 Results

The experiments were performed for selected resolutions (320 × 240, 640 × 480, 1280 × 960) of hardness-field layers and a number of faults between 1000 and 9000 for each resolution. The process was repeated 100x times for the defined resolution and fault count. In addition, the required method attributes (the range and the fault position) were defined by a pseudo-random number generator. All simulations processes were completed on the standard nVidia GeForce GTX 560 Ti GPU and Intel Core i7-860 CPU. The prototype simulation program was implemented in C++ with nVidia CUDA technology support.

Like in the classical step-based function variant, the numerical complexity of the method depends on the total number of generated faults (circles). The experimental results of our simulations also confirm linear nature of the method in the circle-based fault function variant. Performance diagrams (Fig. 6) are derived from the

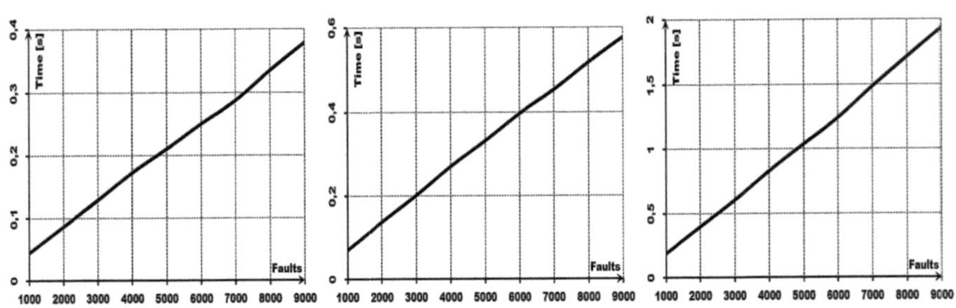

Fig. 6 Averaged performance diagram of hardness layer generation (*left*—320 × 240; *middle*—640 × 480; *right*—1280 × 960)

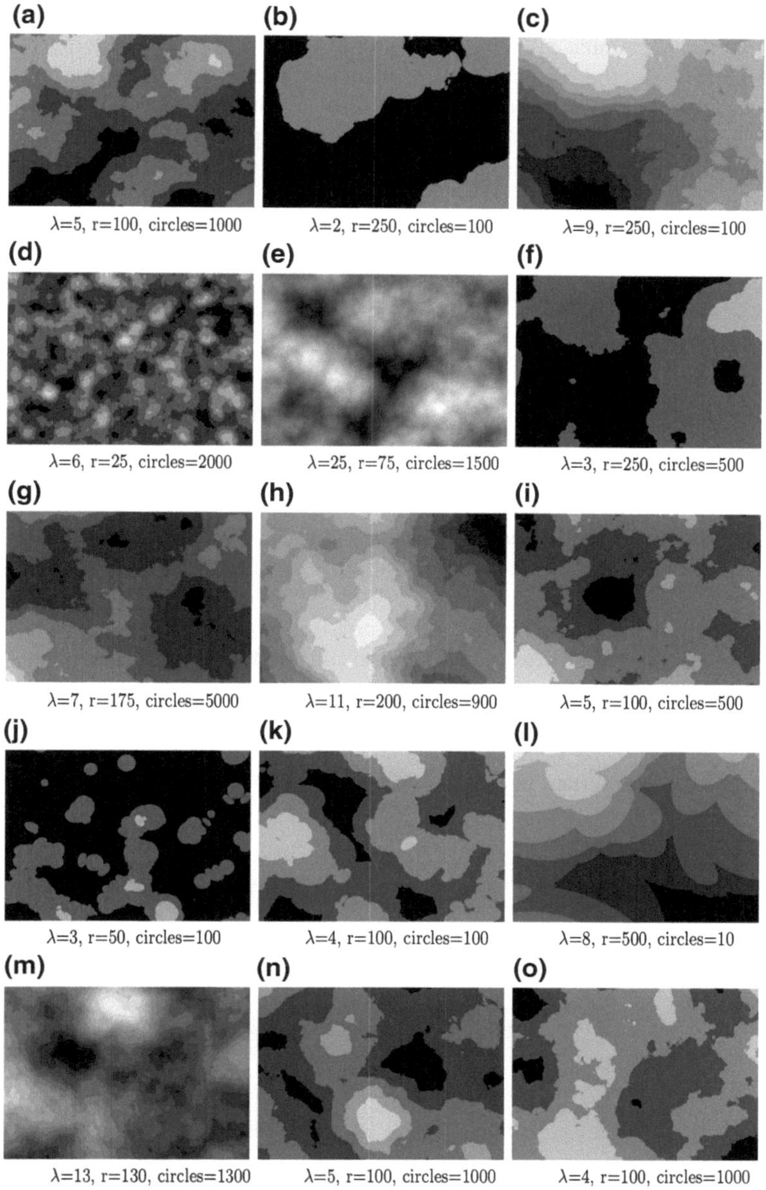

Fig. 7 Examples of generated hardness layer in grey-scale representations (λ—class of hardness, r—circle radius)

simulations. They show that performance of proposed technique depends mostly on the number of generated faults (circles) and resolution of the underlying model.

The terrain structure achieved as result of hardness-field generation algorithm, shows structures similar to those observed as horizontal cross-sections of soil. The model can be used to further simulations of erosion processes and its influence to landscape shape and structure. The proposed circle-based function as faulting technique, supports production of more regular forms than in step-based one. Figure 7 presents several hardness layers generated with the proposed method.

4 Conclusions

The main advantage of the hardness model for terrain modelling is the dynamic approach where we can model the final results of erosion processes i.e. canyons by the river flow, depicted at Fig. 8, or weathering of the mountains, shown in Fig. 9 (all samples was rendered in Terragen). We can control the surface synthesis process indirectly, by the data/underlying structure, and not by the parameterization typical to classical fractal surface generation algorithms (Warszawski and Nikiel 2014b).

The performance of the method offers possibility to make many generations of different hardness structures for a single terrain model in interactive rates, which can be the basis for further simulations of erosion forces affecting on the landscape shape and consistency. The use of Poisson Faulting algorithm with circular fault function, enables us to generate clustered hardness data, which is comparable to natural structures observed in cross-sections of terrain. The hardness-fields obtained

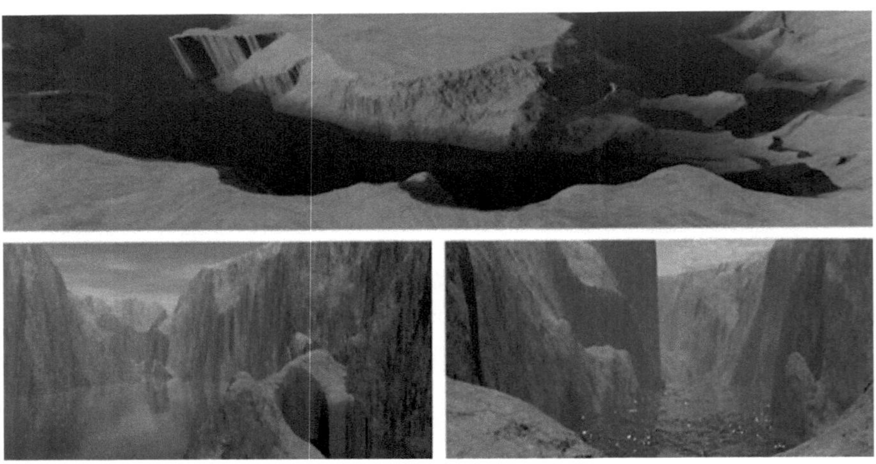

Fig. 8 Sample canyon generated by particle flow algorithm using hardness layer

Fig. 9 Sample weathered peak of a mountain generated by erosion algorithm using hardness layer

with the method are suitable for simulation of more regular and smooth shapes geological structure than in the classical step-based variant.

The further research will be concentrated on adoption of the model for landscapes based on the voxel representation. It will be potentially helpful in the studies on effect of erosion in three-dimensional models, i.e. caves, rock shelves or cliffs. Moreover, this would be very helpful in development of interactive landscape modelling tools for virtual terrain design.

Acknowledgments

HUMAN CAPITAL
HUMAN – BEST INVESTMENT!
Lubuskie
Worth your while
EUROPEAN UNION
EUROPEAN
SOCIAL FUND

The author is a scholar within Sub-measure 8.2.2 Regional Innovation Strategies, Measure 8.2 Transfer of knowledge, Priority VIII Regional human resources for the economy Human Capital Operational Programme co-financed by European Social Fund and state budget.

References

Barnsley MF (1993) Fractals Everywhere. Academic Press, San Diego, USA

Benes B (2007) Real-time erosion using shallow water simulation. In: Proceedings of workshop in virtual reality interactions and physical simulation (VRIPHYS '07), Dublin, Ireland, 43–50

Bezin R, Peyrat A, Crespin B, Terraz O, Skapin X, Meseure P (2011) Intearctive hydraulic erosion using CUDA. Mach Graph Vis 2(20):157–172

Bezin R, Crespin B, Skapin X, Terraz O, Meseure P (2014) Generalized maps for erosion and sedimentation simulation. Comput Graph 45:1–16

Crilly AJ, Earnshaw RA, Jones H (1993) Applications of fractals and chaos. Springer, New York, USA

Dixon AR (1994) The synthesis of fractal terrain and its application to flight and space simulation. Ph.D. Thesis, University of Hull, Hull, England

Dixon AR, Kirby GH, Wills DMP (2009) Artificial planets with fractal feature specification. Vis Comput 3(15):219–228

Fournier A, Fussell D, Carpenter L (1982) Computer rendering of stochastic models. Commun ACM 6(25):371–384

Kristof P, Benes B, Krivanek J, Stava O (2009) Hydraulic erosion using smoothed particle hydrodynamics. Comput Graph Forum 2(28):219–228

Mandelbrot BB (1982) The fractal geometry of nature. H.W. Freeman and Co., San Francisco, USA

Musgrave FK, Kolb CE, Mace RS (1989) The synthesis and rendering of eroded fractal terrains. In: Proceedings of annual conference on computer graphics and interactive techniques (SIGGRAPH '89), Boston, USA, pp 41–50

Nagashima K (1997) Computer generation of eroded valley and mountain terrains. Vis Comput 13:456–464

Nikiel S, Kirby GH (2000) Iterated function systems for terrain synthesis. Mach Graph Vis 1/2 (9):47–55

Peitgen H-O, Saupe D (1988) The science of fractal images. Springer, New York, USA

Warszawski K, Nikiel S, Zawadzki T (2012) Hardness data synthesis for height-field based landscape models. In: Proceedings of surface models for geosciences (GIS '12), Ostrava, Czech Republic, pp 275–284

Warszawski KK, Nikiel SS (2014a) Particle systems-based riverbed modelling over a terrain with hardness layer. In: Poster proceedings of International conferences in central Europe on computer graphics, visualization and computer vision (WSCG '14), Plzen, Czech Republic, pp 49–53

Warszawski KK, Nikiel SS (2014b) A proposition of erosion algorithm for terrain models with hardness layer. J Theor Appl Comput Sci 8(1):76–84

Yokoya N, Yamamoto K, Funakubo N (1989) Fractal based analysis and Interpolation of 3D natural surface shapes and their application to terrain modelling. Comput Vis Graph Image Process 46(3):289–302

Photogrammetric Point Clouds for GIS-Based High-Resolution Estimation of Solar Radiation for Roof-Top Solar Systems

Roland Zink, Luis Ramirez Camargo, Patrick Reidelstürz and Wolfgang Dorner

Abstract Solar radiation is the main factor that must be considered when predicting solar systems energy output. Especially in built-up areas the use of Digital Surface Models (DSM) and analysis with Geographic Information Systems (GIS) has gained popularity for these predictions in recent years. However, the identification of small objects and their impact in the shadowing of solar systems is a very important factor that cannot be entirely acquired by models from common available Light Detection and Ranging (LiDAR) data. Thus higher resolutions DSMs are required. These can be generated from photogrammetric point clouds derived from Unmanned Aerial System (UAS) based large scale image flights. This paper presents a comparison between a common available resolution DSM generated from standardized LiDAR data and a DSM with very high resolution from photogrammetric point clouds. The added value of using high-resolution DSMs for solar radiation estimation for PV rooftop-systems is discussed with a focus on elevation, slope, aspect and on the suitability for selection of usable roof-top areas. Although DSMs with resolutions up to 1 cm have been generated, the results show that a DSM of 25 cm is enough for considering all small roof-top objects that are relevant for estimation of solar radiation.

Keywords LiDAR · Point clouds · UAS · Photogrammetry · GIS · Solar systems · Photovoltaic

R. Zink · L. Ramirez Camargo (✉) · P. Reidelstürz · W. Dorner
Applied Energy Research Group Technologie Campus Freyung,
Deggendorf Institute of Technology, Grafenauer Straße 22, 94078 Freyung, Germany
e-mail: luis.ramirez-camargo@th-deg.de

R. Zink
e-mail: roland.zink@th-deg.de

L. Ramirez Camargo
Institute of Spatial Planning and Rural Development, University of Natural Resources and Life Sciences, Vienna, Austria

K. Růžičková and T. Inspektor (eds.), *Surface Models for Geosciences*,
Lecture Notes in Geoinformation and Cartography,
DOI 10.1007/978-3-319-18407-4_24

1 Introduction: DSM for Solar Radiation Calculation

One of the most promising renewable technologies is the use of solar based systems, including photovoltaic and solar thermal systems. Thus, in recent years—founded by government subsidies—especially photovoltaic panel installation was experiencing a strong upswing in many countries like the USA, China, Germany, Spain or Italy (Lenardič 2011; Jäger-Waldau 2013). As a consequence of the focus on state subsidies, the search for suitable sites is dominated by the economic profitability. To indicate areas or roofs which provide the maximum amount of energy per year or promise economic profitability, solar potential maps and solar cadastre maps are a popular information base (Jakubiec and Reinhart 2013; Kanters et al. 2014). These data often are available as WebGIS where roof surfaces suitable for PV are shown in a clear way (Agugiaro et al. 2012; Theodoridou et al. 2012; Ludwig et al. 2009). Calculation basis of these maps are GIS tools like r.sun in GRASS GIS (Hofierka and Suri 2002) or Solar Analyst in ESRI ArcGIS (Fu and Rich 1999) and DSMs. The maps also provide information at city and municipal scale about the annual solar radiation on a surface (kWh/year/m^2), the average sunshine duration (hours/year), the estimation of electricity production from roof-top photovoltaic systems in kWh per year or only the roofs where photovoltaic or solar thermal modules can be installed. This information seems to be acceptable for the investor or building owner, but not to ensure security of energy supply.

Although roof-top installations are interesting from an ecological perspective, the fluctuating power output is a serious disadvantage. Consequently, more detailed information is necessary, if the system is used as stand-alone photovoltaic power system or is intended to be integrated in a virtual power plant. In both cases corresponding storage systems or other renewable energies must be installed to compensate daily and seasonal fluctuations. As part of one of Germany's largest electro-mobility research projects, "E-WALD", a virtual power plant based on renewable energies for the provision of electricity for electric mobility is developed (http://e-wald.eu/). The research area is located in the southeast of Bavaria and is spanning about 7,000 km^2 of the Bavarian Forest and some of its neighbour-districts. For this rural region low mountains with elevation differences up to 500 m and challenging weather-conditions, including cold winters and a yearly average temperature of 4–6 °C are characteristic.

The challenge for supply systems like virtual power plants is to integrate the fluctuating photovoltaic technology. Therefore the aim is to improve the estimation of solar irradiance using a high resolution DSM, which guarantees the integration of detailed shadowing in the modelling and appropriate recognition of roof-top structures (Ramirez and Zink 2014). Ramirez et al. (2014) utilize the aspect maps with 1 m resolution from LiDAR data to automatically recognize homogeneous areas of the roofs for a certain number of orientations to generate synthetic time series for potential PV plants. The result of this recognition process is very important since a time series is generated for every single surface. This implies that the identification of several surfaces for describing a part of a roof that is actually one single homogeneous surface would increase the computation time and the storage requirements.

In this paper a comparison between a common available resolution and a very high resolution DSM is presented for a study object in Löwendorf, a village in the E-WALD region. The first DSM is generated from standardized LiDAR data and the second from photogrammetric point clouds derived from UAS based large scale image flights. In order to recognize the impact of different pixel resolutions on the most important parameters for solar radiation estimation on roof-tops—elevation, slope and aspect—several DSMs ranging 1 cm to 1 m were generated from the photogrammetric point clouds. The analysis is performed by generating aspect and slope maps from the LiDAR-DSM and also from the photogrammetric point cloud with resolutions 1 cm, 7.5 cm, 10 cm, 25 cm, 50 cm and 1 m. These are compared between each other and the slope maps are used for a classification of usable roof top areas. The results of the comparison and the classification are discussed and a desirable resolution for the DSM, which allows considering all small roof-top objects that are relevant for estimation of solar radiation, is defined.

2 Digital Surface Models

2.1 LiDAR

The acronym LiDAR corresponds to Light Detection and Ranging. It is an active optical remote sensing technology that emits laser pulses. The position and height values derived from the reflected laser pulses are stored as 3D coordinates in order to constitute dense point clouds. If there are several reflection surfaces for the same pulse, for example in the case that the pulse hits vegetation or buildings, it is splitted in an amount of returns that coincides to the number of reflective surfaces. The Bavarian agency for surveying provides data sets of first pulse points (FPL) and last pulse points (LPL). These are also divided in defined categories for the LPL (certain ground point, uncertain ground point, object point and building point) and variable categories for the FPL (the classification is not the same for all data sets). The density of the used data reaches four points per square meter and covers one square kilometre around the study object. The FPL and the spline approximation method implemented in GRASS GIS by Mitasova et al. (2005) are used to generate a DSM with 1 m^2 (DSM1).

2.2 High Resolution Photogrammetric Point Clouds

To generate a very high resolution DSM an UAS based photogrammetric point cloud is derived. The images are taken from a low flying multicopter platform. For this a “falcon 8” from ascending technologies is used as sensor carrier. As sensor the 14 megapixel digital camera “Panasonic DMC LX3” is applied. Images are taken from different heights, starting from 30 m up to 60 m above ground. The flight controller stabilizes the whole system and holds the flight altitude while a gimbal (stabilized camera platform) reduces residual movements on the camera. As flight strategy, in manual flight mode parallel lines and crossed lines are flown. While flying, a GPS tagger is logging positioning data. These data are written to the EXIF image header using the software "GPS photo tagger" in post-processing. Also satellite based augmentation data are received. In addition the system time of the digital camera is synchronized with GPS UTM time allowing the GPS image tagging.

To build the point cloud the images with the best resolution (height of 30 meters above ground) are selected. A photogrammetric bundle adjustment software tool is applied for the 3D reconstruction. Such structure from motion (SFM) tools differ from the classic aerial photogrammetric ones (aerotriangulation). The image orientation is not reconstructed using stereo pairs but the image block itself will be triangulated as a whole bundle. This procedure was implemented in the software "rolleimetric" for analytical terrestrial photogrammetric 3D reconstructions already in the 1990s (Bondar et al. 1997; Riedel 2002). In recent years the computer vision community developed such bundle adjustment systems as high performance full automatic digital 3D image orientation systems. These systems were also complemented with forward intersection high density 3D point cloud adjustment. More detailed and precise results would be able using "Semi-Global Matching" (Hirschmueller 2011). The use of SFM methods has become possible with the increasing performance of consumer PCs. Meanwhile, there are some providers of 3D photogrammetric reconstruction software based on this principles like "aSPECT 3D" or "Agisoft PhotoScan Professional" as well as some cloud based solutions e.g. "Autodesk 123D Catch". Open Source alternatives are "Visual SFM", "Python Photogrammetry Toolbox" or "ArcheOS".

While classical aerotriangulation strongly depends on requirements concerning scale-invariant and small attitude differences taking aerial images, much less restriction is given using bundle adjustment. Therefore such SFM concepts are very relevant for UAS based photogrammetric applications, since these problems are quite common if small and light UAS platforms are employed. A feature point matching tool identifies homologous points in different images and triangulates them. At the end the inner orientation of the image (camera calibration) is found and the images are properly orientated. This means, that the camera positions (x, y, z) and rotation angles (omega, phi, kappa) are known for each image. With such tools plenty of UAS based photogrammetric 3D reconstructions under challenging flight conditions could be done (Drauschke et al. 2014; Reidelstuerz 2012; Zink et al. 2012). Forward intersection is applied to build a dense cloud i.e. each pixel will be intersected with the homologous pixels of the overlapping images. Computing the backward intersection, 256,336 homologous points are found to orientate 40 images. On this basis a dense point cloud with 48,627,743 points is reconstructed. Finally, for each of them the three dimensional coordinates are known (Fig. 1). All points cover a study area of 2,000 m^2 resulting in a point density of 24,313 points/m^2.

The software "Agisoft PhotoScan Professional" is used for image orientation and dense point cloud construction using forward intersection triangulation. Therefore subpixel precision is required. Since photogrammetry is using digital images,

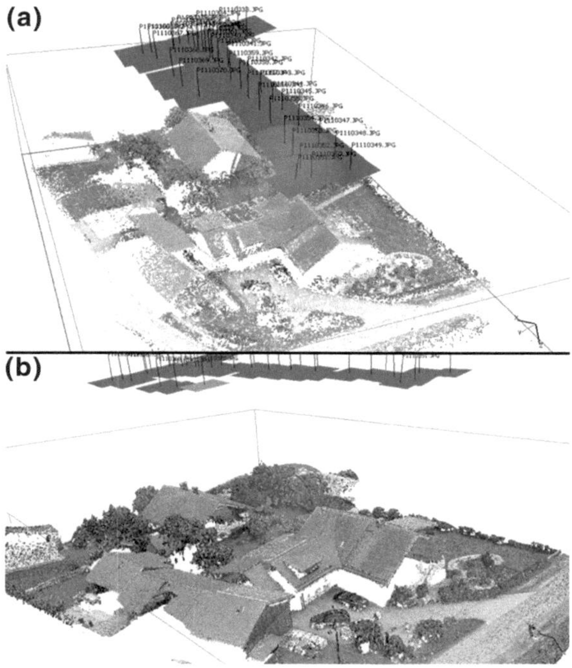

Fig. 1 **a** Image orientation using 256,336 homologous points. **b** Point cloud with 48,627,743 3D points

residuals are defined using "pixel" as units i.e. results of photogrammetric image orientations can be compared without considering different image formats and focal length combinations. The mean error of the orientation in this study amounts 0.461 pixels. The worst error is 0.538 and the best 0.396 pixels. Thus the precision of the reconstruction can be named as very good i.e. there is only minor noise in pixel positioning. As result a 3D point cloud with a resolution from 0.6 mm is available (Fig. 2). From this data DSMs with the following resolutions are derived: 1 cm, 7.5 cm, 10 cm, 25 cm, 50 cm and 1 m. These DSMs are characterized by very low

Fig. 2 Detailed point cloud with resolution of 0.6 mm shows 3D details

roughness and noise. Missing values in the DSMs obtained from the high resolution photogrammetric point clouds are filled with the same approximation method applied for the generation of the DEM1 from LiDAR.

3 Comparison of DSMS: Elevation, Aspect and Slope

The DSMs from both data sources are the basis for the comparison. Slope and aspect maps are calculated from the DSMs using the terrain analysis tool r.slope. aspect of GRASS GIS. Since the areas of interest are only the roof-tops of the buildings, the contours of the roof-tops are drawn and stored in a vector map. This vector map is utilized to clip the DSMs, aspect and slope maps in all available resolutions. The comparison of the DSMs is performed calculating the arithmetic differences between the maps with 1 cm resolution and each one of the other resolutions, as well as the differences between the maps with 1 m resolution from LiDAR data and each one of the other resolutions. Aspect maps cannot be properly compared in the same way since these include angular data (Gaile and Burt 1980). For instance consider two surfaces with aspect values 359° and 1° with 0° as east. The arithmetic difference is 358° although the actual difference is only 2°. Therefore, the comparison of the aspect maps is performed using Eq. 1 for the same resolutions considered in the comparison of the DSMs.

$$\text{AngleDiff} = \arccos(\sin(\text{map1}) * \sin(\text{map2}) + \cos(\text{map1}) * \cos(\text{map2})) \quad (1)$$

where map1 and map2 are the maps to be compared.

In the case of the slope maps both procedures for the comparison can be used. We opted for the second alternative using Eq. 1 also for these maps to be coherent with the handling of angular data. Figure 3 presents a selection of maps, namely, the DSMs, aspect and slope with resolutions 1, 25, 50 cm from the high resolution photogrammetric point clouds and 1 m from LiDAR as well as the comparison between the 1 m and the 1 cm maps in 1 cm resolution. Three neighbour buildings of different complexity and size are presented in every map, named from the west to the east with the letters A, B and C (Fig. 3).

4 Results

4.1 Elevation

While in the original 1 cm point cloud the structures of the roof tiles can be identified (Fig. 2), this seems not to be possible in the DSM0.01. However, the corresponding slope and aspect maps in Fig. 3 show that this fine information about the structures is not lost. Instead it is the resolution of the grey-scale range including

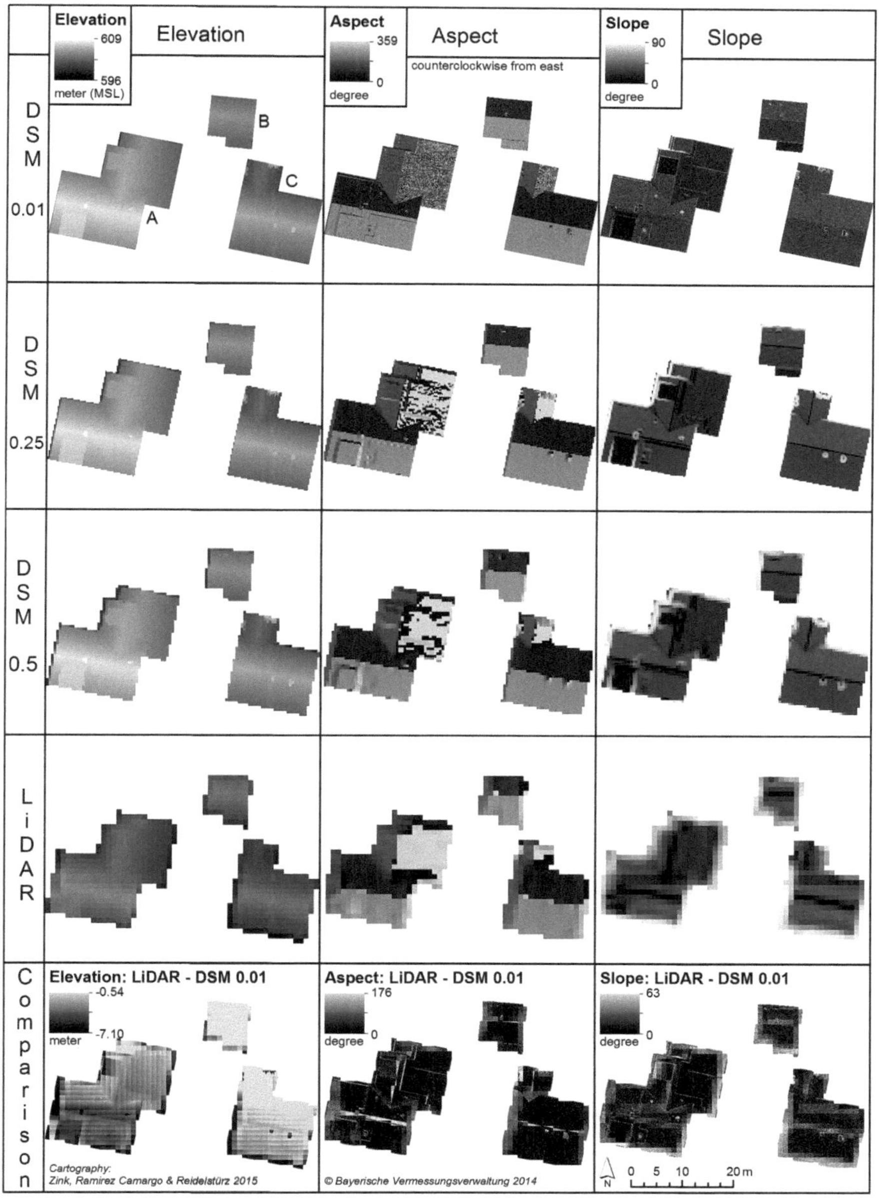

Fig. 3 Comparisons of DSMs: elevation, aspect and slope

256 digital numbers in the illustration of the DSM0.01 that is not high enough to show these elements. Nevertheless the resolution is enough to recognize every detail of small objects as smokestacks or skylights. The roof-edges are sharp and the

ridge is well defined. At the northern side of building C, tree branches over the roof can be recognized. At the southern roof part of building A, the circumference of a dormer is well defined. Similarly, a roof-top PV power plant can be seen in the southeast of building A. DSM0.075, DSM0.1 and DSM0.25 are only marginally discrepant from the DSM0.01, even though the fine resolution of the structures does decrease.

The DSM0.5 represents a threshold in terms of resolution for the elevation parameter. On the one hand none of the objects that can be visually recognized in the DSM0.01 is lost. On the other hand the fine structures are rasterized in a way that the definition of their edges and dimensions is no longer obvious. Furthermore, this is confirmed with the comparison maps, where the largest differences appear in the edges of the roofs and roof objects. With decreasing resolution over 0.5 m, small objects are no longer clearly recognizable. Information about the existence, the position or the dimension of small smokestacks is lost in the DSM1. The tree branches in the northern part of Building C are still influencing the DSM1, but they cannot be recognized as objects beside the roof. Roof-edges and ridges are now rasterized and not well defined. The dormer cannot be recognized as dormer and could be confused with other roof parts as any roof fittings. Moreover, the approximately 20 cm gap between the roof and the roof-top power-plant is too narrow to be represented in the DSM1. The elevation values in the LiDAR based DSM1 are in general slightly lower than the values in all DSMs based on photogrammetric UAS flights (Fig. 4). Automatic positioning and remotely acquired ground control points are not enough for a perfect matching. However, the only additional observable difference between the DSMs with 1 m resolution from both sources is that the large smokestacks, which are hardly recognizable from DSM1 from the photogrammetric source, disappear entirely from the LiDAR based DSM1. Differences are summarized in Table 1.

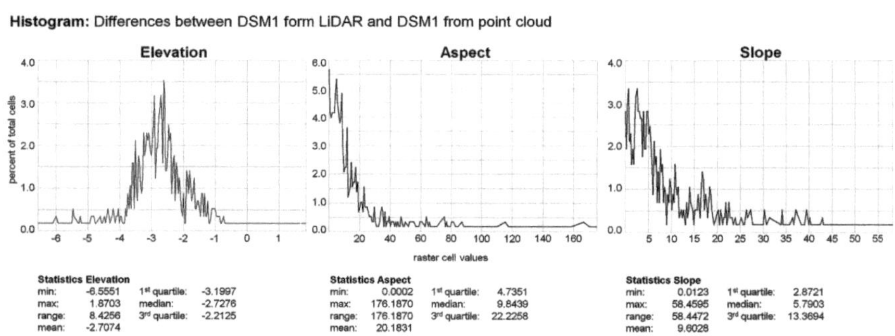

Fig. 4 Histogram of elevation, aspect and slope differences between DSM1 LiDAR and DSM1 from photo-grammetric point cloud as percentage of total cells

Table 1 Summary of the comparison of DSMs

	DSM 0.01	DSM 0.25	DSM 0.5	LiDAR
UAS flight 30–50 m altitude (same flight)	X	X	X	
Land surveying office LiDAR data				X
Mean residual of image orientation (pixel)	0.461	0.461	0.461	–
Maximum resolution of source (m)	0.006	0.006	0.006	–
Classification of perceptibility				
Differentiation of single tile				
Roof edge				
Normal dimensioned skylight of a room in the roof				
Small skylights as at storage rooms in the roof				
Small details like antennas				
Smokestack				

Sharp recognizable	Recognizable	Not recognizable

4.2 Slope

As in the slope map from the DSM0.01 it is possible to recognize the slope of the different faces of a single tile, in the slope map of the DSM1 (independently of the data source) the roof structure disappears almost entirely. The roof structure and objects are still observable in the slope map of the DSM0.5. However the differences in the edges (already with 50 cm resolution) are more evident than in the case of the DSMs. The majority of the slope map values from the DSM1 are similar to the values of the slope map from the DSM1 from LiDAR data (see Fig. 4). 90 % of the cells present differences less than 22.48°. The differences in the values are distributed around 5.79° (median) and 50 % of all values are concentrated between 2.87° (first quartile) and 13.36° (third quartile). The highest differences in the slope maps, as could be expected from the differences in the DEMs, appear in the edges of the roofs and the edges of the roof's objects. These are up to 59° for all slope comparison maps.

4.3 Aspect

In the case of the aspect maps it seems to be similar to the slope and elevation maps. Again the high resolution of the DSM0.01 allows the identification of the different aspects of the surface of a single tile and the small roof objects remain recognizable until the 50 cm resolution. Moreover, the 1 m aspect maps from the two different sources (LiDAR and photogrammetric point cloud) look also very similar. The median of the differences is 9.84° and 50 % of the cells range between 4.73° and

22.22° degrees (Fig. 4). The 90th percentile is already 50.36°. These are very good numbers considering that differences of 179.99° are possible. Furthermore, in contrast to the slope maps, the roof structure is preserved also in the aspect maps from the 1 m DSMs.

4.4 Classification of Usable Roof-Top Areas for PV Systems

In order to determine in how far a higher resolution could contribute to a better automatic recognition of suitable surfaces, the following procedure is implemented: First, the aspect maps are reclassified from all possible values between 0° and 360° to eight cardinal directions (east, north-east, north, north-west, west, south-west, south and south east). Second, the resulting raster map is smoothed using r.neighbors in GRASS GIS by selecting the statistical mode in a radius of r cell neighbours. r is the closest integer of the factor between 100 and the resolution of the aspect map in cm. In the cases where r is not an odd number, one unit is added to r. Finally, the resulting raster map is transformed into a vector map that represents the homogeneous parts of the roofs.

The aspect vector maps generated with the aspect map from the DSM0.01 (Fig. 5) present in a proper and very detailed way different parts of the roofs. Surfaces smaller than 6 m^2 (the average area of 1 kWp PV capacity) are removed. It is guessed, that the remaining areas are the ones, where a PV plant can be installed. In sum the total usable roof-area identified with the aspect map of the DSM0.01 is

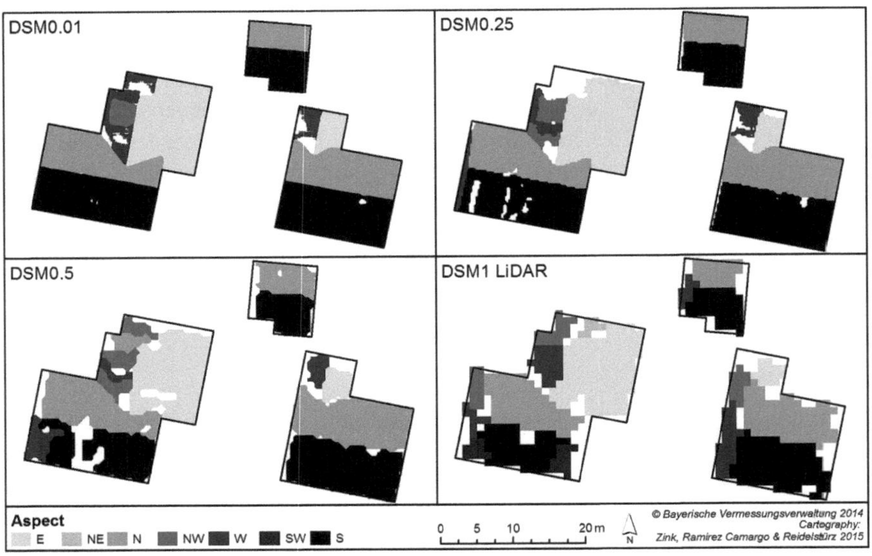

Fig. 5 Result of the aspect classification and usable roof-top areas from different DSMs

694.98 m^2, corresponding to 12 different areas ranging from 7.18 to 124.52 m^2. These values do not change strongly for the aspect maps from the DSM0.075, DSM0.1 and DSM0.25. In the aspect vector map from the DSM0.25 the total area is 656.90 m^2 divided in 13 areas with sizes ranging between 7.09 and 118.00 m^2. Reducing the resolution to 50 cm and 1 m also decreases the total potential area to 625.75 m^2 and 613.96 m^2 respectively. Furthermore, the amount of areas increases to 16 for the aspect vector map from the DSM0.5 and 18 for the aspect vector map from the LiDAR data.

5 Conclusion

The comparison of DSMs and their respective aspect and slope maps in different resolutions shows that a 1 m-raster cell size cannot represent relevant roof objects. Thus small objects like dormers, smokestacks or already installed PV plants are invisible and their shadowing effect on surrounding surfaces will be ignored. But precisely these small objects are decisive to estimate solar radiation and to classify usable areas on roof-tops. These parameters have to be taken into account to model photovoltaic power plants, especially if the system is used as stand-alone power system or is intended to be integrated in a virtual power plant.

As indicated photogrammetric point clouds derived from UAS based large scale image flights are a meaningful alternative to LiDAR data in order to generate DSMs with higher resolutions. The point density that can be achieved with the described large scale image flights are thousands times higher than for LiDAR. In the study area the point density per square meter is four respectively 24,313 for LiDAR and photogrammetric point clouds. However, the costs for data collection, the increased required calculation time and storage capacity have to be paid for this level of detail.

The aspect and slope maps derived from the DSM0.01 allow recognizing every single element of the roof: The roof structure is well defined, the edges are sharp, dormers, smokestacks, already installed PV plants and even the curvature of every single tile can be visualized without distortion and noise. Moreover, the use of the aspect map in this resolution for the classification of usable roof-top areas for PV Systems enables results comparable to manual editing. Nevertheless, the high level of detail does not necessarily imply that a solar radiation algorithm will deliver more accurate results. Standard PV panels are flat and the surface of a module will be properly represented by one single slope and one single aspect value. The very high resolution and the consequent changing slope and aspect for every single tile face results in a false estimation. The examination of these differences remains as an open question for further research.

Due to the similarities between the maps with 1 cm, 7.5 cm, 10 cm and 25 cm resolution it is possible to conclude that a resolution of 25 cm presents the best balance between detail and computational requirements for the discussed purpose. This resolution is sufficient to recognize all the elements that can be observed with the best resolution maps. The classification of usable areas provides also results very similar to the ones of the aspect maps from the DSM0.01. Furthermore, the expected additional expenses in terms of calculation time to handle raster maps of 25 cm is considerably lower than for maps of 1 cm resolution. Therefore it should be thought about the use of small fixed wing UAS as camera carrier, concerning UAS based photogrammetric data acquisition to derive DSMs with a resolution of 25 cm. Such UAS in motorglider-configuration can cover up to 100 ha in one flight (20 min). Equipment acquisition is much more inexpensive than using professional multicopter systems. First experiments have shown that also the use of low-budget digital camera is possible with good results. This also allows data acquisition for small regions quite spontaneously, low-cost and periodically, thus enabling multitemporal change detection applications. Exploiting this idea, image flights with enhanced high-resolution digital single lens reflex cameras (SLR) or up to date system cameras integrated in sports airplanes or modern microlight airplanes with flying heights about 300 m above ground could be a very effective method to acquire larger regions.

References

Agugiaro G, Nex F, De Remondino F, Filippi R, Droghetti S, Furlanello C (2012) Solar radiation estimation on building roofs and web-based solar cadaster. ISPRS Ann Photogrammetry Remote Sens Spat Inf Sci 1:177–182

Bondar I, Koch B, Krüger J, Reidelstürz P (1997) Anwendungspotential der photogrammetrischen Mehrbildtriangulation zur Vermessung terrestrisch forstlicher Aufnahmen und zur Raumlagebestimmung implantierter Micro – Multielektroden im Affenhirn aus simulierten Röntgenaufnahmen. Projektbericht Abteilung Fernerkundung und Landschaftsinformations systeme, Institut für forstliche Betriebswirtschaftslehre, Forstliche Fakultät der Albert Ludwigs Universität Freiburg, 60S. Freiburg, Germany

Drauschke M, Bartelsen J, Reidelstürz P (2014) Towards UAV-based forest monitoring. In: Bendig J, Bareth G (eds) Proceedings of the workshop on UAV-based remote sensing methods for monitoring vegetation, Cologne, Germany, Geographisches Institut der Universität zu Köln - Kölner Geographische Arbeiten, Cologne, 9–10 Sept 2013, pp 21–32

Fu P, Rich PM (1999) Design and Implementation of the solar analyst: an ArcView extension for modeling solar radiation at landscape scales. In: Proccedings of the 19th annual ESRI user conference, San Diego, California, July 1999

Gaile GL, Burt JE (1980) Directional statistics. Geo Abstracts, Norwich

Hirschmueller H (2011) Semi-global matching—motivation, developments and applications. http://www.ifp.uni-stuttgart.de/phowo/2011/presentations/180Hirschmueller.pdf. 7 Sept 2014

Hofierka J, Suri M (2002) The solar radiation model for open source GIS: implementation and applications. In: Proceedings of the open source GIS-GRASS user conference 2002, Trento, Italy, 11–13 Sept 2002, pp 1–19

Jakubiec JA, Reinhart CF (2013) A method for predicting city-wide electricity gains from photovoltaic panels based on LiDAR and GIS data combined with hourly Daysim simulations. Sol Energy 93:127–143

Jäger-Waldau A (2013) PV status report 2013. Scientific and Policy Report by the Joint Research Centre of the European Commission, Luxembourg. Publication Office of the European Union, Belgium. http://iet.jrc.ec.europa.eu/remea/sites/remea/files/pv_status_report_2013.pdf. 5 July 2014

Kanters J, Wall M, Kjellsson E (2014) The solar map as a knowledge base for solar energy use. Energy Procedia 48:1597–1606

Lenardič D (2011) Large-scale photovoltaic power plants, annual and cumulative installed power output capacity 2000-2010. Denis Lenardič, s.p. Jesenice, Slovenia. http://www.pvresources.com/Portals/0/Download/AnnualReview2010.pdf. 12 Sept 2014

Ludwig D, Lanig S, Klärle M (2009) Location analysis for solar panels by LiDAR-data with geoprocessing—SUN-AREA. In: Wohlgemuth V, Page B, Voigt K (eds) Environmental informatics and industrial environmental protection: concepts, methods and tools. Shaker Verlag, Aachen, pp 83–89

Mitasova H, Mitas L, Harmon RS (2005) Simultaneous spline approximation and topographic analysis for lidar elevation data in open source GIS. IEEE Geosci Remote Sens Lett 2:375–379

Ramirez Camargo L, Pagany R, Marquardt A (2014) Zeitlich und räumlich hochaufgelöste Modellierung der potentiellen solaren Einstrahlung - ein Methodenvergleich. In: Strobl J, Blaschke T, Griesebner G, Zagel B (eds) Angewandte Geoinformatik 2014, Beiträge zum 26. AGIT-Symposium Salzburg, Wichmann, Berlin, 2–4 July 2014, pp 143–152

Ramirez Camargo L, Zink R (2014) Photovoltaik in virtuellen Kraftwerken zur Versorgung regionaler Elektromobilitätskonzepte. In: Strobl J, Blaschke T, Griesebner G, Zagel B (eds) Angewandte Geoinformatik 2014, Beiträge zum 26. AGIT-Symposium Salzburg, Wichmann, Berlin, 2–4 July 2014, pp 153–158

Reidelstuerz P (2012) Potential von UAV gestützten Multispektralkamerabefliegungen im Forstbereich. UAS - Tagung im Festsaal der Humboldt Graduate School, ausgerichtet durch das Projekt ADLER, gefördert durch die DBU, Berlin, 5 Mar 2012

Riedel A (2002) Einsatz der Mehrbildtriangulation zur Strukturellen Analyse von Baumpflanzen, vol 73. Lehrstuhl für Waldbau und Forsteinrichtung, Technische Universität München, Forstarchiv, pp 55–61

Theodoridou I, Karteris M, Mallinis G, Papadopoulos AM, Hegger M (2012) Assessment of retrofitting measures and solar systems' potential in urban areas using geographical Information Systems: application to a Mediterranean city. Renew Sustain Energy Rev 16 (8):6239–6261

Zink R, Dorner W, Grzemba A, Schlott M, Reidelstürz P (2012) Photovoltaik unter Aufsicht! Thermographische Kontrolle von PV-Analgen mittels GPS-gestützter Drohne. In: Angewandte Geoinformatik 2012. Beiträge zum 24. AGIT-Symposium Salzburg, Wichmann, Heidelberg, 04–06 July 2012, pp 563–568

Author Index

A
Abolmasov, Biljana, 37

B
Bajat, Branislav, 137
Bakoň, Matúš, 165
Bárek, Viliam, 99
Barton, Gábor, 1
Bartoněk, Dalibor, 13, 27
Barvíř, Radek, 45
Basarić, Irena, 37
Benčoková, Mária, 263
Benes, Bedrich, 237
Bódis, Katalin, 1
Bogdanović, Snežana, 37
Brus, Jan, 45
Bureš, Jiří, 13, 27

C
Čada, Václav, 225
Caha, Jan, 61
Chybicki, Andrzej, 151
Çomut, Fatma Canaslan, 73

D
Dorner, Wolfgang, 293
Đurić, Uroš, 37

F
Fárek, Vladimír, 87
Fuska, Jakub, 99

G
Géczi, Róbert, 1
Gürboğa, Şule, 73

H
Hassanien, Aboul Ella, 249
Hlaváčová, Ivana, 165
Holata, Lukáš, 113

J
Jančíková, Andrea, 125
Jankowska, Marlena, 151

K
Kilibarda, Milan, 137
Kolejka, Jaromír, 189
Kolingerová, Ivana, 237
Krunić, Nikola, 137
Kulawiak, Marcin, 151

L
Lazecký, Milan, 165

M
Marjanović, Miloš, 37
Markiewicz, Lukasz, 151

N
Nikiel, Sławomir S., 281

O
Opatřilová, Irena, 27

K. Růžičková and T. Inspektor (eds.), *Surface Models for Geosciences*,
Lecture Notes in Geoinformation and Cartography,
DOI 10.1007/978-3-319-18407-4

P
Pavelka, Karel, 173
Pospíšil, Lubomil, 13

R
Ramirez Camargo, Luis, 293
Rášová, Alexandra, 61
Reidelstürz, Patrick, 293
Reulke, Ralf, 249
Ruda, Aleš, 189
Růžička, Jan, 205
Růžičková, Kateřina, 205

S
Šafář, Václav, 215
Šedina, Jaroslav, 173
Šilhavý, Jakub, 225
Skorkovská, Věra, 237
Švábenský, Otakar, 13
Světlík, Radek, 113

T
Tahoun, Mohamed, 249

U
Unucka, Jan, 87, 125
Üstün, Aydın, 73

V
Vizi, Ladislav, 263

W
Warszawski, Korneliusz K., 281

Z
Zink, Roland, 293

MIX
Papier aus verantwortungsvollen Quellen
Paper from responsible sources
FSC® C105338

If you have any concerns about our products,
you can contact us on
ProductSafety@springernature.com

In case Publisher is established outside the EU,
the EU authorized representative is:
Springer Nature Customer Service Center GmbH
Europaplatz 3, 69115 Heidelberg, Germany

Printed by Libri Plureos GmbH
in Hamburg, Germany